Concepts and Applications of Plant Cell Biology

Concepts and Applications of Plant Cell Biology

Edited by **Chris Frost**

SYRAWOOD
PUBLISHING HOUSE
New York

Published by Syrawood Publishing House,
750 Third Avenue, 9th Floor,
New York, NY 10017, USA
www.syrawoodpublishinghouse.com

Concepts and Applications of Plant Cell Biology
Edited by Chris Frost

International Standard Book Number: 978-1-68286-272-8 (Hardback)

Printed in China.

Contents

Preface

Owing to scientific advancements over the past few decades, the study of plant biology has progressed considerably and its concepts have found applications in many allied disciplines. This book unravels the recent studies in the field of plant cell biology. Concepts such as cell anatomy, types of plant cells, organelles and functions, physiological processes like photosynthesis, transpiration, etc. have been discussed in detail. It presents the case studies of internationally renowned scientists and academicians. This book will serve as a reference to a broad spectrum of readers and will help them in keeping pace with the rapid changes in this field.

The information contained in this book is the result of intensive hard work done by researchers in this field. All due efforts have been made to make this book serve as a complete guiding source for students and researchers. The topics in this book have been comprehensively explained to help readers understand the growing trends in the field.

I would like to thank the entire group of writers who made sincere efforts in this book and my family who supported me in my efforts of working on this book. I take this opportunity to thank all those who have been a guiding force throughout my life.

Editor

1

Cellular Auxin Transport in Algae

Suyun Zhang [1] and Bert van Duijn [1,2,*]

[1] Plant Biodynamics Laboratory, Biology Institute Leiden, Leiden University, Sylviusweg 72, Leiden 2333, BE, The Netherlands; E-Mail: s.zhang@biology.leidenuniv.nl

[2] Fytagoras, Sylviusweg 72, Leiden 2333, BE, The Netherlands

* Author to whom correspondence should be addressed; E-Mail: bert.vanduijn@fytagoras.nl

Abstract: The phytohormone auxin is one of the main directors of plant growth and development. In higher plants, auxin is generated in apical plant parts and transported from cell-to-cell in a polar fashion. Auxin is present in all plant phyla, and the existence of polar auxin transport (PAT) is well established in land plants. Algae are a group of relatively simple, autotrophic, photosynthetic organisms that share many features with land plants. In particular, Charophyceae (a taxon of green algae) are closest ancestors of land plants. In the study of auxin function, transport and its evolution, the algae form an interesting research target. Recently, proof for polar auxin transport in *Chara species* was published and auxin related research in algae gained more attention. In this review we discuss auxin transport in algae with respect to land plants and suggest directions for future studies.

Keywords: auxin; algae; polar auxin transport; *Chara*

1. Introduction

Auxins are a class of plant hormones with morphogen-like characteristics that regulate the rate of cell division, cell elongation and expansion, ethylene biosynthesis, apical dominance, organ differentiation and some other essential processes in plant growth [1,2].

In the late 19th century, Charles Darwin investigated "the power of movement in plants" [3]. However, it took nearly a century before, in 1926, Fritz Went opened a new field of study by successfully isolating "the plant growth substance" auxin [4,5]. Later he introduced the avena curvature test enabling many experimenters to be active in auxin research [6,7]. Now it is known that auxin is locally synthesized in (land) plants, mainly in shoot tissues, from where it is distributed unevenly throughout the whole plant [5]. Concentration differences of auxin can cause different responses in plant development. During the early 20th century, the traditional "cut-off and stuck-on" tests and later the use of ^{14}C-indoleacetic acid showed that auxin is distributed through the plant by energy-requiring polar transport, with three important characteristics: chemically highly specific, oxygen-requiring and pH dependent [8].

The most abundant naturally occurring auxin is indole-3-acetic acid (IAA) which shows characteristic "polar transport" throughout the whole plant [9–12]. Mid 1970s, based on the known properties of auxin movement (saturable, energy- and protein synthesis-dependent and unidirectional), the "chemiosmotic hypothesis" was formulated as a classic model to explain the mechanism of polar auxin transport (PAT) [10,13]. According to this model, IAA can easily enter the cell cytoplasm in an undissociated lipophilic form (IAAH) when in a slightly acidic extracellular environment (pH 5.5). While in the cytosol, with a neutral pH of about 7, most of the IAA will be dissociated into anions (IAA^-) and, therefore, trapped inside the cell. To aid the exit of IAA^- active auxin anion efflux carriers were proposed, and the asymmetric distribution of such carriers was postulated to attribute to the directionality of auxin transport [14]. This hypothesis later has been proven to be suitable to describe in general auxin transport in land plants [13].

Although auxins are shown to be present in early-diverged plant lineages such as mosses, liverworts and algae [15,16], research on auxin transport and signaling remains mainly focused on seed plants, predominantly in the model plant *Arabidopsis thaliana* [17]. However, algae share many specialized characteristics with the land plants while evolutionally earlier and of simpler structures. Hence, they may provide unique features that may help to unravel (new) auxin working mechanisms, auxin transport characteristics and functions. In this paper, we look into auxin as a plant hormone in algae, and specially focus on auxin transport.

2. The Roles of Auxin in Algae

Algae belong to a very large and diverse group of simple, typically autotrophic organisms. Increasingly data appears that all land plants (embryophytes) diverged from ancestral *Charophycean* algae, a class of green algae, about 400–500 million years ago, Table 1 [18,19]. Table 1 also shows that the red algae and brown algae are more distant from the streptophytes. In this respect, green algae especially members in the class of Charophyceae are promising candidates to study evolutionary aspects, (new) functionalities and cellular physiology of auxin.

Bioassays, high-performance liquid chromatography, mass spectrometry and some other physicochemical analysis, together with other circumstantial evidences, proof the existence of auxin or auxin-like compounds in many species of algae [17,20,21]. The measured concentrations in these studies are highly variable. The presence and action of auxins have been shown both in unicellular and multi-cellular algae [9,17,22–24]. For example, in red algae (e.g., *Grateloupia dichotoma*, *Gracilaria vermiculophylla*, *Agardhiella subulata*) and green algae (e.g., *Chlorella pyrenoidosa*, *Micrasterias thomasiana*) auxin stimulates cell division and cell enlargement [25–30] and affects the development and growth of rhizoids as well [31–33]. Although the full-value of plant hormone systems in algae is still under debate, the aforementioned studies about auxins on algal growth and development indicate that its functions likely correspond to its activity in higher land plants [20,34,35]. As studies so far concentrated on land plant corresponding functions other, new and unexpected, roles for auxins in algae may have been overlooked and remain to be discovered.

Table 1. Partial classification of plants and brown algae.

KINGDOM	DIVISION	CLASS	ORDER
Plantae	Chlorophyta (green algae)		
	Streptophyta	**Charophyceae** (green algae)	Zygnematales, Coleochaetales, Charales, Desmidiales, Klebsormidiales, Mesostigmatales
		Embryophyceae (land plants:mosses lato sensu, lycophytes, ferns and horsetails, seed plants)	
	Rhodophyta (red algae)		
Chromalveolata	Heterokontophyta	Phaeophyceae (brown algae)	

3. Transport of Auxin in Algae

Auxin acts both as hormone and morphogen. The role of auxin in apical dominance can be regarded as a classical example of hormonal integration based on hormone distribution in land plants. Similar phenomena have also been described in various seaweeds, suggesting that in algae similar auxin distribution systems and carriers supporting PAT may be present [34,36].

The events in auxin signaling as established in seed plants do not only involve the sensing of auxins at the level of the target cells and their responses but also auxin biosynthesis and metabolism, intracellular compartmentalization, and directional transport through cells facilitated by specific transporters [37,38].

Auxin transport in seed plants is characterized by its polarity, directionality, distance, and transporting cells [17,39]. Polar auxin transport (PAT) is facilitated by influx carriers (AUX1/LAX proteins) and efflux carriers (PIN proteins), together with some other transport proteins (ABCB/PGP transporter family) [11,13]. The asymmetric distribution of efflux-carriers, mainly related to the plasma-membrane-localized PINs, aids to the gradient of auxin concentration through the whole plant, while the differences of auxin concentration in turn regulate the number and location of efflux-carriers [14,40–42]. The endoplasmic reticulum (ER)-localized PIN proteins (like PIN5, PIN8) are thought to charge the intracellular compartmentalization of auxin and homeostasis, in cooperation with members of the recently-found auxin carrier family PIN-LIKEs (PILS) [43–45].

To understand this whole system better, splitting it into several parts and searching them back in the evolutionarily earlier organisms of green or brown algae lineages will be helpful.

In view of some of the apparent morphological similarities between algae and land plants one might ask if the ability of auxin polar transport is required for such differential development? Studies in the large coenocytic (multi-cellular structures without cross walls) green alga *Caulerpa paspaloides* show different results. Although these algae show characteristics like a leaf-like assimilator which grows up, rhizoid clusters that grow down, and a rhizome that grows horizontally, auxin displays an uniform and non-polar distribution, which might be caused by diffusion and cytoplasmic streaming [22,46,47]. This suggests that auxin polar transport and auxin gradients do not participate in, at least, the later development and maintenance of these three different organs. In multi-cellular algae the existence of PAT has also been investigated. In *Chara*, a branched, multi-cellular green alga, a specific carrier system is suggested to mediate the transmembrane auxin transport [46]. This carrier lacked the inhibition by 1-*N*-Naphthylphthalamic acid (NPA, a phytotropin) which is typical for inhibiting efflux carriers in land plants [46]. In the same study, such specific auxin carriers were not found in the simple unicellular green alga *Chlorella vulgaris* [46]. On the other hand, Klämbt and coworkers reported that in growing rhizoids of *Chara*, auxin efflux showed NPA-sensitive activation [32]. These contradictory results are thought to be due to some additional effects of NPA, unrelated to its ability to increase intracellular IAA levels by blocking IAA efflux [17]. Recent experiments, showing NPA sensitive polar transport of radioactive labeled auxin, provide more direct proof of the presence of PAT in *Chara* and suggest the presence of specific auxin efflux carriers as in land plants [48]. In addition, PAT has been shown to exist in brown algae species like *Fucus distichus* and *Ectocarpus siliculosus* [31,49].

Comparing studies in unicellular (micro or coenocytic) green algae with multi-cellular green algae (*Chara*), Dibb-Fuller and Morris proposed that "the appearance of specific auxin carrier systems in the *Chlorophyta* may have been functionally associated with the evolution of multi-cellularity, rather than with the evolution of a plant body which is characterized by distinctly different morphological regions" [46]. This conclusion seems in accordance with the early hypothesis raised by others, that after the development of multicellular organisms, simple diffusion of IAA would not be efficient enough, the movement across cell membranes is required for polar transport of IAA [8,47,50]. Although the relatively lipid-soluble IAA could account for very slow diffusive transport through membranes, for more effective longer distance transport and gradients in tissue, membrane bound auxin transporters and channels are required.

Since PAT in seed plants is largely dependent on the asymmetric distribution of PIN proteins, the PIN proteins seem a key in the investigation of PAT mechanisms. Though the PILS carrier family is quite conserved throughout the evolution of plants and can be found from unicellular algae to highly developed seed plants, the PIN exporter families at the plasma membrane are comparatively young in evolutionarily perspective [37,51]. The plasma membrane-localized PIN proteins are thought to exist only in land plants, while the endoplasmic reticulum (ER)-localized PIN proteins (like PIN5, PIN8 in *Arabidopsis*) are evolutionarily older, and can be traced back to an origin in Streptophyta algae [51].

By using a basic local alignment search tool (BLAST), partial PIN protein sequences can be identified in several species of green algae like *Spirogyra*, *Penium*, and the evolutionarily even earlier lineage *Klebsormidiophyceae* [51,52]. However, the above results are based on the expressed sequence tag (EST) database, since the nuclear genomes sequences database is barely described in the multicellular green algae of the Streptophyta group. The scant annotated genomes are almost all from the unicellar algae of the Chlorophyta group, and there is not yet evidence of the existence of PIN protein sequences [37,51]. With the growing information on nuclear genomes of multicellular green algae groups, a clearer picture is soon expected on the presence of PIN proteins.

Although evidence for functional PIN proteins supporting PAT in algae is (still) scarce this does not rule out the (functional) existence of PAT in algae. In the brown algae *Ectocarpus siliculosus* PAT has been shown while this species lacks any PIN homolog [53]. The possibility of other types of auxin (polar) transporter mechanisms rather than PIN proteins cannot be excluded. Besides, the presence of plasmodesmata or other active (e.g., vesicle based) transport systems may exist. Researches have shown that there exists a special plasma membrane invagination-structure named charasomes in Chara species, but not in Nitella species. Interestingly, PAT also shows up in *Chara corallina*, but not in Nitella [48,54–57]. Charasomes are thought to be closely related to the ability of endocytosis in Chara and the origin of vesicles [54]. Although even if it is possible for auxin to be transported in vesicles, the driving force of these vesicles in these algae species remains a mystery: the block of cytostreaming by cytochalasin H could not stop the auxin transport in Chara while the normal microtubule system seems not fast enough to reach the observed speed of auxin transport. This suggests the existence of an unknown amplification mechanism [48,58]. Taking everything into consideration, it is still a challenge to identify possible PIN protein independent polar auxin transport mechanisms that may have emerged in the early period of evolution, and may or may not have been eliminated during the evolution of land plants. Algae form an attractive model to investigate these potentially early systems and learn more about the basic properties of auxin transport and auxin function.

At present, our knowledge on the role of auxin and its transport in algae is still very limited. However, we are still able to draw some conclusions based on the available studies. With regard to (polar) auxin transport, Figure 1 illustrates a summary and comparison between land plants and multi-cellular algae, like *Chara corallina.* In both models PAT is present based on auxin influx carriers and NPA sensitive efflux carriers, and a chemiosmotic mechanism. Although the presence of NPA sensitive PAT in *Chara* in combination with our understanding of PAT in land plants justifies this hypothetical model, it must be emphasized that the hypothetical presence of these specific transport proteins in algae is only based on indirect evidence. In addition, the membrane potential difference across the plasmalemma should also be taken into consideration as this plays a significant role in the chemiosmotic mechanism, and complicates the auxin transport model [59–61]. Compared with our knowledge the model system of seed plants—Arabidopsis, studies on algae are still facing many challenges, especially with regard to the still insufficient genetic background database and possibilities for application of molecular biology tools.

Figure 1. (**A**) Auxin transport in seed plants; (**B**) The hypothetical auxin transport system in *Chara*.

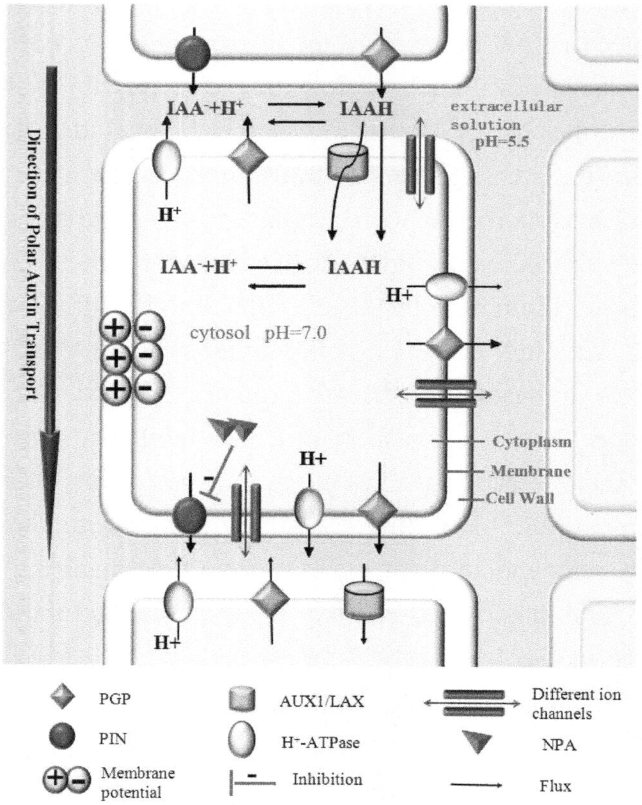

(**A**)

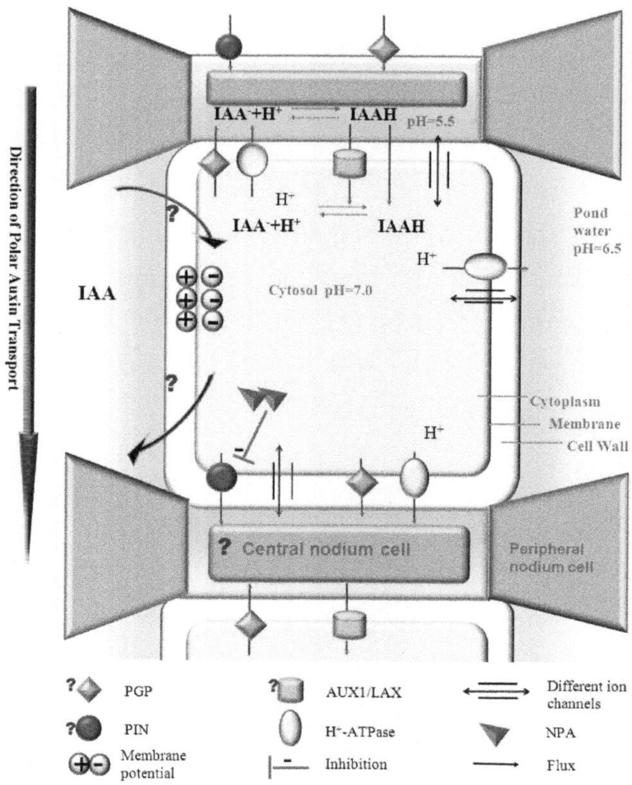

(**B**)

4. Algae as Model System

Research in the field of auxin and algae were mainly on the brown algae when related to the function of auxin, while the green algae were more favorable in studies related to signaling and transport. In view of modern phylogenetics, the division Charophyta contains the closest living relatives to the land plants. In this division, the order Charales has been widely accepted as a sister-relationship to the land plants over the past century, which was based also on the morphology since the Charales develop the most complicated, land-plant-like body structures as compared to all the other orders [62]. Now the phenomenon of PAT is observed in *Chara*, a genera of Charales, showing that, after the discovery of PAT in moss species (*Physcomitrella patens*) [49,63], these green algae species are so far the earliest life forms showing PAT in the non-land plants [46]. Recent research data show that Zynematales and Coleochaetales are more close to the land plants than Charales [64–66] and it hås been proposed that the grade of organismal complexity of Charales may have evolved independently. It remains to be discovered whether the Charales share a similar auxin polar transport mechanism with the land plants or have evolved a different mechanism of PAT. Therefore, from an evolutionary point of view, auxin related research aiming at early mechanisms and functions should be focused on these organisms. Within the order of *Charales*, species from the genera *Chara* have shown a great potential for the future studies in auxin signaling and transport. Species of *Chara* have the giant internodal cells that provide an ideal and unique model for polar auxin transport research at the single cell level while they do show the ability of polar transport of auxin. To some extent, these cells are superior to the model system *Arabidopsis thaliana*, such as for cell biological and cell physiology studies of auxin transport and signaling at the single cell level as advanced microscopy, electrophysiology, and transport studies that can be applied to the intact cell under natural conditions.

Using algae in auxin research requires some precautions and approaches that are less relevant in studies on land plant. For instance, the use of non-sterile algae materials could lead to false conclusions as bacteria or secondary algae may produce auxin-like substances or break down endogenous algal auxin. Some special cautions are also necessary in choice and conditioning of cells for auxin transport studies, especially since the function of auxin in algae remains unclear. In our recent experience, we noticed that in some situations the *Chara* internodes may not show the PAT phenomenon (data not show). This could be an example of auxin feedback regulation on the efflux carriers. The reasons and details need a further study. Also, in higher plants, such effects have been reported. Decapitation of growing shoots can result in the loss of polar auxin transport in segments from internodes subtending the apex [67].

Besides the possibility that "apical dominance" may also be part of the algae development strategies, there are some other unique characteristics of algae that require attention. Unicellular algae have a big surface area in connection with the environment, and also multi-cellular algae cells have a strong exposure to the environment, in contrast to cells in land plants. Despite this strong exposure to the environment algae are able to control environmental parameters around the cells. The internodal cells of *Chara corallina*, for example, can build up several alkaline bands separated by acid bands along the longitudinal axis to facilitate the uptake of inorganic carbon [68–70]. This band-formation ability is quite strong and shows that the cells can manipulate the extracellular micro-environment. Due to this acid-alkaline banding pattern, application of the classical "chemiosmotic mechanisms" for auxin transport should only be applied with great care (Figure 1), as pH balance conditions may vary strongly and abruptly along the cell surface length.

5. Conclusions

Genetic research and development of molecular biological tools, such as genetic transformation, as well as tools for studying the developmental biology for *Chara* and *Nitella* species is still in its infancy. However, based on our detailed knowledge of auxin function, signaling and transport in land plants, combined molecular and cell physiological research in *Chara* and *Nitella* may open a new page on our understanding of auxin's central role in plant growth and development.

Acknowledgements

S.Z. is supported by the Chinese Scholarship Council (2011635073)

Conflicts of Interest

The authors declare no conflict of interest.

Reference

1. Raven, P.H.; Evert, R.F.; Eichhorn, S.E. *Biology of Plants*, 6th ed.; W. H. Freeman: New York, NY, USA, 1999; pp. 673–700.
2. Sslisbury, F.B.; Ross, C.W. *Plant Physiology*, 4th ed.; Wadsworth Publishing Company: Belmont, CA, USA, 1992; pp. 375–378.
3. Darwin, C.; Darwin, F. *The Power of Movement in Plants*; D. Appleton and Company: New York, NY, USA, 1881; pp. 523–545.
4. Went, F.W. Reflections and speculations. *Annu. Rev. Plant Physiol.* **1974**, *25*, 1–26.
5. Taiz, L.; Zeiger, E. *Plant Physiology*; The Benjamin/Cummings Publishing Company, Inc.: Redwood, CA, USA, 1991; pp. 398–425.
6. Went, F.W.; Thimann, K.V. *Phytohormones*; The Macmillan Company: New York, NY, USA, 1937; pp. 21–56.
7. Went, F.W. Growth hormones in the higher plants. *Annu. Rev. Biochem.* **1939**, *8*, 521–540.
8. Jacobs, W.P. Movement of Hormones. In *Plant Hormones and Plant Development*; Cambridge University Press: New York, NY, USA, 1979; pp. 195–243.
9. Lau, S.; Shao, N.; Bock, R.; Jurgens, G.; Smet, I.D. Auxin signaling in algal lineages: Fact or myth? *Trends Plant Sci* **2009**, *14*, 182–188.
10. Friml, J. Auxin transport-shaping the plant. *Curr. Opin. Plant Biol.* **2003**, *6*, 7–12.
11. Friml, J.; Palme, K. Polar auxin transport—Old questions and new concepts? *Plant Mol. Biol.* **2002**, *49*, 273–284.
12. Bartel, B. Auxin biosynthesis. *Annu. Rev. Plant Physiol. Plant Mol. Biol.* **1997**, *48*, 51–66.
13. Vieten, A.; Sauer, M.; Brewer, P.B.; Friml, J. Molecular and cellular aspects of auxin-transport-mediated development. *Trends Plant Sci.* **2007**, *12*, 160–168.

14. Petrasek, J.; Friml, J. Auxin transport routes in plant development. *Development* **2009**, *136*, 2675–2688.
15. Ergun, N.; Topcuoglu, S.F.; Yildiz, A. Auxin (indole-3-acetic acid), Gibberellic acid (GA3), Abscisic Acid (ABA) and Cytokinin (Zeatin) production by some species of mosses and lichens. *Turk. J. Bot*. **2002**, *26*, 13–18.
16. Van Overbeek, J. Auxin in marine algae. *Plant Physol*. **1940**, *15*, 291–299.
17. Cooke, T.J.; Poli, D.B.; Sztein, A.E.; Cohen, J.D. Evolutionary patterns in auxin action. *Plant Mol. Biol.* **2002**, *49*, 319–338.
18. Proseus, T.E.; Boyer, J.S. Identifying cytoplasmic input to the cell wall of growing *Chara coralline*. *J. Exp. Bot*. **2006**, *57*, 3231–3242.
19. Niklas, K.J. The Evolution of plant body plans—A biomechanical perspective. *Ann. Bot.* **2000**, *85*, 411–438.
20. Tarakhovskaya, E.R.; Maslov, Y.I.; Shishova, M.F. Phytohormones in algae. *Russ. J. Plant Physiol.* **2007**, *54*, 163–170.
21. Niemann, D.I.; Dorffling, K. Growth inhibitors and growth promoters in *Enteromorpha compressa* (Chlorophyta). *J. Phycol.* 1**980**, *16*, 383–389.
22. Jacobs, W.P.; Falkenstein, K.; Hamilton, R.H. Nature and amount of auxin in algae. *Plant Physiol.* **1985**, *78*, 844–848.
23. Kenneth, V.T.; Kurt, B. Action of auxins on *Acetabulatia* and the effect of enucleation. *Nature* **1959**, *183*, 946–948.
24. Nowak, J.; Sonaike, B.; Lawson, G.W. Auxin induced stress tolerance in algae. *Environ. Pollut.* **1988**, *51*, 213–218.
25. Yokoya, N.S.; Handro, W. Effects of auxins and cytokinins on tissue culture of *Grateloupia dichotoma* (Gigartinales, Rhodophyta). *Hydrobologia* **1996**, *326/327*, 393–400.
26. Garcia-Jimenez, P.; Rodrigo, M.; Robaina, R.R. Influence of plant growth regulators, Polyamines and Glycerol interaction on growth and morphogenesis of carposporelings of *Grateloupia* cultured *in vitro*. *J. Appl. Phycol.* **1998**, *10*, 95–100.
27. Yokoya, N.S.; Hirotaka, K.; Hideki, O.; Takao, K. Effects of environmental factors and plant growth regulators on growth of the red alga *Gracilaria vermiculophylla* from Shikoku Island, Japan. *Hydrobiologya* **1999**, *398/399*, 339–347.
28. Vance, B.D. Phytohormone effects on cell division in *Chlorella pyrenoidosa* chick (TX-7-11-05) (Chlorellaceae). *J. Plant Growth Regul.* **1987**, *5*, 169–173.
29. Wood, N.L.; Berliner, M.D. Effects of indoleacetic acid on the desmid *Micrasterias thomasiana*. *Plant Sci. Lett.* **1979**, *16*, 285–289.
30. Bradley, P.M.; Cheney, D.P. Some effects of plant growth regulators on tissue cultures of the marine red alga *Agardhiella subulata* (Gigartinales, Rhodophyta). *Hydrobiologia* **1990**, *204/205*, 353–360.
31. Basu, S.; Haiguo, S.; Brian, L.; Quatrano, R.L.; Muday, G.K. Early embryo development in *Fucus distichus* is auxin sensitive. *Plant Physiol.* **2002**, *130*, 292–302.
32. Klambt, D.; Knauth, B.; Dittmann, I. Auxin dependent growth of rhizoids of *Chara globularis*. *Physiol. Plant.* **199**2, *85*, 537–540.

33. Provasoli, L.; Carlucci, A.F. Vitamins and Growth Regulators. In *Algal Physiology and Biochemistry*; Stewart, W.D.P., Ed.; Blackwell Scientific Publications Ltd: Oxford, UK, 1974; pp. 741–787.
34. Bradley, P.M. Plant hormones do have a role in controlling growth and development of algae. *J. Phycol.* **1991**, *27*, 317–321.
35. Evans, L.V.; Trewavas, A.J. Is algal development controlled by plant growth substances? *J. Phycol.* **1991**, *27*, 322–326.
36. Buggeln, R.G. Morphogenesis and Growth Regulators. In *The Biology of Seaweeds*; Lobban, C.S., Wynne, M.J., Eds.; University of California press: Berkeley, CA, USA, 1981; pp. 627–660.
37. Smet, I.D.; Voß, U.; Lau, S.; Wilson, M.; Shao, N.; Timme, R.E.; Swarup, R.; Kerr, I.; Hodgman, C.; Bock, R.; *et al*. Unraveling the evolution of auxin signaling. *Plant Physiol.* **2011**, *155*, 209–221.
38. Feraru, E.; Vosolsobe, S.; Ferar1, M.I.; Petrášek, J.; Kleine-Vehn, J. Evolution and structural diversification of PILS putative auxin carriers in plants. *Front. Plant Sci.* **2012**, *3*, 1–13.
39. Lomax, T.L.; Muday, G.K.; Rubery, P.H. Auxin Transport. In *Plant Hormones: Physiology, Biochemistry, and Molecular Biology*, 2nd ed.; Davies, P.J., Ed.; Kluwer Academic Publishers: Dordrecht, The Netherlands, 1995; pp. 509–530.
40. Morris, D.A. Transmembrane auxin carrier systems—Dynamic regulators of polar auxin transport. *Plant Growth Regul.* **2000**, *32*, 161–172.
41. Paciorek, T.; Zazimalova, E.; Ruthardt, N.; Petrasek, J.; Stierhof, Y.D.; Kleine-Vehn, J.; Morris, D.A.; Emans, N.; Jurgens, G.; Geldner, N.; *et al*. Auxin inhibits endocytosis and promotes its own efflux from cells. *Nature* **2005**, *435*, 1251–1256.
42. Friml, J. Subcellular trafficking of PIN auxin efflux carriers in auxin transport. *Eur. J. Cell Biol.* **2010**, *89*, 231–235.
43. Barbez, E.; Kubes, M.; Rolcik, J.; Beziat, C.; Pencik, A.; Wang, B.; Rosquete, M.R.; Zhu, J.; Dobrev, P.I.; Lee, Y.; *et al*. A novel putative auxin carrier family regulates intracellular auxin homeostasis in plants. *Nature* **2012**, *485*, 119–122.
44. Mravec, J.; Skupa, P.; Bailly, A.; Hoyerova, K.; Krecek, P.; Bielach, A.; Petrasek, J.; Zhang, J.; Gaykova, V.; Stierhof, Y.; *et al*. Subcellular homeostasis of phytohormone auxin is mediated by the ER-localized PIN5 transporter. *Nature* **2009**, *495*, 1136–1140.
45. Sauer, M.; Robert, S.; Kleine-Vehn, J. Auxin: Simply complicated. *J. Exp. Bot.* **2013**, *64*, 2565–2577.
46. Dibb-Fuller, J.E.; Morris, D.A. Studies on the evolution of auxin carriers and phytotropin receptors: Transmembrane auxin transport in unicellular and multicellular Chlorophyta. *Planta* **1992**, *186*, 219–226.
47. Brennan, T.; Jacobs, W.P. Polarity and the movement of [14C]-indol-3-ylacetic acid in the coenocyte, *Caulerpa prolifera*. *Ann. Bot.* **1980**, *46*, 129–131.
48. Boot, K.J.M.; Libbenga, K.R.; Hille, S.C.; Offringa, R.; van Duijn, B. Polar auxin transport: An early invention. *J. Exp. Bot.* **2012**, *63*, 4213–4218.
49. Sun, H.; Basu, S.; Brady, S.R.; Luciano, R.L.; Muday, G.K. Interactions between auxin transport and the actin cytoskeleton in developmental polarity of *Focus distichus* embryos in response to light and gravity. *Plant Physiol.* **2004**, *135*, 266–278.

50. Goldsmith, M.H.M. The polar transport of auxin. *Ann. Rev. Plant Physiol.* **1977**, *28*, 439–478.
51. Viaene, T.; Delwiche, C.F.; Rensing, S.A.; Friml, J. Origin and evolution of PIN auxin transporters in the green lineage. *Trends Plant Sci.* **2013**, *18*, 5–10.
52. Leliaert, F.; Smith, D.R.; Moreau, H.; Herron, M.D.; Verbruggen, H.; Delwiche, C.F.; de Clerck, O. Phylogeny and molecular evolution of the green algae. *CRC Crit. Rev. Plant Sci.* **2012**, *31*, 1–46.
53. Bail, A.L.; Billoud, B.; Kowalczyk, N.; Kowalczyk, M.; Gicquel, M.; Panse, S. L.; Stewart, S.; Scornet, D.; Cock, J. M.; Ljung, K.; *et al*. Auxin metabolism and function in the multicellular Brown alga *Ectocarpus siliculosus*. *Plant Physiol.* **2010**, *153*, 128–144.
54. Franceschi, V.R.; Lucas, W.J. Structure and possible function(s) of charasomes; complex plasmalemma-cell wall elaborations present in some Characean species. *Protoplasma* **1980**, *104*, 253–271.
55. Lucas, W.J.; Franceschi, V.R. Characean charasome-complex and plasmalemma vesicle development. *Protoplasma* **1981**, *107*, 255–267.
56. Schmolzer, P.M.; Hoftberger, M.; Foissner, I. Plasma membrane domains participate in pH banding of *Chara* internodal cells. *Plant Cell Physiol.* **2011**, *52*, 1274–1288.
57. Boot, K.J.M.; Libbenga, K.R.; Offringa, R.; van Duijn, B. Institute Biology Leiden, Leiden, The Netherlands. Unpublished work, 2013.
58. Raven, J.A. Polar auxin transport in relation to long-distance transport of nutrients in the Charales. *Exp. Bot.* **2013**, *64*, 1–9.
59. Raven, J.A. Transport of indoleacetic acid in plant cells in relation to pH and electrical potential gradients, and its significance for polar IAA transport. *New Phytol*. **1975**, *74*, 163–172.
60. Rubby, P.G.; Sheldrake, A.R. Effect of pH and surface charge on cell uptake of auxin. *Nat. New Biol*. **1973**, *139*, 285–288.
61. Fisahn, J.; Hansen, U.P.; Lucas, W.J. Reaction kinetic model of a proposed plasma membrane two-cycle H^+-transport system of *Chara corallina*. *Proc. Natl. Acad. Sci. USA* **1992**, *89*, 3261–3265.
62. Karol, K.G.; McCourt, R.M.; Cimino, M.T.; Delwiche, C.F. The closest living relatives of land plants. *Science* **2001**, *294*, 2351–2353.
63. Fujita, T.; Hasebe, M. Convergences and divergences in polar auxin transport and shoot development in land plant evolution. *Plant Signal. Behav.* **2009**, *4*, 313–315.
64. Finet, C.; Timme, R.E.; Delwiche, C.F.; Marletaz, F. Multigene phylogeny of the green lineage reveals the origin and diversification of land plants. *Curr. Biol*. **2010**, *20*, 2217–2222.
65. Wodniok, S.; Brinkmann, H.; Glockner, G.; Heidel, A.J.; Philippe, H.; Melkonian, M.; Becker, B. Origin of land plants: Do conjugating green algae hold the key? *BMC Evol. Biol*. **2011**, *11*, e104.
66. Timme, R.E.; Bachvaroff, T.R.; Delwiche, C.F. Broad phylogenomic sampling and the sister lineage of land plants. *PLoS One* **2012**, *7*, e29696.
67. Morris, D.A.; Johnson, C.F. The role of auxin efflux carriers in the reversible loss of polar auxin transport in the pea (*Pisum sativum* L.) stem. *Planta* **1990**, *181*, 117–124.
68. Lucas, W.J. Alkaline band formation in *Chara corallina*. *Plant Physiol.* **1979**, *63*, 248–254.
69. Lucas, W.J. Plasmalemma transport of OH^- in *Chara corallina III*. Further studies on transport substrate and directionality. *Plant Physiol.* **1980**, *66*, 46–50.

70. Shimmen, T.; Wakabayashi, A. Involvement of membrane potential in alkaline band formation by internodal cells of *Chara corallina*. *Plant Cell Physiol.* **2008**, *49*, 1614–1620.

2

The Half-Size ABC Transporter FOLDED PETALS 2/ABCG13 Is Involved in Petal Elongation through Narrow Spaces in *Arabidopsis thaliana* Floral Buds

Seiji Takeda [1,2,†,*], Akira Iwasaki [3,4,†], Kiyoshi Tatematsu [3] and Kiyotaka Okada [3,‡]

1 Graduate School of Life and Environmental Sciences, Kyoto Prefectural University, Sakyo-ku, Kyoto 606-8522, Japan
2 Biotechnology Research Department, Kyoto Prefectural Agriculture Forestry and Fisheries Technology Center, Seika, Kyoto 619-0244, Japan
3 Laboratory of Plant Organ Development, National Institute for Basic Biology, Okazaki, Aichi 444-8585, Japan; E-Mails: iwaaki0123@gmail.com (A.I.); ktatem@nibb.ac.jp (K.T.); kiyo@nins.jp (K.O.)
4 Department of Botany, Graduate School of Science, Kyoto University, Kyoto 606-8502, Japan

† These authors contributed equally to this work.

‡ Present address: National Institute of Natural Sciences, Tokyo 105-0001, Japan.

* Author to whom correspondence should be addressed; E-Mail: seijitakeda@kpu.ac.jp

---

Abstract: Flowers are vital for attracting pollinators to plants and in horticulture for humans. Petal morphogenesis is a central process of floral development. Petal development can be divided into three main processes: the establishment of organ identity in a concentric pattern, primordia initiation at fixed positions within a whorl, and morphogenesis, which includes petal elongation through the narrow spaces within the bud. Here, we show that the *FOLDED PETALS 2* (*FOP2*) gene, encoding a member of the half-size ATP binding cassette (ABC) transporter family ABCG13, is involved in straight elongation of petals in *Arabidopsis thaliana*. In *fop2* mutants, flowers open with folded petals, instead of straight-elongated ones found in the wild type. The epicuticular nanoridge structures are absent in many abaxial epidermal cells of *fop2* petals, and surgical or genetic generation of space in young *fop2* buds restores the straight elongation of petals, suggesting that the physical contact of sepals and petals causes the petal folding. Similar petal folding has been reported in the *fop1* mutant, and the petals of *fop2 fop1* double mutants resemble those of both the *fop1 and fop2* single mutants, although the epidermal structure and permeability of the petal surface is more affected in *fop2*. Our results suggest that synthesis and transport of cutin or wax in growing petals play an important role for their smooth elongation through the narrow spaces of floral buds.

Keywords: *Arabidopsis thaliana*; floral organs; folded petals; petal elongation; ABC transporter; ABCG13

---

1. Introduction

Petals have evolved to have various colors, shapes, and fragrances, and, thus, their morphogenesis is critical in attracting pollinators for plants and in horticulture for humans. Flowers usually develop three other distinct organs, each with a specialized function. In *Arabidopsis thaliana*, the floral primordia initiate at the flanks of the shoot apical meristem as floral meristems, subsequently sepals arise, first in the outermost concentric region, and cover the floral meristem, protecting the other inner organs as they develop. Gynoecia and stamens are reproductive organs: gynoecia arise in the center of the flower, surrounded by a number of stamens.

Petal development can be divided into three main processes. First is the identity determination, described by the floral ABCE model, which explains that the organ identity is established in a concentric pattern by combined functions of the floral homeotic genes [1–4]. In addition to these main factors, miRNA172 and other transcription factors such as *AINTEGUMENTA*, *LEUNIG*, *SEUSS*, *RABBIT EARS* (*RBE*), and *STERILE APETALA*, are involved in fine-tuning of expression of the floral ABCE genes in *Arabidopsis thaliana* [5–13]. The second process is primordia initiation at fixed positions within each concentric region or whorl. In many flowering plants, such as Brassicaceae, petals initiate at the inner site of the sepal boundary region, thus, the number of petals and sepals is the same. Usually, positions where floral organ primordia arise within a whorl are fixed relative to the adaxial-abaxial and lateral axes of floral meristem. Although the molecular mechanism for determination of initiation position remains unclear, some genes that are expressed in a position-dependent manner, such as *PETAL LOSS* (sepal boundary), *RBE* (petal primordia), *ROXY1* (young floral organ primordia), and *UNUSUAL FLORAL ORGANS* (which changes its expression pattern as a flower develops), seem to be involved in position-dependent primordia development [14–18]. The third process is petal morphogenesis after initiation, which is regulated by cell division, cell expansion, and endoreduplication, regulated by *JAGGED*, *BIGPETALp*, and *FRILL* genes [19–26].

Previously we showed a mutant, *folded petals 1* (*fop1*), which in *Arabidopsis thaliana* has a defect in the petal elongation process [27]. In the wild type, petals elongate straight through the narrow space between sepals and anthers in the floral bud. In *fop1*, petals become stuck in the bud during elongation, resulting in the formation of folded petals in the open flower. *FOP1* encodes a member of the WAX ESTER SYNTHASE/DIACYLGLYCEROL ACYLTRANSFERASE (WS/DGAT) family WSD11 and is expressed in growing petals, suggesting that FOP1 is involved in production of fatty acids that makes petal elongation smooth in narrow space in floral buds, and that petal morphogenesis is affected by the physical interaction of floral organs.

Here, we describe the characterization of another mutant, *folded petals 2* (*fop2*), which has a similar petal folding phenotype to *fop1*. Our results suggest that FOP2 regulates petal elongation at distinct stages of the same pathway of FOP1, and that export of cutin and/or wax to the surface of the petal epidermis is important for petal morphogenesis.

2. Results and Discussion

2.1. Folded Petals Phenotype in the Fop2 Mutant

During a screen for floral organ defective mutants in *Arabidopsis thaliana*, we found a couple of mutants defective in the petal elongation process. Their flowers open with folded petals instead of wild-type straight-elongated ones, so we named these mutants *folded petals* (*fop*). One of these mutants, named *fop2-1*, produces flowers with folded petals with the shape of 'N' letter, instead of wild-type straight elongated petals (Figure 1a,b). About half of petals were unfolded (Table 1), but most of them were wrinkled, suggesting the trace of physical friction during petal elongation in flower buds (Supplementary Figure S1). After identification of the *FOP2* gene by map-based cloning (see below) we analyzed a SALK T-DNA insertion line (SALK_046735), which we named *fop2-2*, and found that this line showed the weaker but similar petal defects to *fop2-1* (Figures 1c and S1). Since the *fop1* mutant showed similar petal defects [27], we examined the relationship between *fop1* and *fop2* by generating the double mutant. Flowers in the *fop2-1 fop1-1* double mutant showed similar petal folding to single mutants, and this phenotype was not enhanced in the double mutant as compared to the *fop2* single mutant (Figure 1d and Table 1), suggesting that *FOP1* and *FOP2* are involved in the same pathway of petal elongation.

To examine the petal defects in more detail, we analyzed the structure of the petal epidermis using scanning electron microscopy. In wild-type, petal epidermal cells are conical-shaped and epicuticular nanoridges cover their surface (Figure 1e). In *fop2-1*, many epidermal cells are flattened and lack the epicuticular nanoridges at their tip (Figure 1f). This defect was also observed on the *fop1* petal epidermis, suggesting that *fop2* petals make physical contact with sepals, as do the *fop1* petals [27]. In several regions on the *fop2-1* petal surface, a number of long cells lacking epicuticular nanoridges seemed to be aggregated at one site (Figure 1f). This phenotype was not detected in the wild type or *fop1*, suggesting that *FOP1* and *FOP2* have different roles in determining the petal epidermal character.

Some other mutants showing the petal-folding phenotype are sensitive to dye immersion due to the lowered repellency of petal surface [28–30]. We examined the surface repellency of *fop2* by immersing petals in toluidine blue solution [31]. Petals from wild-type flowers were not stained, but the *fop2* petals were stained in the blade region (Figure 1g,h). Interestingly, the *fop1* petals were not stained except for the claw edge (Figure1g,h), probably because this region contacts to the sepal surface and thus has reduced repellency [27]. These results indicate that the petal repellency is lower in *fop2* than in *fop1*, suggesting that *FOP2* and *FOP1* have distinct functions.

Together, our results suggest that *FOP1* and *FOP2* regulate petal elongation in two distinct stages of the same pathway.

Figure 1. (**a**) Wild-type (WS) flower with straight petals; (**b**) *fop2-1* flower with folded petals; (**c**) *fop2-2*/SALK_046735 flower with folded petals; (**d**) *fop2-1 fop1-1* flower; (**e**,**f**) Scanning electron micrographs of wild-type (**e**) and *fop2-1* (f) petals; (**g**,**h**) Toluidine blue-stained flowers (**g**) and petals (**h**) of wild-type (WS), *fop1-1*, and *fop2-1*. Arrowhead indicates the edge staining of the *fop1-1* petal. Bars: a, b, c, d, g, h, 1 mm; e, f, 20 μm.

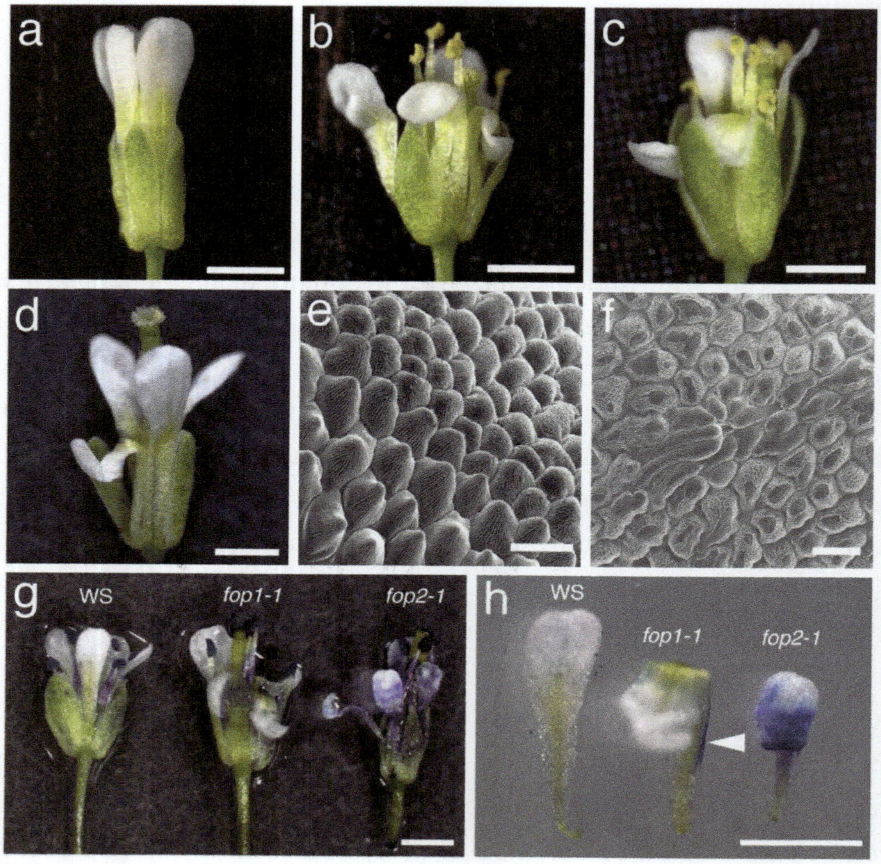

Table 1. Floral organ number of the *fop2-1* single and double mutants.

Genotype	Sepal	Petal		Stamen	Carpel
		Unfolded	Folded		
WS	4 ± 0	4 ± 0	0 ± 0	5.55 ± 0.60	2 ± 0
fop2-1	4 ± 0	2.15 ± 0.75	1.85 ± 0.75 [a]	5.90 ± 0.45	2 ± 0
fop2-1 fop1-1	4 ± 0	1.85 ± 1.27	2.15 ± 1.27 [a]	5.75 ± 0.44	2 ± 0
prs-1 fop2-1	4 ± 0	3.95 ± 0.22	0.55 ± 0.22	5.95 ± 0.22	2 ± 0

Twenty flowers were examined in each line. Numbers indicate the average organ number ± SD. [a] No significant difference each other (Student's *t*-test, $p > 0.05$).

2.2. Surgical and Genetic Generation of Space in a Floral Bud Restored Petal Elongation

We hypothesized that petal folding of *fop2* is caused by the narrow space in the floral buds, as shown in the *fop1* mutant [27]. To elucidate this, we removed the sepals from the *fop2-1* floral buds prior to flower opening (Figure 2a, around floral stage 10 to 12 as defined in [32]). After three days, we found that all examined flowers restored the straight elongation of petals (Figures 2b and S2). We also examined the *fop1 fop2* double mutant in the same way, with similar results (Figure S2). Next, we investigated the *pressed flower-1 (prs-1) fop2-1* double mutant. *prs-1* is defective in the development of lateral domains of organs, and raises flowers with narrower sepals, resulting in "open" floral buds [33]. The floral buds of *prs-1 fop2-1* double mutants generate space in these open buds, compared to the closed buds in *fop2-1* (Figure 2d and inset), and almost all petals elongated straight in the double mutant (Figure 2c and Table 1). These results indicate that surgical or genetic generation of space in a floral bud restores the straight elongation of *fop2* petals. Taken together, the data suggest that the *fop2* lacks lubricant-like materials that allow petals to elongate smoothly in the narrow space in a floral bud.

We also investigated flowers of the *ap3-5 fop2-1* double mutant, whose petals were transformed to sepals, and found that the second whorl organ elongated normally (Figure 2e). This suggests the possibility that *FOP2* function is restricted to the organs with petal identity, and that sepals may have other ABCG transporters involved in cuticle wax deposition.

Figure 2. (**a**) *fop2-1* flower whose sepal is removed at stage 11. Petals are indicated by white dotted line. p, petal; st, stamen; g, gynoecium; (**b**) Same flower as (**a**) after three days. Note the straight elongation of petals; (**c**) *prs-1 fop2-1* flower; (**d**) Inflorescence of *prs-1 fop2-1*. Inset is inflorescence of *fop2-1*; (**e**) *ap3-5 fop2-1* flower. Bars: 1 mm.

2.3. Mapping and Characterization of the FOP2 Gene

To characterize the molecular function of *FOP2*, we mapped the *FOP2* gene on chromosome 1 and identified a mutation in *At1g51460* (Figure 3). The *fop2-1* mutation is a nucleotide substitution of cytosine to thymine in the third exon, causing the change of a glutamine at the position of 194 to a premature stop in the amino acid residues (Figures 3b and S3). We analyzed the SALK_046735 line, in which the T-DNA is inserted in the 4th exon of *At1g51460*. Sequencing of border regions revealed that the T-DNA caused a 15 bp deletion in the 4th exon (Figures 3b and S3). This line showed the same petal-folding phenotype as shown above (Figure 1c).

FOP2 encodes a member of the ATP-binding cassette (ABC) transporter family protein, ABCG13, which has been already reported by Panikashvili *et al.* [34]. They showed the folded-petal phenotype in SALK_046735 (*abcg13*) and RNAi-based knockdown lines, and several species of cutin monomers were reduced in these flowers. We examined the *FOP2* expression pattern by generating *FOP2 genomic:GUS* translational fusion lines, and found that *FOP2* was expressed in petals and ovule integuments, as well as the L1 layer of peduncles (Figure S4). The expression pattern is largely consistent with *promoter:GUS* transcriptional fusion results from Panikashvili *et al.* [34].

Figure 3. (**a**) Diagram of *FOP2* mapping. DNA markers and number of recombinants are indicated; (**b**) Genomic structure of the *FOP2* gene. Point mutation of *fop2-1* and T-DNA insertion, which causes the 15 bp deletion, in *fop2-2*/SALK_046735 are indicated. LB, left border of T-DNA. See Figure S2 for *FOP2* genomic sequence.

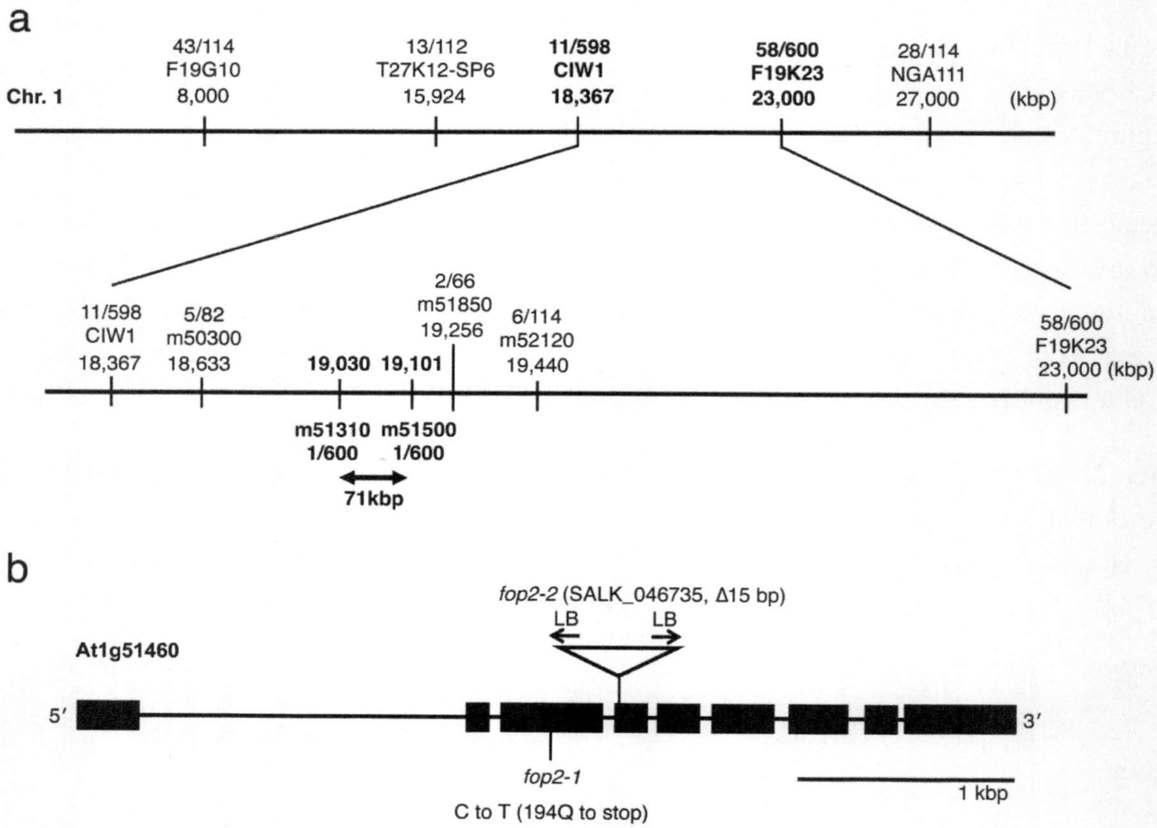

2.4. Molecular Function of FOP2 in Petal Elongation Process

FOP2/ABCG13 belongs to the WHITE-BROWN COMPLEX (WBC) subfamily, members of which are half-transporter proteins [35,36]. ABCG11/AtWBC11/DSO and ABCG12/CER5 in the same subfamily function as homodimers or heterodimers on plasma membranes *in planta* [37], thus, FOP2/ABCG13 would interact with itself or the other WBC members on plasma membrane of petal epidermal cells. Our results from petal epidermis characterization, petal surface repellency tests and the phenotype of the double mutant suggest that production and deposition of the lubricant-like substrates, mediated by FOP1/WSD11 and FOP2/ABCG13, respectively, are a critical process for smooth elongation of petals in the narrow space in floral buds. Since the known half-size ABCG transporters are involved in secretion of both cutin and wax [34,38–42], it is possible that FOP2 transports the FOP1 products to the petal surface. Interestingly FOP1/WSD11 is located to the plasma membrane, suggesting that FOP2 and FOP1 may co-localize on the plasma membrane of petal epidermal cells. Identification of substrates and products of FOP1 and FOP2 will unveil the molecular mechanism of petal elongation in detail.

3. Experimental Section

3.1. Plant Lines and Growth Conditions

The Wassilewskija (WS), Columbia (Col), and Landsberg *erecta* (L*er*) ecotypes of *Arabidopsis thaliana* were used as wild-types. The *fop2-1* mutant was isolated from an M2 population of ethyl methanesulfonate-mutagenized *long hypocotyl 5* (*hy5*) in which the *HY5* genomic fragment was transformed [43]. *fop2-2*/SALK_046735, a T-DNA insertion mutant, was obtained from the SIGnAL Website [44] and the Arabidopsis Biological Resource Center [45]. Plants were grown on MS media or vermiculite in small pots under long-day conditions (16 h light and 8 h dark) with white light at 22 °C to 24 °C.

3.2. Histology and Microscopes

Images were captured using an S8AP0 binocular microscope or DM2500 light microscope equipped with an EC3 digital camera system (Leica, Wetzlar, Germany). For scanning microscopy, samples were directly observed with JSM-5800 (JEOL). For toluidine blue staining, flowers were immersed in TB solution (0.05% toluidine blue (w/v, Waldeck, Münster, Germany) and 0.01% Tween 20 (v/v)) for 5 min and washed in water. At least fifteen (WS and *fop2-1*) and five (*fop1-1*) flowers were examined, respectively.

3.3. Cloning and Expression Analysis of the FOP2 Gene

F2 plants generated by crossing the *fop2-1* with Col and L*er* were used for mapping. DNA marker information was obtained from The Arabidopsis Information Resource [46] and generated using polymorphisms between ecotypes. For genomic-GUS translational fusion, we fused the 5.7 kb genomic fragments including FOP2 5'-promoter and coding regions (without stop codon), β-glucuronidase (GUS), and 1.7 kb 3' region, and cloned into the pGWB1-NB1 binary vector, and transformed to plants by a vacuum infiltration procedure with the *Agrobacterium* strain ASE1.

4. Conclusions

Our results suggest that a half-size ABC transporter, FOP2/ABCG13, is involved in the transport of the possible lubricant materials on the developing petal surface. It is possible that FOP2/ABCG13 transports the products of FOP1/WSD11 on the plasma membrane of petal epidermis, resulting in the smooth elongation of petals through the narrow spaces in floral buds. Main future works would be (1) examining the molecular relation between FOP2 and FOP1, including spatiotemporal expression/localization analysis *in planta*, and enzymatic activity assay; and (2) identifying the substances that function as lubricants during petal elongation. Our results propose the possibility to control the petal morphogenesis by modification of enzymes or transporters, or by exogenous application of these products/substances.

Acknowledgments

We thank the Model Plant Research Facility, the NIBB Bioresource Center, the Salk Institute, and the Arabidopsis Biological Resource Center for providing materials, Rebecca Horn for critical reading of the manuscript, and lab members for helpful discussion. This work was supported by grants for young scientists from Kyoto Prefectural Public University Corporation to ST, a Grant-in-Aid for Scientific Research on Priority Areas (No. 19060004), and a Grant-in-Aid for Creative Scientific Research from the Japan Society for the Promotion of Science (No. 19GS0315) from the Ministry for Education, Culture, Sports, Science, and Technology (MEXT) of Japan to KO.

Author Contributions

S.T. and A.I. performed experiments, S.T. wrote papers, and K.T. and K.O. supervised this work.

Conflicts of Interest

The authors declare no conflict of interest.

References

1. Bowman, J.L.; Smyth, D.R.; Meyerowitz, E.M. Genes directing flower development in *Arabidopsis*. *Plant Cell* **1989**, *1*, 37–52.
2. Krizek, B.A.; Fletcher, J.C. Molecular mechanisms of flower development: An armchair guide. *Nat. Rev. Genet.* **2005**, *6*, 688–698.
3. Weigel, D.; Meyerowitz, E.M. The ABCs of floral homeotic genes. *Cell* **1994**, *78*, 203–209.
4. Theissen, G.; Saedler, H. Plant biology. Floral quartets. *Nature* **2001**, *409*, 469–471.
5. Byzova, M.V.; Franken, J.; Aarts, M.G.; de Almeida-Engler, J.; Engler, G.; Mariani, C.; van Lookeren Campagne, M.M.; Angenent, G.C. *Arabidopsis* STERILE APETALA, a multifunctional gene regulating inflorescence, flower, and ovule development. *Genes Dev.* **1999**, *13*, 1002–1014.
6. Conner, J.; Liu, Z. LEUNIG, a putative transcriptional corepressor that regulates AGAMOUS expression during flower development. *Proc. Natl. Acad. Sci. USA* **2000**, *97*, 12902–12907.
7. Franks, R.G.; Wang, C.; Levin, J.Z.; Liu, Z. SEUSS, a member of a novel family of plant regulatory proteins, represses floral homeotic gene expression with LEUNIG. *Development* **2002**, *129*, 253–263.
8. Grigorova, B.; Mara, C.; Hollender, C.; Sijacic, P.; Chen, X.; Liu, Z. LEUNIG and SEUSS co-repressors regulate miR172 expression in *Arabidopsis* flowers. *Development* **2011**, *138*, 2451–2456.
9. Krizek, B.A.; Lewis, M.W.; Fletcher, J.C. RABBIT EARS is a second-whorl repressor of AGAMOUS that maintains spatial boundaries in *Arabidopsis* flowers. *Plant J.* **2006**, *45*, 369–383.
10. Krizek, B.A.; Prost, V.; Macias, A. AINTEGUMENTA promotes petal identity and acts as a negative regulator of AGAMOUS. *Plant Cell* **2000**, *12*, 1357–1366.
11. Liu, Z.; Meyerowitz, E.M. LEUNIG regulates AGAMOUS expression in *Arabidopsis* flowers. *Development* **1995**, *121*, 975–991.

12. Sridhar, V.V.; Surendrarao, A.; Gonzalez, D.; Conlan, R.S.; Liu, Z. Transcriptional repression of target genes by LEUNIG and SEUSS, two interacting regulatory proteins for *Arabidopsis* flower development. *Proc. Natl. Acad. Sci. USA* **2004**, *101*, 11494–11499.
13. Chen, X. A microRNA as a translational repressor of APETALA2 in *Arabidopsis* flower development. *Science* **2004**, *303*, 2022–2025.
14. Takeda, S.; Matsumoto, N.; Okada, K. RABBIT EARS, encoding a SUPERMAN-like zinc finger protein, regulates petal development in *Arabidopsis thaliana*. *Development* **2004**, *131*, 425–434.
15. Griffith, M.E.; da Silva Conceicao, A.; Smyth, D.R. PETAL LOSS gene regulates initiation and orientation of second whorl organs in the *Arabidopsis* flower. *Development* **1999**, *126*, 5635–5644.
16. Lee, I.; Wolfe, D.S.; Nilsson, O.; Weigel, D. A LEAFY co-regulator encoded by UNUSUAL FLORAL ORGANS. *Curr. Biol.* **1997**, *7*, 95–104.
17. Samach, A.; Klenz, J.E.; Kohalmi, S.E.; Risseeuw, E.; Haughn, G.W.; Crosby, W.L. The UNUSUAL FLORAL ORGANS gene of *Arabidopsis thaliana* is an F-box protein required for normal patterning and growth in the floral meristem. *Plant J.* **1999**, *20*, 433–445.
18. Xing, S.; Rosso, M.G.; Zachgo, S. ROXY1, a member of the plant glutaredoxin family, is required for petal development in *Arabidopsis thaliana*. *Development* **2005**, *132*, 1555–1565.
19. Schiessl, K.; Muino, J.M.; Sablowski, R. *Arabidopsis* JAGGED links floral organ patterning to tissue growth by repressing Kip-related cell cycle inhibitors. *Proc. Natl. Acad. Sci. USA* **2014**, *111*, 2830–2835.
20. Sauret-Gueto, S.; Schiessl, K.; Bangham, A.; Sablowski, R.; Coen, E. JAGGED controls *Arabidopsis* petal growth and shape by interacting with a divergent polarity field. *PLoS Biol.* **2013**, *11*, e1001550.
21. Dinneny, J.R.; Yadegari, R.; Fischer, R.L.; Yanofsky, M.F.; Weigel, D. The role of JAGGED in shaping lateral organs. *Development* **2004**, *131*, 1101–1110.
22. Ohno, C.K.; Reddy, G.V.; Heisler, M.G.; Meyerowitz, E.M. The *Arabidopsis* JAGGED gene encodes a zinc finger protein that promotes leaf tissue development. *Development* **2004**, *131*, 1111–1122.
23. Szecsi, J.; Joly, C.; Bordji, K.; Varaud, E.; Cock, J.M.; Dumas, C.; Bendahmane, M. BIGPETALp, a bHLH transcription factor is involved in the control of *Arabidopsis* petal size. *EMBO J.* **2006**, *25*, 3912–3920.
24. Varaud, E.; Brioudes, F.; Szecsi, J.; Leroux, J.; Brown, S.; Perrot-Rechenmann, C.; Bendahmane, M. AUXIN RESPONSE FACTOR8 regulates *Arabidopsis* petal growth by interacting with the bHLH transcription factor BIGPETALp. *Plant Cell* **2011**, *23*, 973–983.
25. Hase, Y.; Fujioka, S.; Yoshida, S.; Sun, G.; Umeda, M.; Tanaka, A. Ectopic endoreduplication caused by sterol alteration results in serrated petals in *Arabidopsis*. *J. Exp. Bot.* **2005**, *56*, 1263–1268.
26. Hase, Y.; Tanaka, A.; Baba, T.; Watanabe, H. FRL1 is required for petal and sepal development in *Arabidopsis*. *Plant J.* **2000**, *24*, 21–32.
27. Takeda, S.; Iwasaki, A.; Matsumoto, N.; Uemura, T.; Tatematsu, K.; Okada, K. Physical interaction of floral organs controls petal morphogenesis in *Arabidopsis*. *Plant Physiol.* **2013**, *161*, 1242–1250.

28. Li-Beisson, Y.; Pollard, M.; Sauveplane, V.; Pinot, F.; Ohlrogge, J.; Beisson, F. Nanoridges that characterize the surface morphology of flowers require the synthesis of cutin polyester. *Proc. Natl. Acad. Sci. USA* **2009**, *106*, 22008–22013.
29. Panikashvili, D.; Shi, J.X.; Schreiber, L.; Aharoni, A. The *Arabidopsis* DCR encoding a soluble BAHD acyltransferase is required for cutin polyester formation and seed hydration properties. *Plant Physiol.* **2009**, *151*, 1773–1789.
30. Oshima, Y.; Shikata, M.; Koyama, T.; Ohtsubo, N.; Mitsuda, N.; Ohme-Takagi, M. MIXTA-like transcription factors and WAX INDUCER1/SHINE1 coordinately regulate cuticle development in *Arabidopsis* and *Torenia fournieri*. *Plant Cell* **2013**, *25*, 1609–1624.
31. Tanaka, T.; Tanaka, H.; Machida, C.; Watanabe, M.; Machida, Y. A new method for rapid visualization of defects in leaf cuticle reveals five intrinsic patterns of surface defects in *Arabidopsis*. *Plant J.* **2004**, *37*, 139–146.
32. Smyth, D.R.; Bowman, J.L.; Meyerowitz, E.M. Early flower development in *Arabidopsis*. *Plant Cell* **1990**, *2*, 755–767.
33. Matsumoto, N.; Okada, K. A homeobox gene, PRESSED FLOWER, regulates lateral axis-dependent development of *Arabidopsis* flowers. *Genes Dev.* **2001**, *15*, 3355–3364.
34. Panikashvili, D.; Shi, J.X.; Schreiber, L.; Aharoni, A. The *Arabidopsis* ABCG13 transporter is required for flower cuticle secretion and patterning of the petal epidermis. *New Phytol.* **2011**, *190*, 113–124.
35. Kang, J.; Park, J.; Choi, H.; Burla, B.; Kretzschmar, T.; Lee, Y.; Martinoia, E. Plant ABC Transporters. *Arabidopsis Book* **2011**, *9*, e0153.
36. Samuels, L.; Kunst, L.; Jetter, R. Sealing plant surfaces: Cuticular wax formation by epidermal cells. *Annu. Rev. Plant Biol.* **2008**, *59*, 683–707.
37. McFarlane, H.E.; Shin, J.J.; Bird, D.A.; Samuels, A.L. *Arabidopsis* ABCG transporters, which are required for export of diverse cuticular lipids, dimerize in different combinations. *Plant Cell* **2010**, *22*, 3066–3075.
38. Bird, D.; Beisson, F.; Brigham, A.; Shin, J.; Greer, S.; Jetter, R.; Kunst, L.; Wu, X.; Yephremov, A.; Samuels, L. Characterization of *Arabidopsis* ABCG11/WBC11, an ATP binding cassette (ABC) transporter that is required for cuticular lipid secretion. *Plant J.* **2007**, *52*, 485–498.
39. Panikashvili, D.; Savaldi-Goldstein, S.; Mandel, T.; Yifhar, T.; Franke, R.B.; Hofer, R.; Schreiber, L.; Chory, J.; Aharoni, A. The *Arabidopsis* DESPERADO/AtWBC11 transporter is required for cutin and wax secretion. *Plant Physiol.* **2007**, *145*, 1345–1360.
40. Pighin, J.A.; Zheng, H.; Balakshin, L.J.; Goodman, I.P.; Western, T.L.; Jetter, R.; Kunst, L.; Samuels, A.L. Plant cuticular lipid export requires an ABC transporter. *Science* **2004**, *306*, 702–704.
41. Luo, B.; Xue, X.Y.; Hu, W.L.; Wang, L.J.; Chen, X.Y. An ABC transporter gene of *Arabidopsis thaliana*, AtWBC11, is involved in cuticle development and prevention of organ fusion. *Plant Cell Physiol.* **2007**, *48*, 1790–1802.
42. Ukitsu, H.; Kuromori, T.; Toyooka, K.; Goto, Y.; Matsuoka, K.; Sakuradani, E.; Shimizu, S.; Kamiya, A.; Imura, Y.; Yuguchi, M.; *et al.* Cytological and biochemical analysis of COF1, an *Arabidopsis* mutant of an ABC transporter gene. *Plant Cell Physiol.* **2007**, *48*, 1524–1533.

43. Oyama, T.; Shimura, Y.; Okada, K. The *Arabidopsis* HY5 gene encodes a bZIP protein that regulates stimulus-induced development of root and hypocotyl. *Genes Dev.* **1997**, *11*, 2983–2995.
44. The SIGnAL Website. Available online: http://signal.salk.edu (accessed on 19 July 2014).
45. The Arabidopsis Biological Resource Center. Available online: http://abrc.osu.edu (accessed on 19 July 2014).
46. The Arabidopsis Information Resource. Available online: https//www.arabidopsis.org (accessed on 19 July 2014).

3

Vacuolar Na^+/H^+ NHX-Type Antiporters Are Required for Cellular K^+ Homeostasis, Microtubule Organization and Directional Root Growth

Tyler McCubbin, Elias Bassil, Shiqi Zhang and Eduardo Blumwald *

Department of Plant Sciences, University of California, Davis, CA 95616, USA;
E-Mails: tjmccubbin@ucdavis.edu (T.M.); esbassil@ucdavis.edu (E.B.); shqzhang@ucdavis.edu (S.Z.)

* Author to whom correspondence should be addressed; E-Mail: eblumwald@ucdavis.edu

---

Abstract: Na^+/H^+ antiporters (NHXs) are integral membrane transporters that catalyze the electroneutral exchange of K^+ or Na^+ for H^+ and are implicated in cell expansion, development, pH and ion homeostasis and salt tolerance. *Arabidopsis* contains four vacuolar NHX isoforms (NHX1–NHX4), but only the functional roles for NHX1 and NHX2 have been assessed thus far. Colocalization studies indicated that NHX3 and NHX4 colocalize to the tonoplast. To investigate the role of all vacuolar NHX isoforms, a quadruple knockout *nhx1nhx2nhx3nhx4*, lacking all vacuolar NHXs, was generated. Seedlings of *nhx1nhx2nhx3nhx4* displayed significantly reduced growth, with markedly shorter hypocotyls. Under high K^+, but not Na^+, pronounced root skewing occurred in *nhx1nhx2nhx3nhx4*, suggesting that the organization of the cytoskeleton might be perturbed. Whole mount immunolabeling of cortical microtubules indicated that high K^+ caused significant microtubule reorganization in *nhx1nhx2nhx3nhx4* root cells of the elongation zone. Using microtubule stabilizing (Taxol) and destabilizing (propyzamide) drugs, we found that the effect of K^+ on *nhx1nhx2nhx3nhx4* root growth was antagonistic to that of Taxol, whereas elevated K^+ exacerbated the endogenous effect of propyzamide on root skewing. Collectively, our results suggest that altered K^+ homeostasis leads to an increase in the dynamics of cortical microtubule reorganization in *nhx1nhx2nhx3nhx4* root epidermal cells of the elongation zone.

Keywords: vacuole; potassium; salt; homeostasis; pH; root; growth; microtubule; cytoskeleton; curling; skewing; NHX

---

1. Introduction

Ion homeostasis and pH regulation are essential cellular processes necessary for plant growth, development and responses to environmental stress. Na^+/H^+ antiporters (NHXs) are integral membrane proteins localized to the plasma membrane and endomembrane compartments that facilitate the electroneutral exchange of Na^+ or K^+ for H^+, thereby maintaining both luminal pH, as well as cation homeostasis. The *Arabidopsis* genome encodes eight NHX homologs that have been grouped based on sequence similarity and function into three distinct classes: plasma membrane (NHX7/SOS1 and NHX8), endosomal/vesicular (NHX5, NHX6), as well as four that are vacuolar (NHX1, NHX2, NHX3, NHX4) [1,2]. We previously reported the vacuolar localization and function of two of the vacuolar NHX isoforms, NHX1 and NHX2 [3,4], that bear strong sequence homology to the putatively vacuolar NHX3 and NHX4. Other work has shown colocalization of AtNHX3 with vacuolar proteins δ-TIP and γ-TIP, as well as SYP22 in tobacco protoplasts, suggesting a vacuolar and PVC localization of AtNHX3 [5]. Similarly, constitutive expression of AtNHX3 in sugar beet (*Beta vulgaris* L.) has been shown to confer increased salt tolerance, but also has led to high K^+ accumulation in roots when grown under high salt [6]. In addition, Liu *et al.* found that *Arabidopsis nhx3* null mutants were hypersensitive to K^+ deficiency, further implicating NHX3 in K^+ homeostasis [5]. AtNHX4 localization was suggested to be similar to NHX1 at the tonoplast [7], but additional experiments are required to demonstrate localization convincingly. Given that little is known about the roles of NHX3 and NHX4 in vacuolar ion and pH homeostasis, further investigations are warranted.

Potassium transport into the vacuole has been shown to be a key regulator of cell expansion [8]. Both NHX1 and NHX2 are necessary for the accumulation of vacuolar K^+ and vacuolar alkalization, because the double knockout *nhx1nhx2* accumulated 70% less vacuolar K^+ and had significantly more acidic vacuoles than comparable wild-type (WT) plants [3]. *nhx1nhx2* plants also exhibited reduced overall growth, had smaller cells in all tissues examined, as well as aberrant flower development, such as incomplete filament elongation and anther dehiscence, which resulted in significantly reduced seed set. Others studies also showed that *nhx1nh2* double knockouts have high cytosolic K^+ [9], altered vacuole morphology, compromised turgor, severely impaired stomatal conductance and less daytime transpiration due to the lack of K^+ accumulation in guard cell vacuoles [10]. One intriguing phenotype of *nhx1nhx2* was their high sensitivity to additional K^+ in the growth media, with a severe reduction in overall growth [3]. This phenotype was not evident when *nhx1nhx2* plants were grown in the presence of equimolar concentrations of Na^+ ions. The presence of high K^+ also led to profound curling of seedling roots, suggesting that altered K^+ homeostasis, due to the lack of NHX1 and NHX2 antiporters, altered directional root growth [3].

Root growth patterning is controlled by microtubule and cytoskeleton reorganization [11]. Cellular Na^+ homeostasis can affect cytoskeleton-mediated root growth, and the effects of salinity on directional root growth have already been reported [12,13]. Little is known about the role(s) of cellular K^+

homeostasis in microtubule organization. Here, we show that vacuolar NHX1 to NHX4 reside at the tonoplast. Using the quadruple knockout *nhx1nhx2nhx3nhx4*, lacking all four vacuolar NHXs, we show that vacuolar NHXs impart essential functions in K^+ homeostasis that affect directional root growth and cytoskeletal dynamics. The quadruple knockout exhibits more pronounced phenotypes compared to the *nhx1nhx2* double knockout, including increased sensitivity to K^+, suggesting that NHX3 and NHX4 have additional roles in cell expansion and K^+ homeostasis.

2. Results and Discussion

We sought to assess the contribution of all vacuolar NHXs to overall plant growth and development by generating a quadruple knockout lacking all vacuolar NHX antiporters, NHX1, NHX2, NHX3 and NHX4. Using available T-DNA insertion lines (Figure A1), we generated the quadruple knockout *nhx1nhx2nhx3nhx4* through successive crosses of single, double and triple knockouts, as described in the Methods. The level of expression of all four vacuolar NHX transcripts in the *nhx1nhx2nhx3nhx4* knockout indicated that no full-length transcripts exist (Figure A1).

2.1. NHX3 and NHX4 Are Colocalized at the Tonoplast

To determine the subcellular localization of NHX3 and NHX4, we generated constitutively expressed translational fusion proteins of both NHX3 and NHX4 with CFP and RFP, respectively. Transient co-expression of NHX3-CFP and NHX4-RFP in *Arabidopsis* cotyledons displayed overlapping expression at the periphery of the cell, as well as at transvacuolar strands (Figure 1A,C). Because NHX1 and NHX2 were shown previously to localize at the tonoplast [3], we transiently expressed NHX4-CFP in plants stably expressing NHX2-YFP and observed an overlapping expression pattern between NHX2 and NHX4 that was characteristic of a vacuolar signal, including trans-vacuolar strands and vacuolar bulbs (arrowheads in Figure 1D–F). We then expressed NHX3-RFP and NHX4-CFP in aquaporin γ-TIP [14] and the R-SNARE VAMP711 [15] expressing lines, which were previously shown to colocalize with NHX1 [3]. In cotyledon epidermal cells, an overlapping expression between NHX3-RFP and γ-TIP-GFP, including in vacuolar bulbs that are characteristic of γ-TIP, was seen (Figure 1G–I). Overlapping expression was also seen with NHX4-CFP and VAMP711-RFP (Figure 1J–L). Taken together, these results indicated that both NHX3 and NHX4 localize to the tonoplast, as suggested by other heterologous expression studies [5,7] and sequence similarity to NHX1 and NHX2 [1].

2.2. Growth Phenotypes of nhx1nhx2nhx3nhx4 and Responses to Na^+, K^+

We used the etiolation response of dark-grown seedlings to examine the contribution of *NHX1*, *NHX2*, *NHX3* and *NHX4* to cell expansion and seedling growth. Previously, we reported that etiolated *nhx1nhx2* double knockout seedlings had profound hypocotyl and root elongation phenotypes [3]. Etiolated *nhx1nhx2nhx3nhx4* seedlings displayed significantly reduced growth of both roots and hypocotyls (Figure 2). In control media, hypocotyl lengths of *nhx1nhx2nhx3nhx4* were less than 50% the length of WT after 11 days of etiolated growth (Figure 2M). The hypocotyl elongation response of *nhx1nhx2nhx3nhx4* was highly responsive to both added Na^+ and K^+. For example, hypocotyls of *nhx1nhx2nhx3nhx4* grown in 30 mM NaCl were 31% longer than those of seedlings grown on control media (1 mM NaCl, 1 mM KCl), but hypocotyls of *nhx1nhx2nhx3nhx4* seedlings grown in 30 mM KCl were 49% shorter (Figure 2D,H,K,L) than those grown in control media. By contrast, little change in hypocotyl length was observed in the WT seedlings grown on either 30 mM NaCl or KCl (Figure 2B,F,J,L). Root growth was also markedly different between *nhx1nhx2nhx3nhx4* and WT vertically-grown seedlings, but only in response to either K^+ or Na^+. Similar to the K^+ and Na^+-dependent growth response of hypocotyls, *nhx1nhx2nhx3nhx4* roots were also significantly shorter (26%) under K^+ and longer (13%) under Na^+ as compared to *nhx1nhx2nhx3nhx4* grown in control media. In addition, the *nhx1nhx2nhx3nhx4* seedlings began to display prominent left-handed skewing (as viewed through the agar, following the convention assigned by Rutherford and Masson [16]) after four days of growth on media supplemented with 30 mM K^+ (Figure A2). Seedlings grown for another seven days on the same media displayed complete curling of roots (Figure 2H). By contrast, WT root growth remained unchanged under 30 mM K^+ even after 11 days.

Figure 1. Intracellular localization of *NHX3* and *NHX4*. Transient co-expression, in *Arabidopsis* cotyledon epidermal cells, of *35S::NHX3-CFP* (**A**) with *35S::NHX4-RFP* (**B**) colocalized as shown in the merge fluorescent panel (**C**). Transient expression of *35S::NHX4-CFP* (**D**) in stable *35S::NHX2-YFP* background (**E**) also colocalized as shown in the merged image (**F**). Transient expression of *35S::NHX3-RFP* (**G**) in stable expressing *γTIP::γTIP-GFP* (**H**) colocalized as shown in the merged panel (**I**). Transient expression of *35S::NHX4-CFP* (**J**) in the stable expressing *Ub10::VAMP711-RFP* (**K**) colocalized as shown in the merged image (**L**). Arrowheads in (C,F,I,L) point to tonoplast invaginations and transvacuolar cytoplasmic strands.

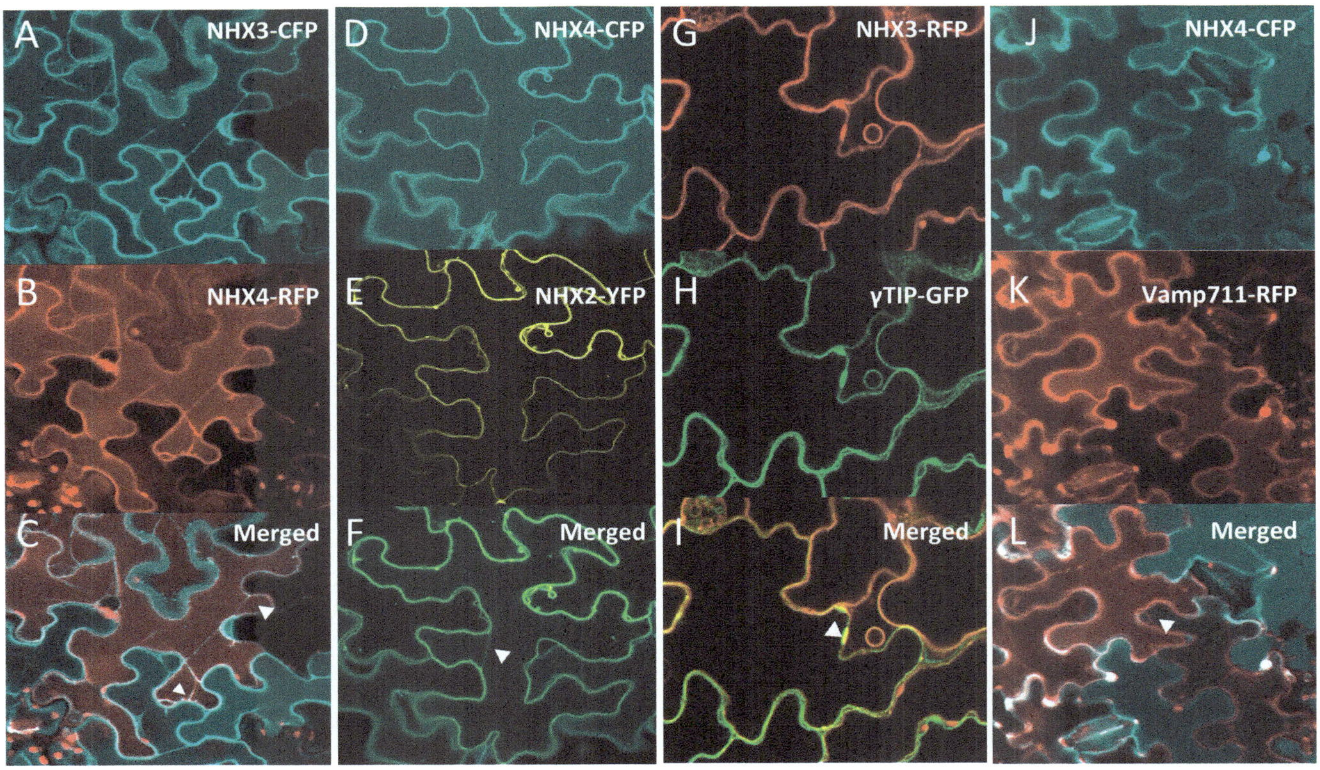

Figure 2. Etiolation response of the knockout *nhx1nhx2nhx3nhx4*. Etiolated wild-type (**A**,**C**,**E**) and *nhx1nhx2nhx3nhx4* (**B**,**D**,**F**) seedlings grown for 11 days in the dark on either 30 mM K^+ (**C**,**D**) or 30 mM Na^+ media (**E**,**F**). No significant difference in hypocotyl or root growth was noted in the wild-type; however, in *nhx1nhx2nhx3nhx4*, root and hypocotyl growth was significantly inhibited in seedlings grown in 30 mM K^+ and increased in those grown in 30 mM Na^+, as compared to control media. Quantification of hypocotyl (**G**) and root (**H**) growth normalized relative to growth on control media (1 mM Na^+ and 1 mM K^+). *nhx1nhx2nhx3nhx4* displayed left-handed root skewing when grown on media containing 30 mM K^+ (**D**), but not 1 mM K^+ (**B**). Values represent the mean ± SE ($n = 29$). * $p \leq 0.05$, *** $p \leq 0.001$; p-values reflect significant differences between control and either K^+ or Na^+ treatments within each genotype. The yellow bar is 0.5 cm.

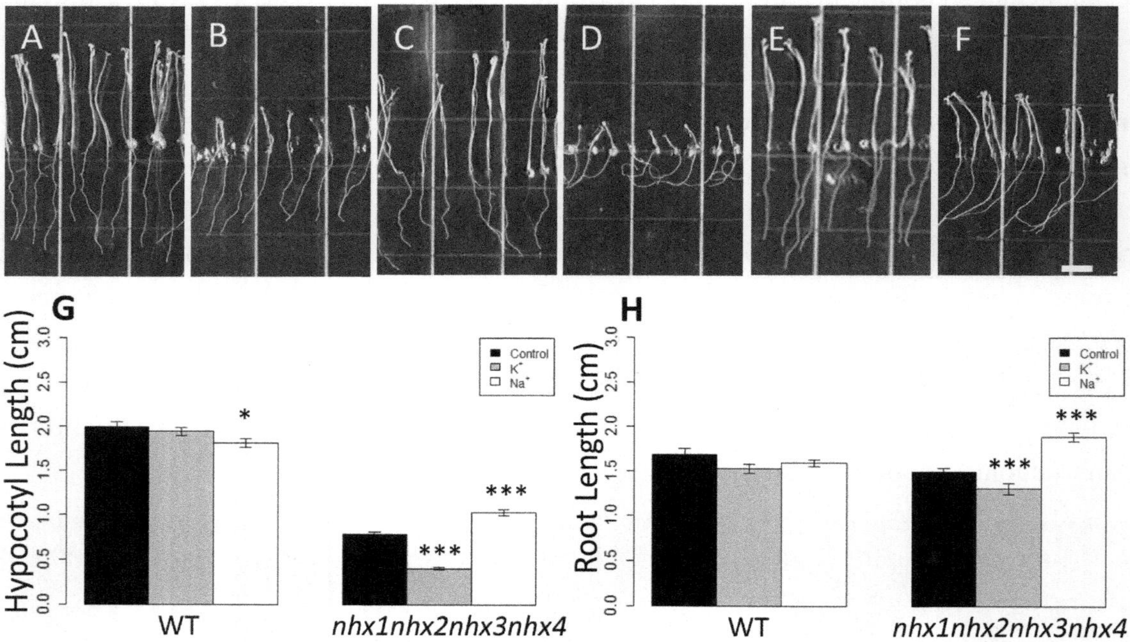

Previously, we reported that etiolated *nhx1nhx2* plants also displayed reduced hypocotyl growth, that was 70% the length of WT hypocotyls on media containing 1 mM K^+ and 50% of WT on 30 mM K^+ [3], while in etiolated *nhx1nhx2nhx3nhx4* seedlings, the average hypocotyl length was 40% and 20% of WT on 1 and 30 mM K^+, respectively. A comparison of root growth responses between *nhx1nhx2nhx3nhx4* and *nhx1nhx2* to high K^+ [3] indicates that root curling is also more pronounced and root growth more sensitive to high K^+, similar to that observed in hypocotyls. The dramatic differences in root and hypocotyl elongation between *nhx1nhx2* and *nhx1nhx2nhx3nhx4* under the higher K^+ suggest that NHX3 and NHX4 have additional roles in K^+ homeostasis and cell expansion.

To determine if the root curling observed in *nhx1nhx2nhx3nhx4* roots was specific to K^+, we compared the root growth of *nhx1nhx2nhx3nhx4* on media supplemented with KCl, KNO_3, NaCl or mannitol. Mannitol was included in order to determine whether the root curling response of *nhx1nhx2nhx3nhx4* under high K^+ was due to an osmotic effect; thus, 60 mM mannitol (60 mOsm, equivalent to the osmolarity of the added salt) was used. We found that *nhx1nhx2nhx3nhx4* root growth on mannitol was not different from the control (Figure A3J,B), and therefore not likely due to an osmotic effect. As shown in Figure A3, both 30 mM KCl and 30 mM KNO_3 supplemented media resulted in similar left-handed root skewing and curling of the root tip that was not seen when *nhx1nhx2nhx3nhx4* was grown on mannitol, suggesting that the response of *nhx1nhx2nhx3nhx4* was specific to K^+. Collectively, experiments comparing *nhx1nhx2nhx3nhx4* seedling growth (*i.e.*, Figures 2 and A3) indicate that the knockouts' root curling response was caused by aberrant K^+ homeostasis.

Previous work has provided a strong link between ion stress and directional root growth. For example, *sos1*(*nhx7*) null mutants displayed a loss of normal gravitropic root growth when grown vertically on media supplemented with high (50–150 mM) Na^+, with young roots exhibiting skewing and eventual upward directed growth [17]. Moreover, roots of both WT and *sos1*(*nhx7*) seedlings grown on high Na^+ showed delayed gravitropic responses, thereby suggesting that aberrant intracellular ion homeostasis can specifically affect directional root growth. The *sos1* skewing phenotype is remarkably similar to *nhx1nhx2nhx3nhx4* grown on K^+ supplemented media, and although effects of high cytosolic Na^+ have been suggested, the effects of K^+ on root skewing remain unclear. Furthermore, *sos1*(*nhx7*) mutants were more sensitive to microtubule reorganization, with whole mount immunostaining experiments in elongating root epidermal cells demonstrating that *sos1* (*nhx7*) in the presence of 50 mM Na^+ had helically-orientated microtubules, while WT plants maintained the normal transverse array [12]. Although microtubule remodeling has been shown to be induced by high external Na^+, reports of K^+-induced microtubule reorganization are sparse [12,13,18]. In KCl (350 mM)-treated *Zea mays*, a reorganization of the cortical microtubule array from the transverse to oblique orientation in root cells of the maturation zone occurred within 3 min after application. By contrast, treatment with sorbitol did not result in the reorganization of microtubules [19]. These results suggested that cortical microtubule organization in roots was affected by high concentrations of Na^+ and K^+. Given the phenotypes of *nhx1nhx2nhx3nhx4*, the knockout represents a useful tool to study the role of K^+ in cytoskeletal organization. The positive effect of Na^+ and the negative effect on K^+ on *nhx1nhx2nhx3nhx4* growth is well supported by similar results obtained in *nhx1nhx2* in which Na^+ was proposed to substitute for the lack of K^+ accumulation in vacuoles, providing the osmoticum needed to generate the turgor for cell expansion [3,10]. The sensitivity of both *nhx1nhx2* and *nhx1nhx2nhx3nhx4* to added K^+, as well as the aberrantly high cytosolic K^+ concentrations of *nhx1nhx2*, together with previously reported effects of Na^+ on cytoskeleton disassembly and reorganization, suggests that aberrant K^+ homeostasis in the *nhx1nhx2nhx3nhx4* might affect the organization of the cytoskeleton.

2.3. nhx1nhx2nhx3nhx4 Lacks Accumulation of K^+ in the Vacuole

Previously, we quantified the concentration of K^+ in vacuoles of the knockout *nhx1nhx2* using the K^+ binding dye, PBFI [3]. Using the same approach, we quantified the vacuolar [K^+] of *nhx1nhx2nhx3nhx4* in root tip cells of the elongation zone, where K^+-induced root curling was observed (Figure 2D). In *nhx1nhx2nhx3nhx4*, vacuolar K^+ was ~19 mM (±7 mM) and significantly lower than WT, which was 77 mM (±13 mM). These values are similar to those obtained in *nhx1nhx2*, suggesting that the lack of *NHX3* and *NHX4* in addition to *NHX1* and *NHX2* did not significantly alter root tip cell vacuolar K^+. In the same *nhx1nhx2* knockout, the cytosolic K^+ was also reported to be aberrantly high compared to WT [9]. Therefore, given that the electrochemical potential of K^+ favors uptake into the cell [1,3], it is possible that in *nhx1nhx2nhx3nhx4*, the lack of K^+ accumulation into the vacuole might lead to an aberrantly high K^+ accumulation in the cytosol.

2.4. Cells of the Root Transition Zone Lack Expansion in nhx1nhx2nhx3nhx4

To investigate the significance of NHX1 to NHX4-mediated Na^+ and K^+ transport on cell expansion and root skewing, we stained roots of *nhx1nhx2nhx3nhx4* seedlings with propidium iodide and visualized cell size. Compared with WT, cells in the *nhx1nhx2nhx3nhx4* root transition zone appeared to be much shorter when grown on 30 mM KCl-supplemented media (Figure 3E,F). Cell size was similar between WT and *nhx1nhx2nhx3nhx4* in the apical meristem, but epidermal cells in the transition zone were markedly smaller in *nhx1nhx2nhx3nhx4*. Whereas in the WT, a gradual increase in cell size from the meristem to the elongation zone (Figure 3E, arrows) was apparent, the length of epidermal cells in the transition zone of *nhx1nhx2nhx3nhx4* was aberrantly small, and as a consequence, the transition zone appeared "compressed" (Figure 3F, white arrows). This phenotype was not evident in *nhx1nhx2nhx3nhx4* grown on 30 mM NaCl-supplemented media, where no difference in cell size or shape was observed between WT and *nhx1nhx2nhx3nhx4* in any root zones (Figure 3C,D).

Figure 3. Root cell expansion in the *nhx1nhx2nhx3nhx4* root transition zone is reduced. WT (**A**,**C**,**E**) or *nhx1nhx2nhx3nhx4* (**B**,**D**,**F**) six-day-old dark-grown seedlings were stained with 10 $\mu g \cdot mL^{-1}$ propidium iodide. Seedlings were grown on control media (1 mM K^+, 1 mM Na^+) (**A**,**B**) or on media supplemented with 30 mM NaCl (**C**,**D**) or 30 mM KCl (**E**,**F**). The transition zone is bounded by the dashed lines. Arrowheads point to cells that show differences in cell size and are discussed in the Results. Bar is 50 μm.

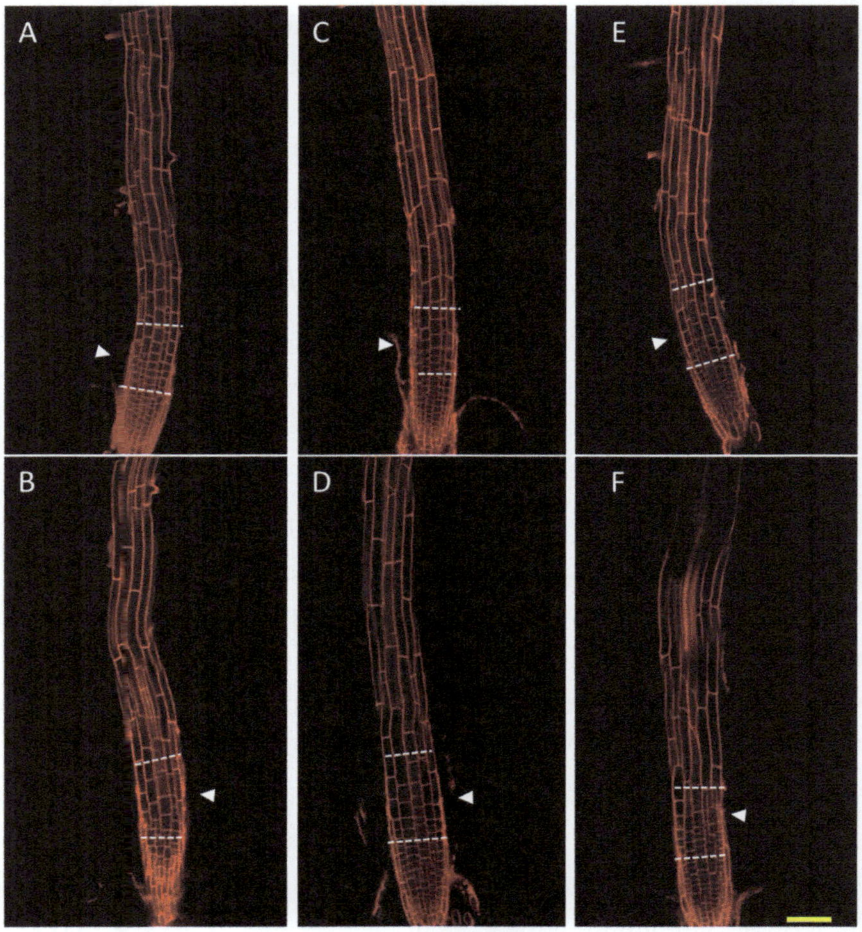

2.5. K^+ Dependent Root Skewing Depends on the Dynamic Organization of the Cytoskeleton

In order to investigate whether the root growth response of *nhx1nhx2nhx3nhx4* to elevated K^+ concentrations was associated with dynamic changes in the organization of the cytoskeleton, we tested the effect of microtubule perturbing drugs on root morphology and directional root growth in etiolated *nhx1nhx2nhx3nhx4* seedlings as above. Propyzamide is a microtubule destabilizing drug that is known to induce right-handed root skewing (when viewed through the agar) without inhibiting overall growth and root elongation [12,20,21]. When applied at 3 μM, anisotropic root growth of wild-type *Arabidopsis* is moderately impaired without causing swelling of epidermal roots cells [22]. We measured root growth angles for both WT and *nhx1nhx2nhx3nhx4* vertically-grown seedlings on minimal media (containing 1 mM KCl) and an identical media supplemented with an additional 30 mM KCl. As shown in Figure 4, both the WT and *nhx1nhx2nhx3nhx4* displayed the characteristic right-handed skewing (Figure 4E,F) when grown in the presence of 3 μM propyzamide, but only *nhx1nhx2nhx3nhx4* displayed left-handed skewing in response to 30 mM K^+ (Figure 4D). When both elevated K^+ and 3 μM propyzamide were included in the media, roots from WT seedlings continued to skew towards the right, unlike the roots from *nhx1nhx2nhx3nhx4*, which displayed less pronounced skewing (Figure 4G,H).

Figure 4. Propyzamide alters potassium-induced root skewing in the *nhx1nhx2nhx3nhx4* knockout. Wild-type (**A**,**C**,**E**,**G**) and *nhx1nhx2nhx3nhx4* (**B**,**D**,**F**,**H**) seedlings were germinated on minimal media and transferred to plates with either 30 mM K^+ (**C**,**D**), 3 μM propyzamide (**E**,**F**) or both 30 mM K^+ and propyzamide (**G**,**H**) when four day-old and grown for an additional seven days. Root skewing is quantified as the angle of root growth in degrees from vertical (**I**). Values represent the mean ± SE (n = 28). The bar is 0.5 cm.

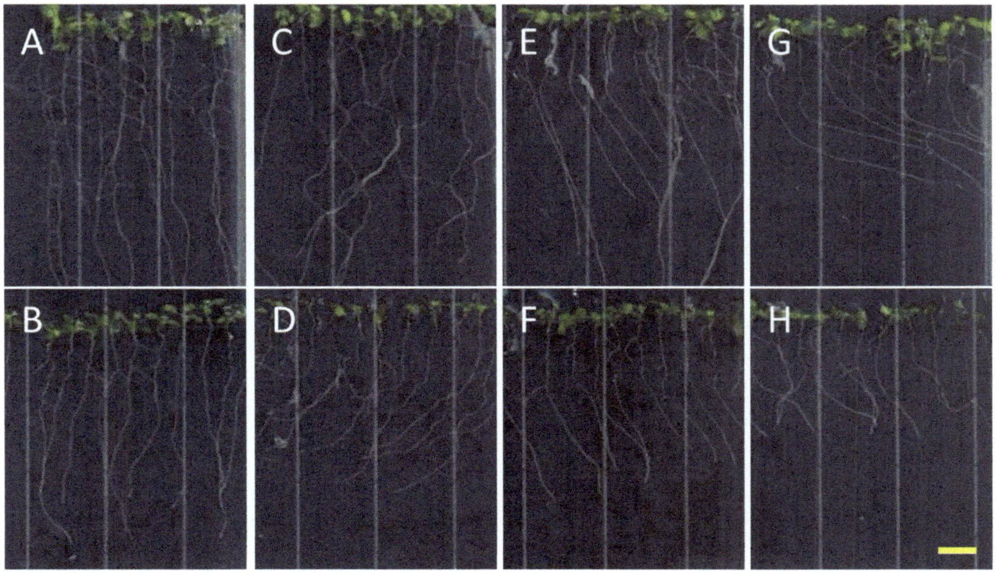

Figure 4. *Cont.*

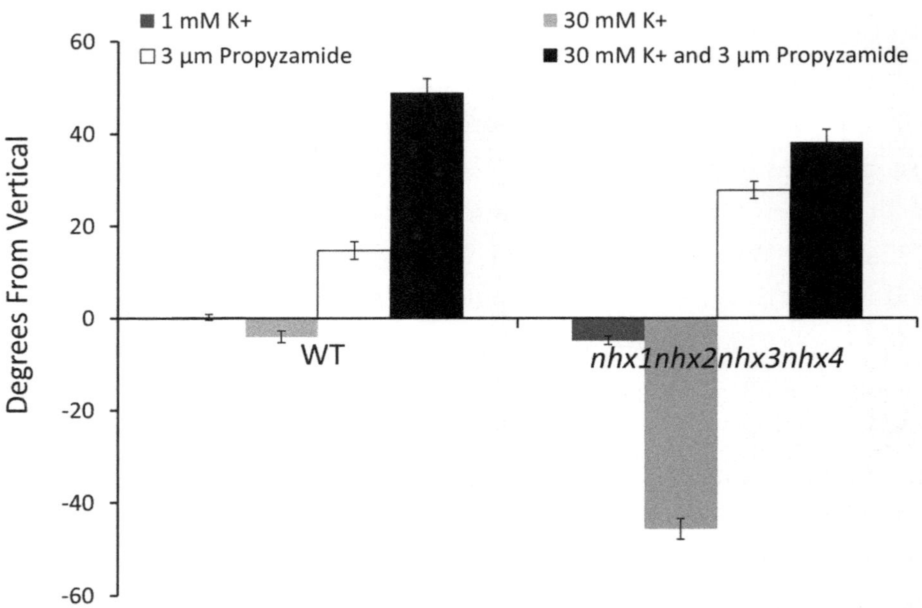

Wang and colleagues demonstrated that WT and *sos1*(*nhx7*) seedlings exhibited left-handed root skewing when grown vertically on NaCl-enriched MS media, similar to the root skewing of *nhx1nhx2nhx3nhx4* in the presence of high KCl. Their results indicated that 50 mM Na^+ was sufficient to initiate root skewing that eventually resulting in strong root curling and the formation of root "loops". Interestingly, Na^+-dependent root skewing was exacerbated in *sos1*(*nhx7*) mutants, which displayed more prominent root skewing than the WT at lower Na^+ treatments, suggesting that cytosolic ion accumulation played an important role in directional root growth. NaCl-induced salt stress resulted in seedling death, which occurred more frequently in *sos1*(*nhx7*) than WT seedlings, but was partially prevented by propyzamide treatment, suggesting that microtubule reorganization was required for salt tolerance [18].

We also tested the effect of Taxol, a microtubule stabilizing drug, on K^+-induced root skewing in *nhx1nhx2nhx3nhx4* plants. A characteristic effect of Taxol on roots is a right-handed skewing response that is dose dependent [23]. We observed that 0.5 µM Taxol was sufficient to induce right-handed root skewing in the WT (Figure 5E), but the same concentration did not significantly change the root growth direction of *nhx1nhx2nhx3nhx4* plants (Figure 5F). It is unlikely that the lack of *nhx1nhx2nhx3nhx4* root skewing observed in the 0.5 µM Taxol treatment was caused by a lack of overall root growth, because *nhx1nhx2nhx3nhx4* roots did not vary significantly in length between 0.5 µM or 1.0 µM Taxol treatments. Indeed, the addition of 30 mM KCl was sufficient to induce left-handed skewing in *nhx1nhx2nhx3nhx4*, despite the presence of 0.5 µM Taxol (Figure 5J). In contrast, the added KCl had no effect on WT root growth, with seedlings continuing to skew to the right (Figure 5I). We observed that a threshold of 1.0 µM Taxol was sufficient to overcome the K^+-induced left-handed skewing in *nhx1nhx2nhx3nhx4* plants and to cause right-handed skewing, as shown (Figure 5L). Previously, it was shown that 0.5 µM Taxol counters the effect of Na^+-induced root skewing, which is dose dependent [18]; a result that is strikingly similar to what we observed in Taxol- and K^+-treated *nhx1nhx2nhx3nhx4* (Figure 5).

Figure 5. Taxol reverses directional root growth in *nhx1nhx2nhx3nhx4* grown on potassium-supplemented media. Wild-type (**top**) and *nhx1nhx2nhx3nhx4* (**bottom**) seedlings were germinated on minimal media containing 1 mM K^+ for three days and transferred to plates containing two concentrations of Taxol and K^+ (indicated below) and grown for an additional four days. (**A,B)** 1 mM K^+; (**C,D**) 30 mM K^+; (**E,F**) 0.5 μM Taxol and 1 mM K^+; (**G,H**) 1 μM Taxol and 1 mM K^+; (**I,J**) 0.5 μM Taxol and 30 mM K^+; (**K,L**) 1 μM Taxol and 30 mM K^+. (**M**) The angle of root growth was quantified as the degrees from vertical; values represent the mean ± SE ($n = 24$). The bar is 0.5 cm.

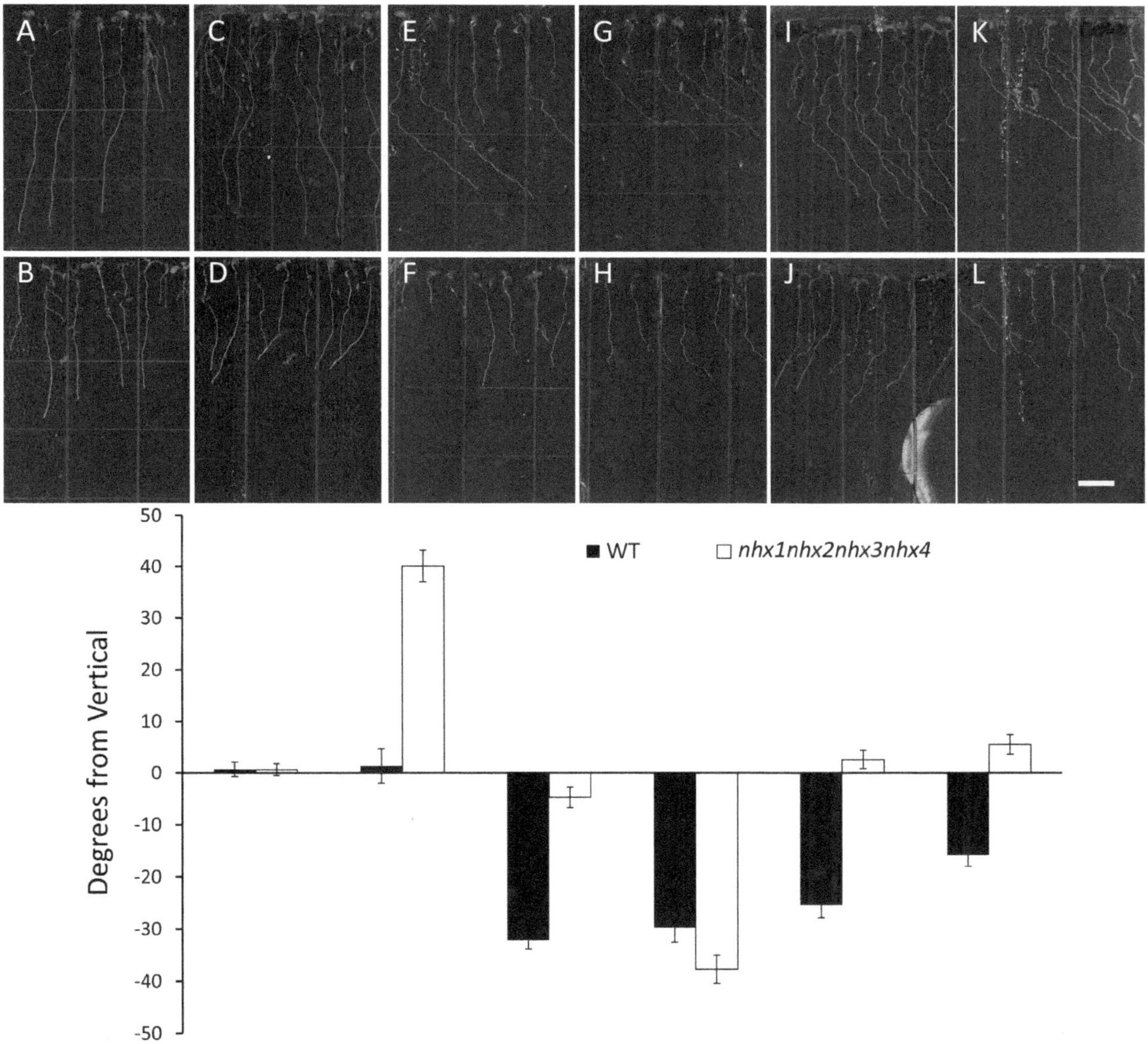

To better understand the possible effect of high K^+ on cytoskeletal organization, we used a whole mount immunostaining approach to visualize the organization of cortical microtubules in cells of the root elongation zone, where pronounced skewing and curling occurred.

As depicted in Figure 6, no differences were observed between the microtubule organization of roots from *nhx1nhx2nhx3nhx4* and the WT when grown on media containing 1 mM K^+ (Figure 6A,B). By contrast, *nhx1nhx2nhx3nhx4* seedlings grown on media containing 30 mM K^+ displayed a helical orientation of the cortical microtubule array in elongating *nhx1nhx2nhx3nhx4* root cells (Figure 6D) that was observed only when roots exhibited left-handed skewing. Both WT and *nhx1nhx2nhx3nhx4* displayed the characteristic Taxol-induced "bundling" of microtubules [24,25] with a highly parallel microtubule organization (Figure 6E,F). However, when seedlings were grown on media with both 0.5 μM Taxol and 30 mM KCl, the *nhx1nhx2nhx3nhx4* microtubules displayed a similar spiraling pattern that was seen when grown on 30 mM KCl plates (Figure 6H,D). Higher K^+ had no observable effect on microtubule organization in the WT, even in the presence of Taxol (Figure 6C,G).

Figure 6. Potassium induces helical reorganization of cortical microtubules of *nhx1nhx2nhx3nhx4* epidermal cells in the root elongation zone. Immunolabeled microtubules in WT (top) and *nhx1nhx2nhx3nhx4* (bottom). Seedlings were germinated on minimal media containing 1 mM K^+ for three days, then transferred to plates with two concentrations of Taxol and grown for an additional five days before fixation. Normal microtubule orientation is observed in WT (**A**) and *nhx1nhx2nhx3nhx4* (**B**) when grown on media containing 1 mM K^+. WT seedlings grown on 30 mM K^+ displayed normally-orientated microtubules (**C**), whereas 30 mM K^+ induced a helical orientation of the cortical microtubules in *nhx1nhx2nhx3nhx4* (**D**). In the presence of 1 mM K^+ and 0.5 μM Taxol, both WT (**E**) and *nhx1nhx2nhx3nhx4* (**F**) display characteristic bundling of cortical microtubules. Whereas 30 mM K^+ had no effect on microtubule orientation in WT grown with 0.5 μM Taxol (**G**), *nhx1nhx2nhx3nhx4* seedlings grown with 30 mM K^+ and 0.5 μM Taxol (**H**) exhibited a helical microtubule organization similar to that seen when grown with just 30 mM K^+. Arrows point to helically-oriented microtubules. The bar is 10 μM. The upper left insets are enlarged 50% to highlight microtubule orientation.

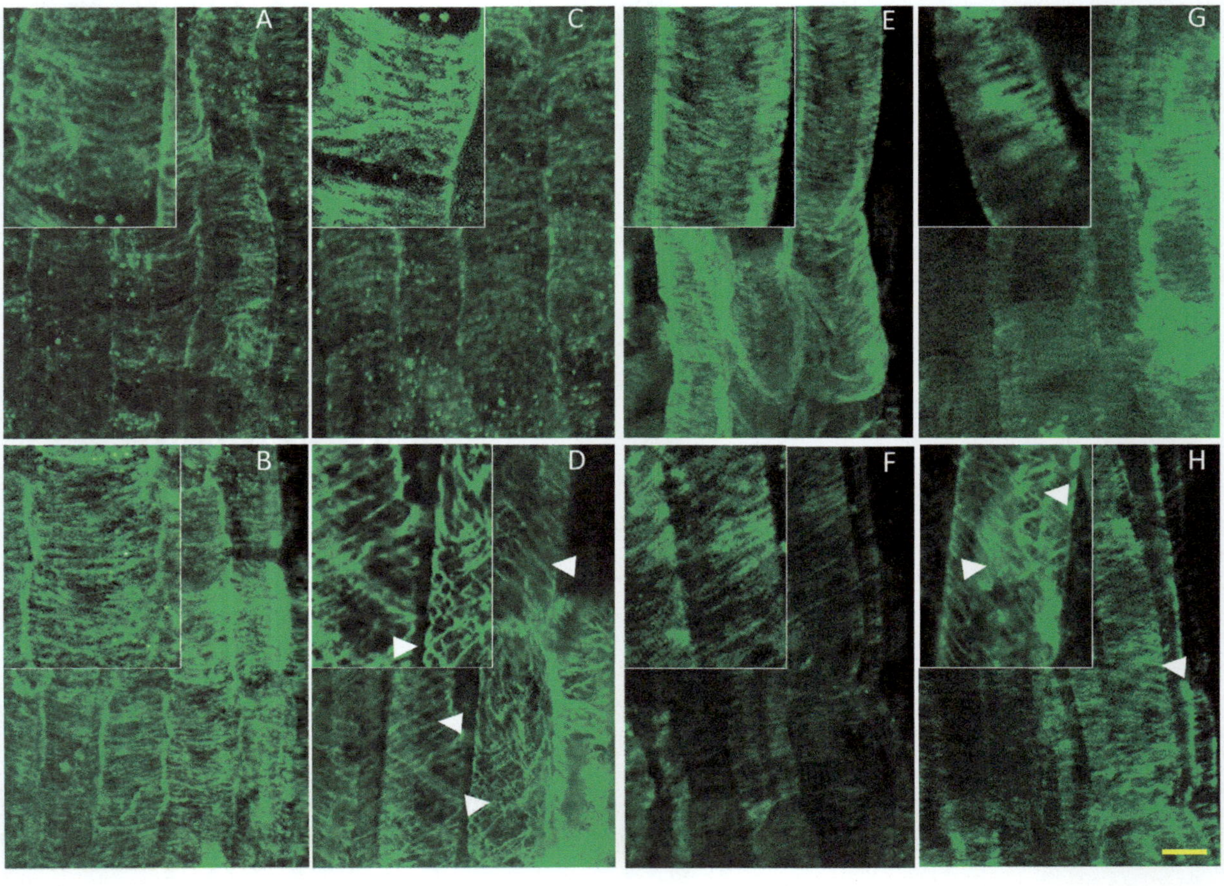

In particular, cortical microtubules depolymerize and reorganize in a Ca^{2+}-dependent manner under salt stress; preventing this response with microtubule stabilizing drugs, such as Taxol, results in seedling death [12,13], suggesting that cytoskeletal reorganization of root cells is important during salt stress. In addition, mutations of the Na^+/H^+ antiporter SOS1 and associated SOS2 kinase were shown to suppress the cortical microtubule disruptions and subsequent helical growth associated with the *Arabidopsis spiral1* mutants. This study also demonstrated that even in the absence of salt stress, loss of function SOS mutants resulted in altered cortical microtubule arrays when seedlings were treated with the same dose of propyzamide as used in the present study [12].

3. Experimental Section

3.1. Plant Materials and Growth Conditions

T-DNA insertions lines for *NHX1* (At5g27150), *NHX2* (At3g05030), *NHX3* (At5g55470) and *NHX4* (At3g06370) were obtained and were *nhx1-1* [SALK_034001], *nhx2-1* [SALK_036114], *nhx3-1* [WISC_345-348F19] and *nhx4-1* [SALK_112901] [26]. All single knockouts were backcrossed twice prior to crosses for generating higher order knockouts. Subsequent crosses generated double knockouts and two triple knockouts, which were then crossed to generate a tetra *nhx* knockout. RT-PCR confirmed no DNA amplification with allele-specific primers from leaf-derived *nhx1nhx2nhx3nhx4* cDNA. Positions of T-DNA insertion sites and RT-PCR results are shown in Figure A1.

Arabidopsis thaliana (ecotype Columbia [Col-0]) and *nhx1nhx2nhx3nhx4* seedlings were grown at 22 °C under diurnal light conditions (8 h light and 16 h dark), unless otherwise specified. Seeds were germinated on vertically-positioned plates containing a modified Murashige and Skoog media [27] (0.5% sucrose, 1 mM Na^+, 1 mM K^+, 0.8% phytagel, pH 5.7) and supplemented as specified for each experiment. For experiments testing stress responses, this media was considered as a control and was modified by the addition of 30 mM NaCl or KCl or 60 mM mannitol. For drug treatment assays, media was supplemented with 3 μM propyzamide (Sigma cat.#45645) or either 0.5 μM or 1 μM Taxol (Sigma cat.#T7191).

3.2. Cloning and Expression of Translational Fusion Proteins

NHX3 and *NHX4* translational fusions were constructed using the Gateway cloning system (Invitrogen, Carlsbad, CA, USA). cDNAs without stop codons were cloned into pDONR207 to generate entry vectors (pDONR207-NHX2, pDONR207-NHX3, pDONR207-NHX4) and recombined into pEarley-Gate101 for YFP fusion and pEarleyGate102 for CFP fusion [28]. For the 35S-NHX4-RFP construct, pDONR207-NHX4 was recombined with pH7RWG2 [29]. All constructs were introduced to *A. tumefasciens* strain GV3101 for both stable and transient transformation of Col-0, VAMP711-RFP [30] and γ-TIP-GFP.

3.3. Transient Transformation of Arabidopsis Seedlings

Transient transformation of *Arabidopsis* was adopted from Li, *et al.* [31] using 4–5-day-old seedlings grown in a sterile 24-well plate. Seedlings were co-cultivated with *Agrobacterium tumefasciens* (strain GV3101) for 18–24 h before observation. Co-transformation using two constructs (Figure 1I,L) was accomplished by co-cultivating seedlings with two bacterial cultures, each diluted to one half of the final optical density used when transforming a single construct (OD 0.5).

3.4. Hypocotyl and Root Length Measurements

Seedlings grown on agar plates were scanned on a flatbed scanner as color TIFF files with 300-dpi resolution. Root and hypocotyl lengths were quantified using the ImageJ [32] plugin SmartRoot [33]. The angles of root growth were measured with the same software suite using gridlines etched on the rear of the petri dishes as horizontal and vertical reference points (see Figure A4).

3.5. Quantification of Vacuolar K^+

Vacuolar K^+ was quantified using the K^+-specific dye PBFI-AM, as described previously [3].

3.6. Propidium Iodide Staining

Dark-grown, six-day-old WT and *nhx1nhx2nhx3nhx4* seedlings were grown as described above. Seedlings were stained with a 10-µg·mL^{-1} solution of propidium iodide (Sigma cat.#81845).

3.7. Whole Mount Immunostaining of Arabidopsis Seedlings

Whole mount immunostaining of root cortical microtubules was modified from the protocol of Hauser *et al.* [34]. Dark-grown seedlings grown for five days after germination were fixed on agar plates in a solution of 4% p-formaldehyde, 5% DMSO, 0.025% glutaraldehyde in 1× microtubule stabilizing buffer (MTSB; 50 mM PIPES, 5 mM EGTA, 5 mM $MGSO_4$, pH 7) vacuum infiltrated at room temperature for 1 h. Seedlings were then washed three times in wash buffer (MTSB with 5% DMSO) three times for ten minutes each, and the solution was exchanged for a cell wall digestion buffer (MTSB with 1% cellulase, 0.1% pectolyase) and incubated at room temperature for 30 min. Following digestion, three washes were performed with MTSB for five minutes each, and the seedlings were incubated for ten minutes in a solution of MTSB with 1 mg/mL $NaBH_4$ for ten minutes. After three more washes with MTSB for five minutes, each seedling was incubated for ten minutes in a solution of MTSB containing 10 mM glycine and then washed once more with MTSB three times for a total of 30 min. Severed roots were transferred to microscope slides (Thermo-Scientific Superfrost plus, cat.#4951PLUS-001) and incubated in a membrane permeabilization solution (MTSB with 1% Triton X-100) at room temperature for 15 min and then in methanol at −20 °C for 20 min.

Immunostaining was carried out by an incubation with DM1A primary antibody (Sigma cat.#T9026) diluted 1:400 in a blocking solution (PBS containing 3% BSA, 0.05% Tween-20, and 0.1% thimerosal) for one hour at 37 °C. After washing, roots were incubated with FITC goat anti-mouse (Sigma cat. #F027) for one hour at 37 °C.

3.8. Confocal Microscopy

Fluorescence microscopy was performed using a confocal laser scanning microscope (Zeiss LSM 710) equipped with a 40× water immersion objective. The excitation wavelength was 488 nm, with emission wavelengths of 500–535 nm for GFP or FITC, 525–615 nm for YFP/Venus, 450–500 nm for CFP and

565–605 nm for RFP. For imaging of GFP/RFP and CFP/YFP double reporter lines, sequential scanning on multiple channels was used to avoid crosstalk between fluorescence emission spectra.

4. Conclusions

We previously reported that NHX1 and NHX2 colocalize to the vacuole and are essential for vacuolar K^+ accumulation and the regulation of pH [3]. Here, we demonstrate that both NHX3 and NHX4 are also localized to the tonoplast and that vacuolar cation/H^+ exchange by all four vacuolar NHXs are essential for normal growth and development. Compared to both WT and the *nhx1nhx2* double knockout, the quadruple knockout *nhx1nhx2nhx3nhx4* has significantly reduced vacuolar K^+ and growth, as well as high sensitivity to added K^+. *nhx1nhx2nhx3nhx4* also exhibited pronounced root skewing and curling responses not observed under equimolar Na^+ that were correlated with an increase in cytoskeletal dynamics and a reduction in the size of the root cells of the transition zone.

Acknowledgments

We thank Bo Liu and Yuh-Ru Julie Lee (Department of Plant Biology, University of California, Davis) and Aviah Zilberstein, (Department of Molecular Biology and Ecology of plant, Tel Aviv University) for technical assistance and helpful discussions. This work was supported in part by grants from the National Science Foundation (MCB-0343279; IOS-0820112) and the Will W. Lester Endowment, University of California, to Eduardo Blumwald.

Author Contributions

Tyler McCubbin, Elias Bassil and Eduardo Blumwald designed the research. Tyler McCubbin, Elias Bassil and Shiqi Zhang performed the research. Tyler McCubbin, Elias Bassil and Eduardo Blumwald wrote the article.

Conflicts of Interest

The authors declare no conflict of interest.

References

1. Bassil, E.; Coku, A.; Blumwald, E. Cellular ion homeostasis: Emerging roles of intracellular nhx Na^+/H^+ antiporters in plant growth and development. *J. Exp. Bot.* **2012**, *63*, 5727–5740.
2. Reguera, M.; Bassil, E.; Blumwald, E. Intracellular NHX-type cation/H^+ antiporters in plants. *Mol. Plant* **2013**, doi:10.1093/mp/sst091.
3. Bassil, E.; Tajima, H.; Liang, Y.-C.; Ohto, M.-A.; Ushijima, K.; Nakano, R.; Esumi, T.; Coku, A.; Belmonte, M.; Blumwald, E. The arabidopsis Na^+/H^+ antiporters NHX1 and NHX2 control vacuolar pH and K^+ homeostasis to regulate growth, flower development, and reproduction. *Plant Cell* **2011**, *23*, 3482–3497.
4. Apse, M.P.; Sottosanto, J.B.; Blumwald, E. Vacuolar cation/H^+ exchange, ion homeostasis, and leaf development are altered in a T-DNA insertional mutant of atnhx1, the *Arabidopsis* vacuolar Na^+/H^+ antiporter. *Plant J.* **2003**, *36*, 229–239.

5. Liu, H.; Tang, R.; Zhang, Y.; Wang, C.; Lv, Q.; Gao, X.; Li, W.; Zhang, H. AtNHX3 is a vacuolar K^+/H^+ antiporter required for low-potassium tolerance in *Arabidopsis thaliana. Plant Cell Environ.* **2010**, *33*, 1989–1999.
6. Liu, H.; Wang, Q.; Yu, M.; Zhang, Y.; Wu, Y.; Zhang, H. Transgenic salt-tolerant sugar beet (*Beta vulgaris* L.) constitutively expressing an arabidopsis thaliana vacuolar Na^+/H^+ antiporter gene, *AtNHX3*, accumulates more soluble sugar but less salt in storage roots. *Plant Cell Environ.* **2008**, *31*, 1325–1334.
7. Li, H.T.; Liu, H.; Gao, X.S.; Zhang, H.X. Knock-out of arabidopsis *AtNHX4* gene enhances tolerance to salt stress. *Biochem. Biophys. Res. Commun.* **2009**, *382*, 637–641.
8. Maathuis, F.J.M.; Podar, D. Uptake, distribution, and physiological functions of potassium, calcium, and magnesium. In *The Molecular and Physiological Basis of Nutrient Use Efficiency in Crops*; Wiley-Blackwell: Hoboken, NJ, USA, 2011; pp. 265–293.
9. Barragán, V.; Leidi, E.O.; Andrés, Z.; Rubio, L.; de Luca, A.; Fernández, J.A.; Cubero, B.; Pardo, J.M. Ion exchangers NHX1and NHX2 mediate active potassium uptake into vacuoles to regulate cell turgor and stomatal function in arabidopsis. *Plant Cell* **2012**, *24*, 1127–1142.
10. Andrés, Z.; Pérez-Hormaeche, J.; Leidi, E.O.; Schlücking, K.; Steinhorst, L.; McLachlan, D.H.; Schumacher, K.; Hetherington, A.M.; Kudla, J.; Cubero, B.; *et al.* Control of vacuolar dynamics and regulation of stomatal aperture by tonoplast potassium uptake. *Proc. Natl. Acad. Sci. USA* **2014**, doi:10.1073/pnas.1320421111.
11. Bibikova, T.N.; Blancaflor, E.B.; Gilroy, S. Microtubules regulate tip growth and orientation in root hairs of *Arabidopsis thaliana. Plant J.* **1999**, *17*, 657–665.
12. Shoji, T.; Suzuki, K.; Abe, T.; Kaneko, Y.; Shi, H.; Zhu, J.K.; Rus, A.; Hasegawa, P.M.; Hashimoto, T. Salt stress affects cortical microtubule organization and helical growth in arabidopsis. *Plant Cell Physiol.* **2006**, *47*, 1158–1168.
13. Wang, S.; Kurepa, J.; Hashimoto, T.; Smalle, J.A. Salt stress-induced disassembly of *Arabidopsis* cortical microtubule arrays involves 26S proteasome-dependent degradation of spiral1. *Plant Cell* **2011**, *23*, 3412–3427.
14. Saito, C.; Ueda, T.; Abe, H.; Wada, Y.; Kuroiwa, T.; Hisada, A.; Furuya, M.; Nakano, A. A complex and mobile structure forms a distinct subregion within the continuous vacuolar membrane in young cotyledons of arabidopsis. *Plant J.* **2002**, *29*, 245–255.
15. Uemura, T.; Ueda, T.; Ohniwa, R.L.; Nakano, A.; Takeyasu, K.; Sato, M.H. Systematic analysis of snare molecules in arabidopsis: Dissection of the post-golgi network in plant cells. *Cell Struct. Funct.* **2004**, *29*, 49–65.
16. Rutherford, R.; Masson, P.H. *Arabidopsis thaliana* sku mutant seedlings show exaggerated surface-dependent alteration in root growth vector. *Plant Physiol.* **1996**, *111*, 987–998.
17. Sun, F.; Zhang, W.; Hu, H.; Li, B.; Wang, Y.; Zhao, Y.; Li, K.; Liu, M.; Li, X. Salt modulates gravity signaling pathway to regulate growth direction of primary roots in arabidopsis. *Plant Physiol.* **2008**, *146*, 178–188.
18. Wang, C.; Li, J.; Yuan, M. Salt tolerance requires cortical microtubule reorganization in *Arabidopsis. Plant Cell Physiol.* **2007**, *48*, 1534–1547.
19. Blancaflor, E.B.; Hasenstein, K.H. Growth and microtubule orientation of *Zea mays* roots subjected to osmotic stress. *Int. J. Plant Sci.* **1995**, *156*, 774–783.

20. Yuen, C.Y.; Sedbrook, J.C.; Perrin, R.M.; Carroll, K.L.; Masson, P.H. Loss-of-function mutations of root hair defective3 suppress root waving, skewing, and epidermal cell file rotation in arabidopsis. *Plant Physiol.* **2005**, *138*, 701–714.
21. Nakamura, M.; Naoi, K.; Shoji, T.; Hashimoto, T. Low concentrations of propyzamide and oryzalin alter microtubule dynamics in *Arabidopsis* epidermal cells. *Plant Cell Physiol.* **2004**, *45*, 1330–1334.
22. Naoi, K.; Hashimoto, T. A semidominant mutation in an arabidopsis mitogen-activated protein kinase phosphatase-like gene compromises cortical microtubule organization. *Plant Cell* **2004**, *16*, 1841–1853.
23. Sedbrook, J.C.; Ehrhardt, D.W.; Fisher, S.E.; Scheible, W.R.; Somerville, C.R. The *Arabidopsis* SKU6/SPIRAL1 gene encodes a plus end-localized microtubule-interacting protein involved in directional cell expansion. *Plant Cell* **2004**, *16*, 1506–1520.
24. Takahashi, H.; Hirota, K.; Kawahara, A.; Hayakawa, E.; Inoue, Y. Randomization of cortical microtubules in root epidermal cells induces root hair initiation in lettuce (*Lactuca sativa* L.) seedlings. *Plant Cell Physiol.* **2003**, *44*, 350–359.
25. Baskin, T.I.; Wilson, J.E.; Cork, A.; Williamson, R.E. Morphology and microtubule organization in arabidopsis roots exposed to oryzalin or Taxol. *Plant Cell Physiol.* **1994**, *35*, 935–942.
26. Alonso, J.M.; Stepanova, A.N.; Leisse, T.J.; Kim, C.J.; Chen, H.; Shinn, P.; Stevenson, D.K.; Zimmerman, J.; Barajas, P.; Cheuk, R.; *et al.* Genome-wide insertional mutagenesis of *Arabidopsis thaliana. Science* **2003**, *301*, 653–657.
27. Spalding, E.P.; Hirsch, R.E.; Lewis, D.R.; Qi, Z.; Sussman, M.R.; Lewis, B.D. Potassium uptake supporting plant growth in the absence of akt1 channel activity: Inhibition by ammonium and stimulation by sodium. *J. Gen. Physiol.* **1999**, *113*, 909–918.
28. Earley, K.W.; Haag, J.R.; Pontes, O.; Opper, K.; Juehne, T.; Song, K.; Pikaard, C.S. Gateway-compatible vectors for plant functional genomics and proteomics. *Plant J.* **2006**, *45*, 616–629.
29. Karimi, M.; Bleys, A.; Vanderhaeghen, R.; Hilson, P. Building blocks for plant gene assembly. *Plant Physiol.* **2007**, *145*, 1183–1191.
30. Geldner, N.; Devervaud-Tendon, V.; Hyman, D.L.; Mayer, U.; Stierhof, Y-D.; Chory, J. Rapid, combinatorial analysis of membrane compartments in intact plants with a multicolor marker set. *Plant J.* **2009**, *59*, 169–178.
31. Li, J.F.; Park, E.; von Arnim, A.G.; Nebenfuhr, A. The fast technique: A simplified agrobacterium-based transformation method for transient gene expression analysis in seedlings of arabidopsis and other plant species. *Plant Methods* **2009**, *5*, 6.
32. Schneider, C.A.; Rasband, W.S.; Eliceiri, K.W. Nih image to imagej: 25 Years of image analysis. *Nat. Methods* **2012**, *9*, 671–675.
33. Lobet, G.; Pagès, L.; Draye, X. A novel image-analysis toolbox enabling quantitative analysis of root system architecture. *Plant Physiol.* **2011**, *157*, 29–39.
34. Hauser, M.T.; Morikami, A.; Benfey, P.N. Conditional root expansion mutants of Arabidopsis. *Development* **1995**, *121*, 1237–1252.

Appendix

Figure A1. Genetic maps of *NHX* T-DNA insertions used in generating *nhx1nhx2nhx3nhx4* (**A**); rt-PCR shows no expression of *NHX1*, *NHX2*, *NHX3*, *NHX4* in *nhx1nhx2nhx3nhx4* (**B**); primers used for verifying gene expression are shown in the table (**C**).

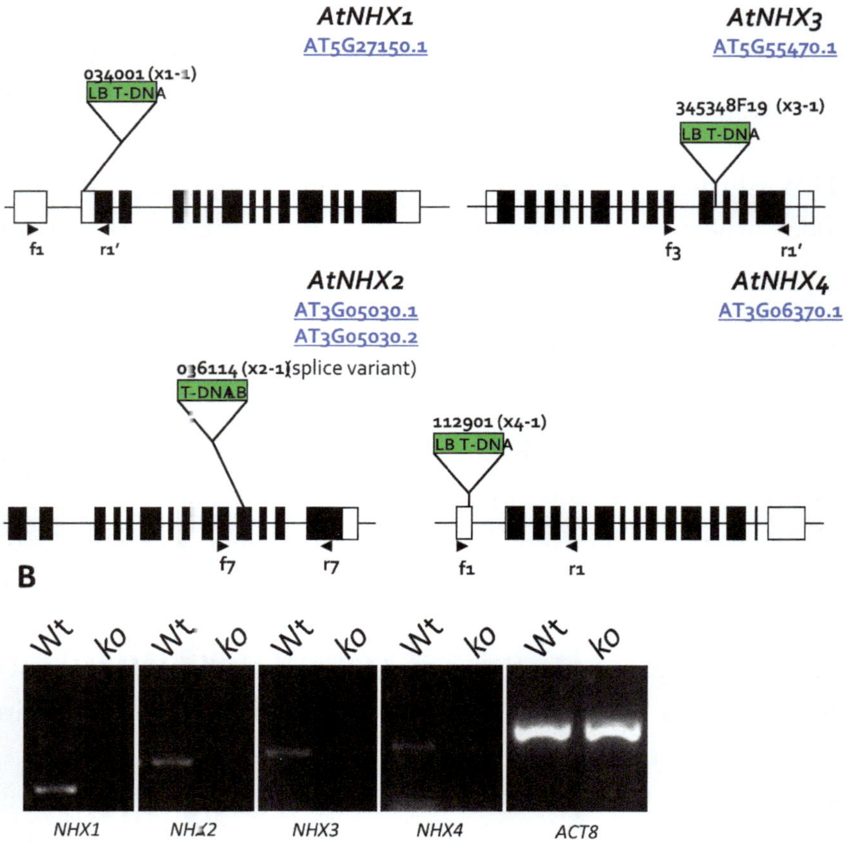

C

Gene name	Forward primer	Reverse primer	Size
NHX1 (f1, r1')	GAATTCGCCTCTCTGTTTCGTTCCTC	GCAACCACAGAAGCGTGATCAGA	357 bp
NHX2 (f7, r7)	AATGGAGATTCGTGAGTGACAG	CAAATGAATCTAGCTGTTCACCATCG	456 bp
NHX3 (f3, r1')	CAGCTTAAGCTTTGGGGGTACT	TTCTGGTTGGTTTTCTCGACG	592 bp
NHX4 (f1, r1)	ATCTTTGGCGGTCTCTCGTCTTC	TCCAATGGCTAGATAGTCCGCAA	610 bp
ACT8 (344F-1042R)	AAGCCAACAGAGAGAAGATGAC	TGGAAAGTGCTGAGGGAAGCAAG	698 bp

Figure A2. Etiolation response of four-day-old *nhx1nhx2nhx3nhx4*. WT (**A**,**C**,**E**) and *nhx1nhx2nhx3nhx4* (**B**,**D**,**F**) grown on 1 mM K^+, 1 mM Na^+ (**A**,**B**), 30 mM K^+ (**C**,**D**) or 30 mM Na^+ (**E**,**F**). Root growth of *nhx1nhx2nhx3nhx4* does not differ between treatments, but hypocotyls are shorter on 30 mM K^+ (**D**) than 1 mM K^+, 1 mM Na^+ (**B**) or 30 mM Na^+ (**F**). Growth of both roots and hypocotyls in WT is unaffected across treatments. The yellow bar is 1 cm.

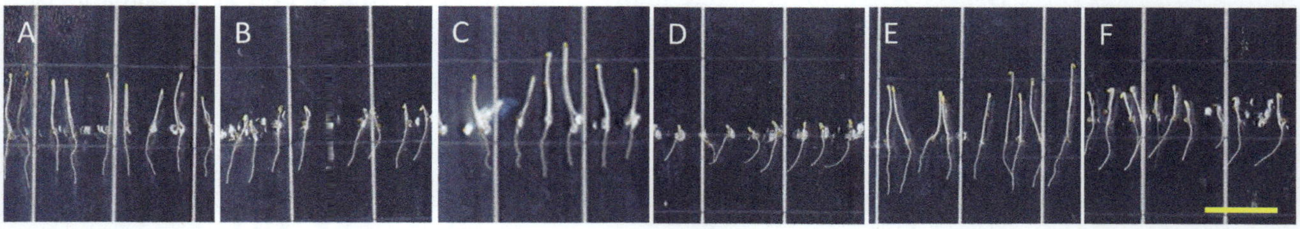

Figure A3. *nhx1nhx2nhx3nhx4* root tip curling is K^+ specific. Six-days-old WT (**A**,**C**,**E**,**G**) and *nhx1nhx2nhx3nhx4* (**B**,**D**,**F**,**H**). Both WT and *nhx1nhx2nhx3nhx4* exhibit straight root growth on media containing 1 mM KCl (**A**,**B**). When grown on media supplemented with either 30 mM KCl (**C**,**D**) or 30 mM KNO_3 (**E**,**F**), WT roots grow normally, whereas roots of *nhx1nhx2nhx3nhx4* exhibit a left-handed curling. Neither 30 mM NaCl (**G**,**H**) nor 60 mM mannitol (**I**,**J**) induced root curling, although the addition of NaCl produced longer root growth in *nhx1nhx2nhx3nhx4* (**H**) compared with other treatments. The yellow bar is 1 cm.

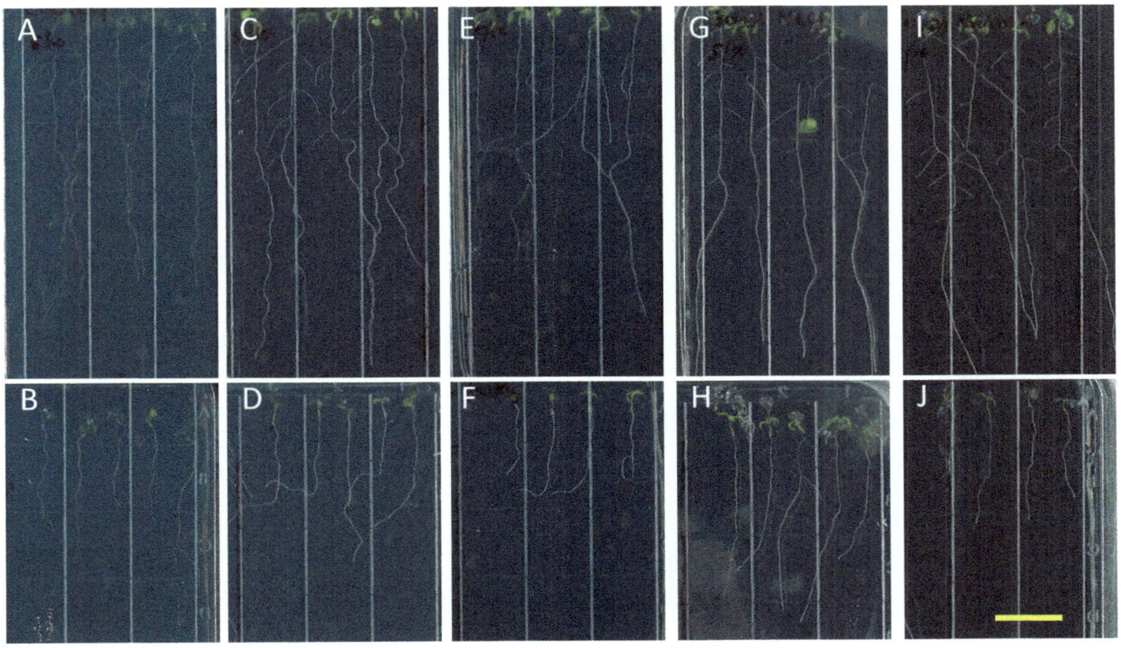

Figure A4. Root skewing measurement methodology. The etched grid on the back of square petri plates served as a reference point for assigning the vertical and horizontal vectors. Growth vectors were measured by drawing a line along the general axis of root growth. For curling roots, the growth vector was determined before the point at which the root begins to curl.

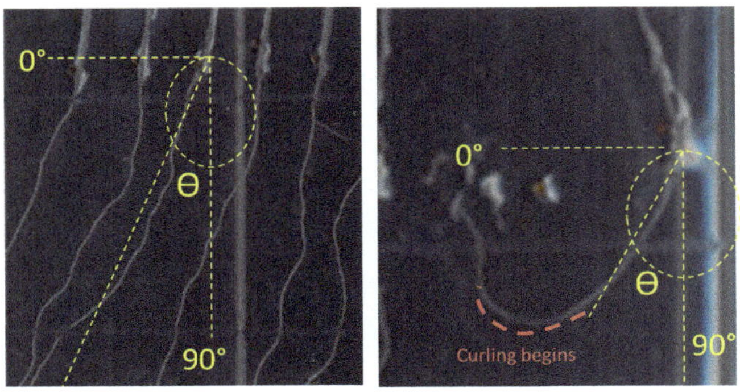

4

Effects of Light and Wounding on Jasmonates in Rice *phyAphyC* Mutants

Rita Brendel [1,†], Katharina Svyatyna [1,†], Yusuke Jikumaru [2], Michael Reichelt [3], Axel Mithöfer [4], Makoto Takano [5], Yuji Kamiya [2], Peter Nick [1] and Michael Riemann [1,*]

1 Botanical Institute, Karlsruhe Institute of Technology, Karlsruhe 76131, Germany; E-Mails: rita.brendel@kit.edu (R.B.); katha.sv@gmx.de (K.S.); peter.nick@kit.edu (P.N.)

2 RIKEN Plant Science Center, Tsurumi, Yokohama, Kanagawa 230-0045, Japan; E-Mails: jikumaru@psc.riken.jp (Y.J.); ykamiya@postman.riken.jp (Y.K.)

3 Department of Biochemistry, Max Planck Institute for Chemical Ecology, Jena 07745, Germany; E-Mail: reichelt@ice.mpg.de

4 Department of Bioorganic Chemistry, Max Planck Institute for Chemical Ecology, Jena 07745, Germany; E-Mail: amithoefer@ice.mpg.de

5 Department of Plant Physiology, National Institute of Agrobiological Sciences, Tsukuba, Ibaraki 305-8602, Japan; E-Mail: mtakano@affrc.go.jp

† These authors contributed equally to this work.

* Author to whom correspondence should be addressed; E-Mail: michael.riemann@kit.edu

---

Abstract: Jasmonates (JA) are lipid-derived plant hormones. They have been shown to be important regulators of photomorphogenesis, a developmental program in plants, which is activated by light through different red and blue light sensitive photoreceptors. In rice, inhibition of coleoptile growth by light is a central event in photomorphogenesis. This growth inhibition is impaired, when jasmonate biosynthesis is knocked out. Previously, we found that *JASMONATE RESISTANT 1* (*OsJAR1*) transcripts were not induced in the phytochrome (phy) mutant *phyAphyC*. Therefore, in the current study we investigated the regulation of JA and its highly bioactive derivative (+)-7-*iso*-jasmonoyl-L-isoleucine (JA-Ile), as well as the transcriptional regulation of several JA-dependent genes both in wild type and *phyAphyC* mutant. JA and JA-Ile levels increased in the mutant seedlings in response to blue light. However, in *phyAphyC* mutant leaves, which were continuously wounded, JA and JA-Ile levels were lower compared to those in the wild type. Hence, the mutation of phyA and phyC has differential effects on jasmonate levels depending on the tissue and developmental stage. Our results suggest that the contribution of JA-Ile to signaling during photomorphogenesis of rice is minor, as coleoptile phenotypes of *phyAphyC* mutants resemble those of jasmonate-deficient mutants despite the fact that induction by blue light leads to higher levels of JA-Ile compared to the wild type. We postulate that phyA and phyC could control the activity of specific enzymes metabolizing JA to active derivatives.

Keywords: *Oryza sativa* L.; jasmonate; jasmonate-isoleucine; JASMONATE RESISTANT 1; phytochrome; photomorphogenesis; mechanical wounding

---

1. Introduction

For plants it is crucial to respond to light fast as they use light as their main energy source. During the first days after germination they are able to sustain growth by consuming the nutrients stored in the seed. Thus, seedlings are able to grow rapidly in complete darkness although they are not photosynthetically active yet. During their passage through the soil, the leaves of rice seedlings are enclosed in an organ called coleoptile, which protects them from mechanical damage. Coleoptiles elongate very quickly in complete darkness, but are extremely light sensitive. Upon detection of light they stop growing and open along a predefined axis to release their leaves. This marks the transition of the seedling to a completely different developmental program, referred to as photomorphogenesis.

In order to perceive light, plants have evolved a sophisticated system of photoreceptors, which are either sensitive to blue (cryptochromes, cry; phototropin, phot [1–4]), or to red/far-red (phytochrome, phy; [5–8]) light. Phys exist in two forms, the red light absorbing P_r- and the far-red light absorbing P_{fr}-form, which both can be interconverted by irradiation of the appropriate wavelength. As rice harbors only three genes coding for the phys phyA, phyB, and phyC [9,10] it is convenient for functional studies of phys as compared to other model systems.

The three phytochromes differ in their kinetic properties after conversion into the P_{fr} form. While the P_{fr} form of phyB is stable in the light, phyA and phyC are degraded after activation, a phenomenon known as photodestruction [10]. For continuous irradiation, light qualities that establish a low photoequilibrium (the proportion of P_{fr} over total phytochrome), such as canopy light depleted in red light, are, therefore, more efficiently activating phyA/phyC signaling. In contrast for light qualities that establish a high photoequilibrium, such as daylight above the canopy, more P_{fr} is established than can be used for signaling and the excessive P_{fr} undergoes photodestruction and therefore is not contributing to sustain signaling [11]. However, very low fluence responses (e.g., germination), which are under control of phyA, also occur preferentially at a low photoequilibrium [12]. Due to these kinetic differences, continuous red light is activating almost exclusively phyB dependent signaling, whereas continuous far-red and blue light act through phyA/phyC.

The inhibition of coleoptile growth during photomorphogenesis is thought to be mediated by phys [13,14], but also an involvement of crys has been discussed recently [15]. Several examples show that the developmental changes during photomorphogenesis are regulated by plant hormones. For example in *Arabidopsis thaliana* brassinosteroids have been demonstrated to be important regulators of photomorphogenesis. When brassinosteroid biosynthesis is impaired, as in the *deetiolation 2 (det 2)* mutant, seedlings undergo photomorphogenesis in darkness [16,17]. In rice, jasmonates (jasmonic acid, JA, and derivatives) have been shown to be very important for the phy-dependent inhibition of coleoptile elongation [18–20]. Jasmonates are signaling compounds, which are biosynthesized from linolenic acid, cleaved from a chloroplast membrane lipid [21]. After conjugation to isoleucine (Ile) by JASMONATE RESISTANT 1 (JAR1) [22], JA is activated and can bind to its co-receptor CORONATINE INSENSITIVE 1 (COI1) in *Arabidopsis* [23,24]. Upon binding of COI1 and JA-Ile the complex can recruit JAZ proteins, repressors of jasmonate signaling [25,26]. Subsequently, JAZ proteins are presumably marked by ubiquitination for degradation in the proteasome.

For rice, it has been shown that the deficiency of jasmonates leads to an altered photomorphogenesis. In the mutants *hebiba* and *coleoptile photomorphogenesis 2 (cpm2)*, the function of ALLENE OXIDE CYCLASE (AOC), a key enzyme of jasmonate biosynthesis, is impaired, leading to a long coleoptile phenotype in light [27]. Moreover, the photodestruction of phyA was found to be much slower in the *hebiba* mutant [28,29], which indicates a function for jasmonate in this process.

JAR1 is important for the conversion of JA to the highly bioactive amino-acid conjugate JA-Ile and this enzymatic step has been shown to be light dependent in lima bean [30]. Therefore, mutants of the rice orthologue of JAR1 were studied. It was found that the long coleoptile phenotype of *osjar1* mutants was especially pronounced under blue light. However, this phenotype of *osjar1* mutants is less severe compared to the rice *aoc* mutants [31]. *OsJAR1* transcripts are inducible by blue light, but this induction is strongly reduced in *phyAphyC* mutants lacking the photodestructible isotypes of phytochrome [32].

The present study was motivated by this observation. We wanted to know whether the blue light induction of jasmonate biosynthesis and signaling is mediated entirely via phyA and phyC. Therefore, we investigated the regulation of the biosynthesis of JA and its derivative JA-Ile in *phyAphyC* mutants, as well as the regulation of JA-responsive genes. We compared the regulation of biosynthesis and signaling of jasmonate in response to light and mechanical wounding, a signal well-known to induce this plant hormone. In seedlings grown in blue light phyA and phyC might either attenuate the synthesis of JA or accelerate its metabolism, as, in mutants, both levels of JA and JA-Ile were elevated. In contrast, both compounds tended to be less abundant in wounded mutant compared to wild type leaves, indicating that the negative effect on JA and JA-Ile accumulation is specific for the seedling stage.

2. Results and Discussion

2.1. The phyAphyC Mutant Displays a Phenotype in Response to Continuous Blue and Far-Red Light

In a recent study, where blue light was irradiated as a pulse of 3 min, it could be shown that the transcript for the JAR1-enzyme is not induced after blue light in etiolated *phyAphyC* mutant seedlings while its transcription is induced in the wild type [32]. Therefore, we compared the response of coleoptile growth of wild type and *phyAphyC* mutants to continuous irradiation with different light qualities. Wild type and *phyAphyC* mutant seedlings were grown in continuous blue (cBL), red (cRL), or far-red (cFRL) light and the coleoptile length were measured after the emergence of the leaves. Seedlings of *phyAphyC* mutants showed a longer coleoptile than the wild type in cFRL and cBL, while there was no such difference visible in cRL (Figure 1).

Figure 1. Length of rice coleoptiles after growth in continuous light of different qualities. cBL: continuous blue light (λ_{max} 470 nm), cRL: continuous red light (λ_{max} 650 nm), cFRL: continuous far red light (λ_{max} 735 nm), fluence rate was adjusted to 10 µmol/m^{2}s for all light qualities. Coleoptile length of wild type (grey bar) and *phyAphyC* mutants (white bar) was measured after the leaves had pierced the coleoptile tip. Statistically significant differences to the wild type are indicated by two asterisks (Student's *t*-test, $p < 0.01$).

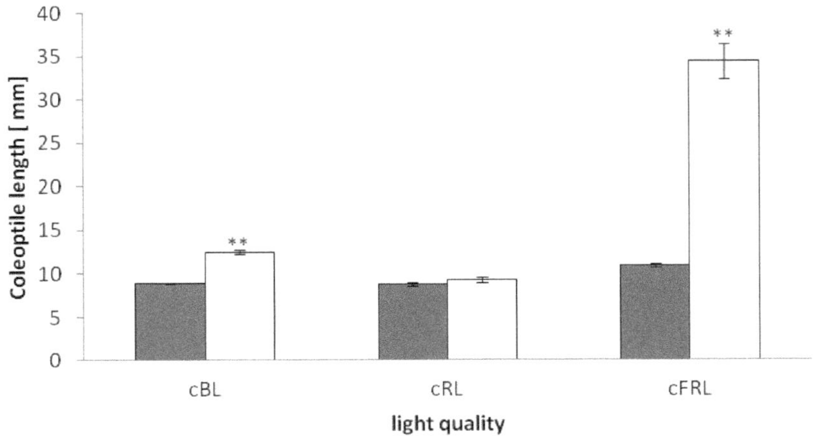

As to be expected from the dependence of phytochrome photodestruction on light quality, the difference between wild type and *phyAphyC* mutant was strongest for far-red light. Under those conditions, mutant coleoptiles were in average about three times longer (34 mm *vs.* 11 mm) than those of the wild type. This result indicates a completely impaired de-etiolation under cFRL, which is in accordance with the findings of Takano *et al.* [10].

Interestingly, the *phyAphyC* mutants also showed a significant blue light phenotype. Grown in cBL, the coleoptile length was in average 12 mm compared to the wild type with only 9 mm. A similar, but stronger phenotype under blue light is also known from mutants, which are impaired in jasmonate biosynthesis, such as *hebiba*, *cmp2*, or *osjar1* [19,27,31]. In cBL, coleoptiles of *phyAphyC* mutants are shorter than those of *osjar1* mutants, which have the same genetic background, while the severely JA-deficient mutants *cpm2* and *hebiba* develop the longest coleoptiles [31].

2.2. The Blue-Light Response of OsJAR1 Is Specifically Suppressed in phyAphyC

As *phyAphyC* and the mutants deficient in JA or bioactive JA amino acid conjugates showed a similar blue light phenotype, we investigated the blue-light response of representative JA-responsive genes, known to be differentially expressed in *osjar1* mutants.

The transcriptional regulation of genes for the biosynthesis of jasmonates (*OsAOC*, *OsJAR1*) and JA signaling (*OsTIFY10c/OsJAZ8*) [31,33] was analyzed in etiolated seedlings exposed to blue light for different incubation times (1 h, 2 h, and 6 h). *OsAOC* is a single-copy gene, while *OsJAR1* and *OsJAZ8* are members of gene families. *OsJAR1* belongs to the GH3 gene family consisting of

12 members in rice [34,35]. *OsJAZ8* belongs to the *TIFY* gene family consisting of 20 members in rice [36]. Out of those, 15 genes are putative *JAZ* genes.

The transcript of *OsJAR1* is induced by blue light treatment in the wild type in a transient manner, peaking 2 h after the onset of irradiation relative to a dark control (six-fold induction in Nipponbare). Although this transcript is also induced in *phyAphyC*, the induction is much weaker (only two-fold, Figure 2A). This very low induction of *OsJAR1* in *phyAphyC* is in line with the results of Riemann *et al*. [32], who found, that *OsJAR1* transcript levels are basically not responsive to blue light in *phyAphyC* mutants.

Figure 2. Transcript levels of the representative JA-biosynthesis genes *OsJAR1* (**A**) and *OsAOC* (**B**) and the JA-signaling gene OsJAZ8 (**C**) in wild type and mutant seedlings in response to blue light treatment. Nipponbare (grey bar) and *phyAphyC* mutant (white bar) seedlings were irradiated at day five with blue light (10 μmol/m^{2}s) for the times indicated. qPCR analysis was carried out using two standard genes, eEF-1α and OsUBI5, for normalization. The fold change induction for each gene was calculated relative to the 0 h control plant, which was defaulted as 1. Data represent the average of three independent experiments. Statistically significant differences to the wild type are indicated by one asterisk (Student's *t*-test, $p < 0.05$).

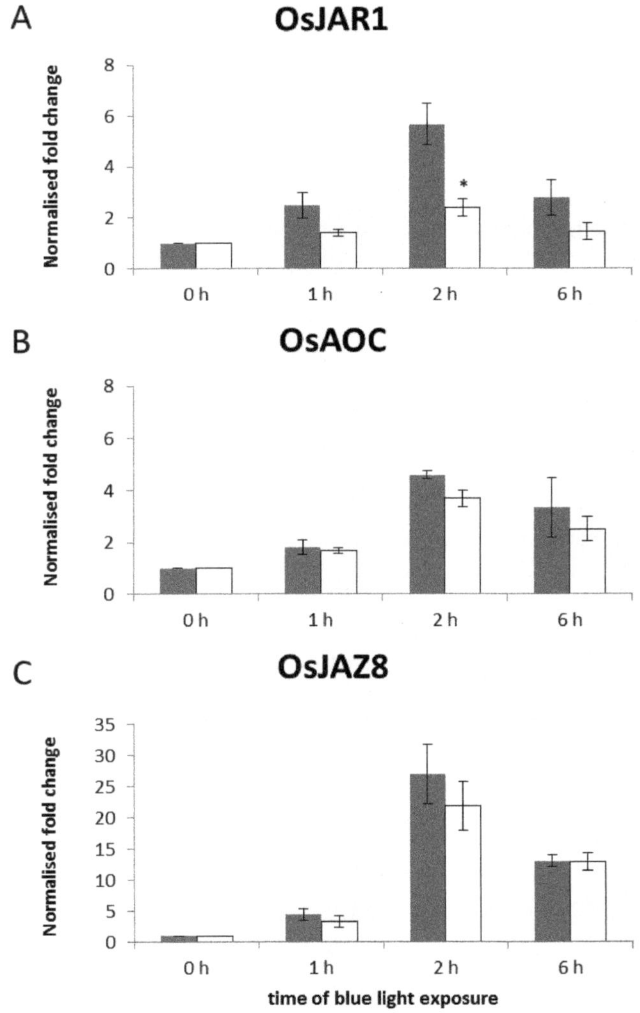

Similar to *OsJAR1*, the JA biosynthesis gene *OsAOC* is also induced transiently by blue light (Figure 2B), in both wild type and mutant *OsAOC* transcript levels are highest at two hours after a blue light pulse.

In the JA-response gene *OsJAZ8* no significant difference between wild type and *phyAphyC* mutants was found (Figure 2C). This gene is induced strongly by blue light, also peaking at 2 h after irradiation (more than 20-fold) in both wild type and mutant.

Thus, amongst the JA-responsive genes we have tested, exclusively the transcriptional regulation of *OsJAR1* was severely affected in *phyAphyC* mutants. It is likely that crys contribute to the transcriptional regulation of JA-dependent genes in blue light. Hirose *et al.* [15] have demonstrated that overexpression of *OsCRY1a* leads to an increased blue-light responsiveness of *OsALLENEOXIDESYNTHASE 1* transcription. The normal induction of *OsAOC* and *OsJAZ8* in the *phyAphyC* mutants might, therefore, be mediated via crys, since photolabile phytochrome is absent. A similar phenomenon has been reported for a mutant in *Arabidopsis* deficient in the chromophore of phytochrome, where, nevertheless, JA-induced genes remained responsive to light [37].

2.3. The phyAphyC Mutant Produces Higher Amounts of Jasmonates in Response to Blue Light

The transcription of *OsJAR1* was only weakly induced by blue light in *phyAphyC* (Figure 2A), and the coleoptiles of this mutant were longer in cBL (Figure 1). However, the two other JA-responsive genes basically retained their responsiveness to blue light in the mutant. Therefore, we examined the hormonal levels of JA and JA-Ile for selected time points during irradiation with blue light (dark, 1 h, 6 h).

From the strongly reduced transcript levels for *OsJAR1* in the *phyAphyC* mutant, the abundance of JA-Ile is expected to be strongly reduced as compared to the wild type, whereas for JA a slight, but significant reduction was predicted. Surprisingly, in contrast to these predictions, *phyAphyC* produces higher levels of JA and JA-Ile after irradiation with blue light (Figure 3). One hour blue light irradiation induces the levels of JA and JA-Ile in both wild type and mutant to the same extent (Figure 3A,B). However, after six hours blue light, JA accumulates to almost three-fold higher (162.6 ng/gFW), and JA-Ile to two-fold higher (5.9 ng/gFW) levels in *phyAphyC* compared to the wild type (53.7 ng/gFW and 2.9 ng/gFW, respectively). An increased JA biosynthesis has also been reported for the already mentioned chromophore-deficient *Arabidopsis* mutant, while in the same study such an effect could not be found in a *phyAphyB* mutant [38].

Although JA and JA-Ile are overproduced in etiolated *phyAphyC* mutants subjected to blue light, these mutants nevertheless develop significantly longer coleoptiles in continuous blue light. On the other hand, mutants with impaired JA biosynthesis develop very long coleoptiles under blue light [31]. We conclude that a jasmonate derivative different from JA or JA-Ile controls coleoptile growth in response to blue light. This could be due to conjugation of JA to an amino acid different from Ile or an entirely different conjugate. In *Arabidopsis* it has been shown that methyl-JA (MeJA) is more effective in inhibiting hypocotyls elongation than JA-Ile [39], which suggests that JA derivatives different from JA-Ile promote growth inhibition in hypocotyls. In addition to JA, auxin may have a positive impact on coleoptile elongation of *phyAphyC* in blue light, as more auxin is present in the mutants (Figure A2). However, similar to the wild type, auxin levels in the mutants are decreasing. Therefore, the observed auxin effect may be a consequence of JA-dependent stimulation of auxin biosynthesis, which has been proposed previously [40]. It is surprising that, not only JA-Ile, but also JA is overproduced in *phyAphyC*, although the mutant develops longer coleoptiles. Hence, it might be possible that there is a phy-dependent, presently unknown step in the bioactivation of jasmonate in young rice seedlings.

Figure 3. JA (**A**) and JA-Ile (**B**) levels of rice seedlings in response to blue light. Etiolated wild type (grey bars) and *phyAphyC* (white bars) seedlings were irradiated with blue light (10 μmol/m^{2}s) at day five for the times indicated. Hormone levels are presented as the average of three independent samples relative to sample fresh weight. Statistically significant differences to the wild type are indicated by two asterisks (Student's *t*-test, $p < 0.01$).

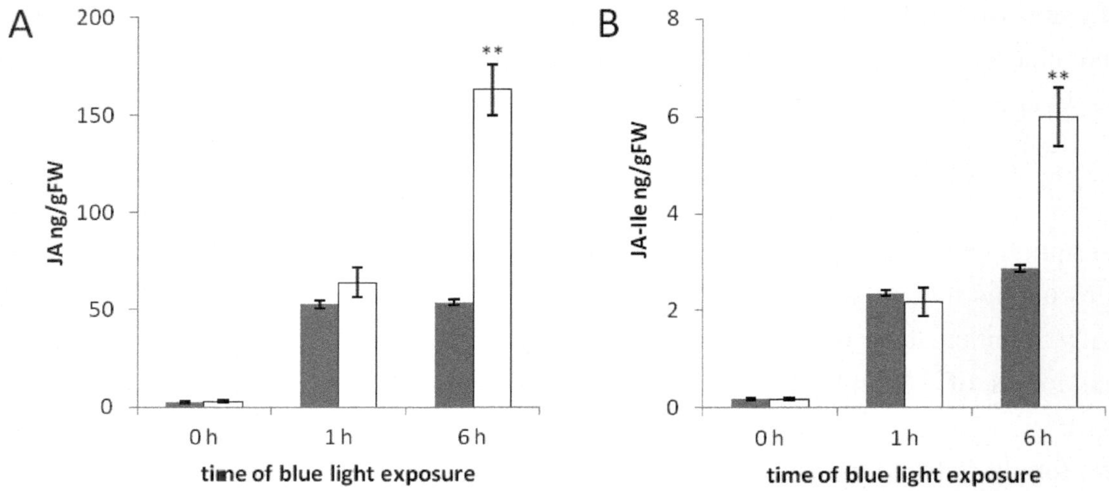

In addition to this unknown JA-dependent branching point responsible for blue-light dependent growth control, there seems to be a branching of the biosynthetic pathway leading to JA-Ile. The higher production of JA-Ile in the phyAphyC mutant contrasts with the severely impaired induction of OsJAR1 transcripts. It might, therefore, be possible that a second enzyme, different from OsJAR1 contributes to the biosynthesis of JA-Ile in young rice seedlings. This assumption would be consistent with our previous observation that JA-Ile still can be synthesized in *osjar1* mutants [31]. It seems to be difficult to identify this enzyme via its transcriptional regulation. At least we found that a putative candidate, *OsJAR2* [31,41], is not regulated on the transcriptional level (Figure A1).

Comparing the temporal regulation of JA levels and gene-expression, we found that *OsJAZ8*, which is not fully induced in *ojsar1* mutants [31], does not respond stronger in *phyAphyC* mutants, although there is a higher amount of active JA present. This finding either means that the induction of *OsJAZ8* transcripts is already saturated by the lower JA levels present in the wild type, or that the expression of this specific *JAZ* gene in blue light is not dependent on phytochrome, or alternatively, that factors different from Jasmonates contribute to the regulation of the gene.

2.4. Less Jasmonates Are Produced in phyAphyC *in Response to Mechanical Wounding*

The results of gene expression analysis and phytohormone levels in *phyAphyC* seedlings raised the question whether the high induction of JA and JA-Ile after blue light might be limited to seedling stage and light treatment or whether other treatments would induce such high JA-amounts as well. We therefore applied a different treatment (wounding), which is known to induce the biosynthesis of JA, at a different developmental stage. In order to wound plants in a reproducible manner, we used the MecWorm system, which mimics the mechanical part of insect feeding [31,42].

We treated rice plants six weeks after germination for specific time intervals using the MecWorm system and measured the hormone levels. In both wild type and *phyAphyC* mutant, JA and JA-Ile were induced due to the wounding treatment (Figure 4A,B). JA and JA-Ile were already strongly induced after 30 min of wounding (wild type: 180.3 ng/gFW and 30.5 ng/gFW, respectively, mutant: 115.2 ng/gFW and 23.6 ng/gFW, respectively). After 1 h of continuous wounding, wild type leaves contained twice the content of JA compared to *phyAphyC* (wild type: 232.8 ng/gFW, mutant: 131.1 ng/gFW; Figure 4A). At the same time point, *phyAphyC* also produced significantly less JA-Ile than the wild type (wild type: 34.2 ng/gFW, mutant: 21.7 ng/gFW; Figure 4B). However, after 6 h of treatment with MecWorm, both Nipponbare and *phyAphyC* showed similar hormone levels, because the wild type response had ceased somewhat decreasing to the levels in *phyAphyC*. In summary, wild type leaves developed a transient peak of JA and JA-Ile after 1 h of wounding, whereas in *phyAphyC*, both hormones remained at the level reached after 30 min.

Figure 4. JA (**A**) and JA-Ile (**B**) levels of rice plants in response to wounding. Plants of wild type (grey bar) and *phyAphyC* (white bar) were wounded with MecWorm six weeks after germination for the indicated time intervals. Hormone levels are depicted relative to sample fresh weight and represented as an average of three independent experiments. C = control plant, W = wounded plant. Statistically significant differences to the wild type are indicated by an asterisk (Student's *t*-test, $p < 0.05$).

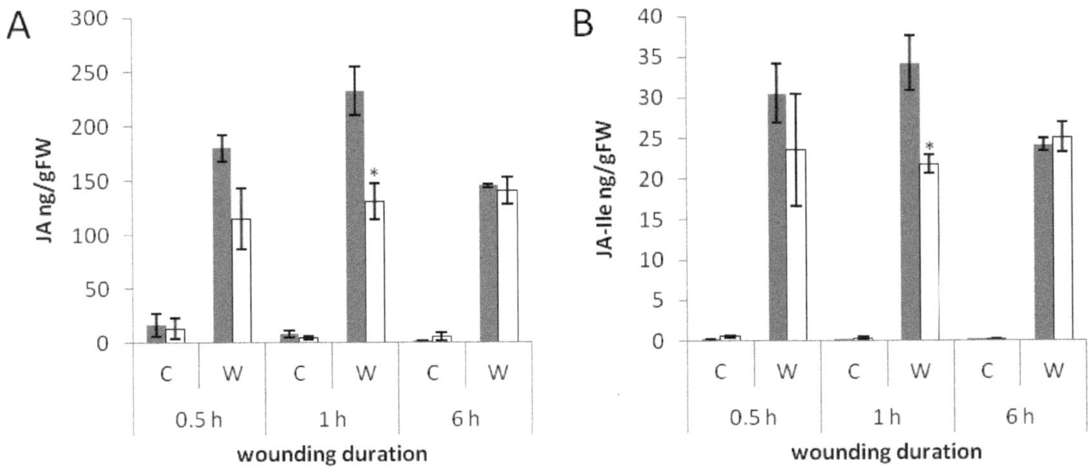

Thus, in contrast to the blue-light response of seedlings, in MecWorm-treated leaves the levels of JA or JA-Ile were not increased in the *phyAphyC* mutant compared to the wild type. We even found that *phyAphyC* produces less JA and JA-Ile than the wild type, when measured after 1 h of MecWorm treatment. Hence, the biosynthesis of JA and JA-Ile is not generally enhanced in *phyAphyC* mutants, but depending on defined developmental stages and stimuli.

2.5. Transcriptional Regulation of JA-Responsive Genes by Continuous Mechanical Wounding

In response to blue light, the induction of *OsJAR1* was impaired in *phyAphyC* while two further JA-responsive genes, *OsAOC* and *OsJAZ8*, were induced as in the wild type (Figure 2). Therefore, we examined the expression of the same genes after wounding with MecWorm.

After 1 h of MecWorm-treatment, two of the genes were not induced at all (*OsJAR1*, *OsAOC*) and the third gene (*OsJAZ8*) was only induced weakly (Figure 5). None of these genes revealed a difference between wild type and *phyAphyC*.

Figure 5. Transcript levels of the representative JA-biosynthesis *OsJAR1* (**A**) and *OsAOC* (**B**), and the JA-signaling gene *OsJAZ8* (**C**) in wild type and mutant plants in response to wounding. Plants of wild type (grey bar) and *phyAphyC* (white bar) were wounded with MecWorm six weeks after germination for the indicated time intervals. qPCR analysis was carried out using two standard genes, eEF-1α and OsUBI5, for normalization. The fold change induction for each gene was calculated relative to a corresponding untreated control plant. Data represent the average of three independent experiments. Statistically significant differences to the wild type are indicated by an asterisk (Student's *t*-test, $p < 0.05$).

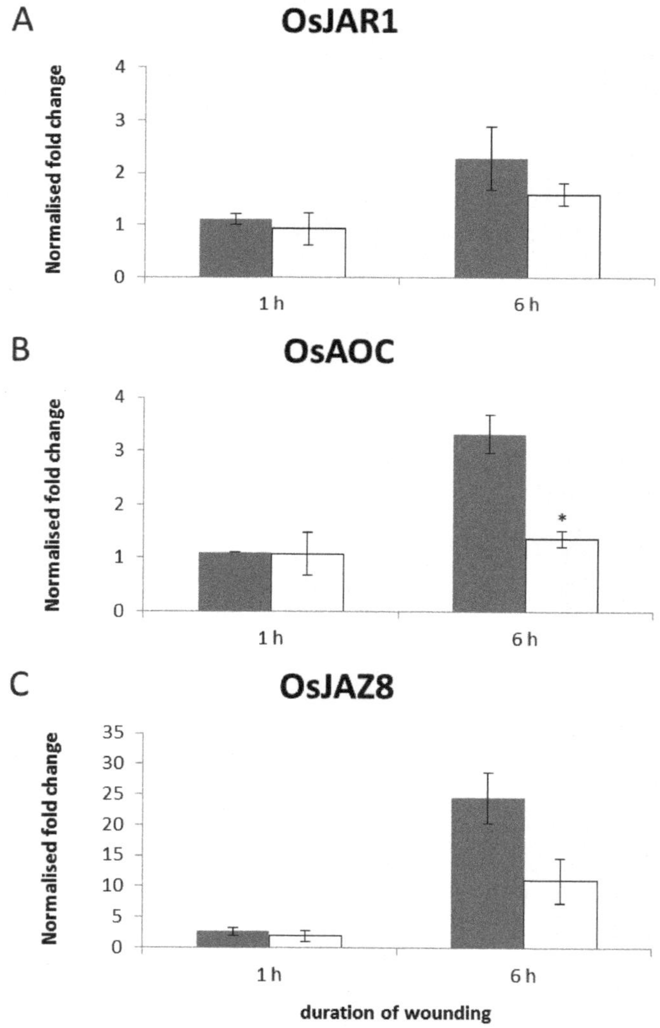

After 6 h of treatment with MecWorm, transcription of *OsAOC* was induced three-fold and *OsJAZ8* 25-fold in the wild type (Figure 5B,C), relative to a non-treated control. The *OsJAR1* transcript was induced only slightly (approximately two-fold; Figure 5A).

In the mutant, the expression of *OsJAR1* in leaves treated for 6 h was only weakly induced with 1.8-fold induction relative to the control, comparable to the results obtained from the wild type (Figure 5A).

The gene expression of *OsAOC*, however, seemed to be affected by the *phyAphyC* mutation. In the mutant, no induction of *OsAOC* could be detected at 6 h treatment (Figure 5B) while in the wild type *OsAOC* was three-fold induced relative to a control plant. The expression of *OsJAZ8* after 6 h MecWorm treatment was reduced in *phyAphyC* by a factor of 2 as compared to the wild type (24.4-fold in the wild type *vs.* 11.0-fold in the mutant, Figure 5C) but the statistical significance of this difference was above the confidence level of 0.05 in a Student's *t*-test.

Despite the impaired *OsAOC*-induction in *phyAphyC*, the JA-level was strongly elevated in the mutant as compared to the control (Figure 4A). This suggests that transcriptional activation of *OsAOC* is not required to sustain the induction of JA in response to wounding.

Compared to the gene expression of blue light irradiated seedlings, the regulation of JA-dependent genes differs in the MecWorm-wounded rice plants. While in the seedlings only *OsJAR1* was expressed differentially in wild type and mutant, in the adult wounded plants it is *OsAOC*.

Whether a given gene of the jasmonate pathway is sensitive to the control by photolabile phytochrome or not, is, thus, not a constant feature of jasmonate signaling, but depends on developmental state and on the quality of stimulation. Trewavas [43] has worked out, that the developmental responses to phytohormones are controlled by the responsiveness of the target tissue rather than by the abundance of the signal. The finding that the phy-dependent changes of jasmonate status depend on developmental state and signal context, *i.e.*, on the competence of the target cells, provides a neat illustration for this point.

3. Experimental

3.1. Plant Material

The *phyAphyC* double mutant (*phyA-2phyC-1* alleles) line was generated by crossing of *phyA* and *phyC* single mutants, which had been obtained by Tos17-mediated mutagenesis. All these mutant lines are described in detail by Takano *et al.* [10].

The wild type background for the *phyAphyC* line is *Oryza sativa* L. ssp. *japonica* cv. Nipponbare, which was used as a control in this study.

All seeds used for the experiments were propagated in the greenhouses of the Botanical Garden of the KIT (Karlsruhe Institute of Technology, Karlsruhe, Germany).

3.2. Plant Cultivation and Treatments

Rice seeds were dehusked and surface sterilized before sowing. Seeds were shaken in 70% ethanol for 1 min and washed with ultrapure H_2O two times. Afterwards, the seeds were incubated in a solution of sodium-hypochlorite (5% w/v, Carl Roth, Karlsruhe, Germany), shaking at 100 rpm for 20 min, and then washed with sterile ultrapure H_2O under sterile conditions.

Subsequently, 20 to 25 seeds were sown into one magenta box (Sigma-Aldrich, St. Louis, MO, USA) on phytoagar (0.6% w/v, Duchefa, Haarlem, The Netherlands).

Seedlings were irradiated with blue light (λ_{max} 470 nm), red light (λ_{max} 650 nm), or far-red light (λ_{max} 735 nm) using custom made LED arrays as described by Qiao *et al.* [44] adjusting to identical fluence rates (10 µmol/m^2s).

For MecWorm treatments, seeds were sown onto 0.4% phytoagar and raised at 25 °C under continuous white light (20 µmol/m^2s). After two weeks, seedlings were transferred to sand and raised for four additional weeks in a phytochamber under short-day conditions (10 h light at 28 °C, 280 µmol/m^2s, 14 h darkness at 22 °C). Once a week plants were fertilized (Wuxal "TopN" and "Super" fertilizers, Manna, Ammerbuch-Pfäffingen, Germany).

The wounding experiments were conducted using the MecWorm system [42]. The youngest, fully developed leaf of a rice plant raised for six weeks was treated for variable time intervals. As a control, the respective leaf of a non-treated plant of the same age, raised in the same phytochamber, was harvested. The details for the MecWorm treatments have been described previously in Svyatyna *et al.* [31].

3.3. Hormone Analysis

For light experiments, plant material was harvested immediately after irradiation and transferred to liquid nitrogen. Fresh weight of each frozen sample was recorded very exactly using an analytical balance, and subsequently the plant material was freeze-dried. Plant hormone extraction and analysis was carried out as described in Svyatyna *et al.* [31].

After wounding experiments, treated and control leaves were harvested directly and frozen in liquid nitrogen. Frozen sample material was ground roughly and fresh weight was measured using an analytical balance. Samples were stored at −80 °C until shipping on dry ice for further analysis. Hormones were extracted by methanol and extracts analyzed using a Zorbax Eclipse XDB-C18 column (50 × 4.6 mm, 1.8 µm, Agilent Technologies, Waldbronn, Germany) installed on an Agilent 1200 HPLC system (Agilent Technologies, Böblingen, Germany) coupled to an API 5000 tandem mass spectrometer (Applied Biosystems, Darmstadt, Germany) as described in Svyatyna *et al.* [31].

3.4. Gene Expression Analysis

Plant material was harvested directly after the end of the respective treatment and frozen in liquid nitrogen. Sample material was stored at −80 °C until RNA extraction.

Samples were homogenized into a frozen powder (TissueLyser, Qiagen, Hilden, Germany) and then RNA was extracted using the innuPREP Plant RNA Kit (Analytik Jena, Jena, Germany) according to the manufacturer's instructions, including the steps to remove genomic gDNA contamination by RNAse-free DNAse (Qiagen, Hilden, Germany).

Only RNA with high quality and integrity, as verified via spectrophotometry and gel electrophoresis, was used for cDNA-synthesis with the DyNAmo cDNA-Sythesis Kit (Thermo Fisher Scientific Inc., Waltham, MA, USA) using oligo-dT primers.

The qPCR analysis was performed using protocol and primers already described in Svyatyna *et al.* [31]. This also applies to the calculation of relative gene expression, which was carried out using the ΔΔCt method, with eEF-1α and OsUBI5 as internal control genes for normalization.

3.5. Statistical Analysis

Data obtained in this study were examined with two different statistical methods, Student's *t*-test and Tukey-Kramer test, with the same results. Confidence levels of Student's *t*-tests were depicted in the respective graphs, one asterisk indicating $p < 0.05$, two asterisks indicating $p < 0.01$.

4. Conclusions

Based on previous findings, that the response to light in rice seedlings depends on jasmonate, and that the transcriptional activation of OsJAR1 is impaired in the *phyAphyC* mutant, we examined the transcriptional and hormonal regulation in this mutant. Surprisingly, JA and JA-Ile were produced at higher levels in *phyAphyC*, and two JA-dependent genes investigated in this study were transcriptionally activated by blue light in the same way as in the wild type. We conclude that the effect of blue light on the activation of JA and JA-Ile biosynthesis proceeds independently of photolabile phytochrome through different blue-light sensitive receptors. Interestingly, although both JA and JA-Ile are present at higher levels after extended exposure to blue light, coleoptiles of *phyAphyC* mutants elongate more compared to the wild type when raised in continuous blue light. Hence, the light response of coleoptile elongation could be mediated by JA derivatives different from JA and JA-Ile. Alternatively, the increased levels of jasmonates in the mutants might affect other hormone pathways such as auxin. The negative regulation of JA biosynthesis by photolabile phytochrome is specific for the light response of coleoptile elongation, and not observed for the wound response of leaves. We conclude that phys in concert with other blue light sensitive receptors regulate the JA-dependent responses to blue light in rice seedlings. Phys might control important enzymatic steps in the metabolism and activation of JA, independent of the canonical positive feedback mechanism of JA on the genes of its own biosynthesis pathway. This model is supported by our observation that the transcription of *OsJAR1* is not induced in *phyAphyC* although this mutant accumulates even more JA than the wild type. Furthermore, we observed increased auxin levels in the *phyAphyC* mutant, in both etiolated and deetiolated seedlings. It is possible that the higher amount of auxin in mutant coleoptiles (Figure A2) contribute to enhanced coleoptile elongation.

Based on our results, we developed a working hypothesis for the role of phyA and phyC in the blue-light induced JA-response. The biosynthesis of JA is induced by blue light, mediated via photoreceptors (including phys, but also other blue-light sensitive receptors). Due to the autocatalytic regulation of JA-synthesis, this would result in a dramatic overshoot of the JA pathway (as characteristic for the wound response). In case of the light response, an equilibrium is formed by a negative feedback, which is, at least partially, mediated by phyA/phyC-dependent signaling via an unknown mechanism (Figure 6). When these photoreceptors are impaired, this negative feedback is removed leading to a severe JA-overproduction. Whether this phy-dependent JA-homeostasis during photomorphogenesis is achieved by inhibition of JA-biosynthesis or by scavenging of JA via induction of synthesis of further JA-derivatives, remains to be elucidated. In order to examine, whether the activity of specific enzymes metabolizing JA to active derivatives is indeed controlled by phyA and phyC we will extend the current investigation to additional photoreceptor mutants, such as cry and phot, as well as combinations of different photoreceptor mutations.

Figure 6. Working model for the modulation of JA-biosynthesis and -metabolism by phyA and phyC during the blue-light response in rice seedlings. Blue light is perceived by photoreceptors (including phy and other blue-light sensitive receptors) leading to initial induction of JA-biosynthesis. PhyA/phyC-dependent signaling constrains JA-synthesis below a certain threshold by either inhibiting further JA-biosynthesis (1) or by inducing JA-depletion via formation of other JA-derivatives apart from JA-Ile (2).

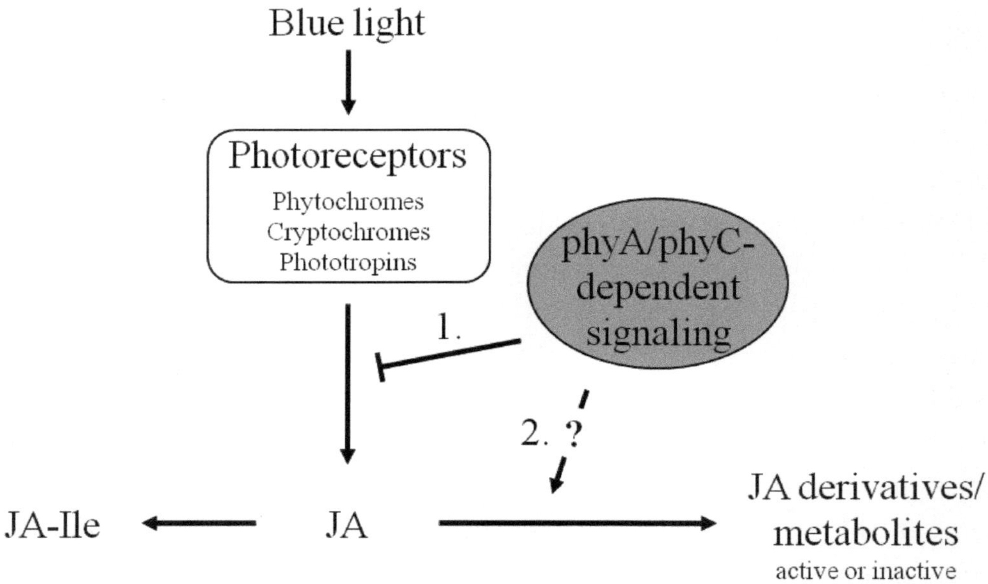

Acknowledgments

R. Brendel and K. Svyatyna were supported by funding from Landesgraduiertenförderungsgesetz of Baden-Württemberg.

Author Contributions

R. Brendel and K. Svyatyna established and performed experiments and statistical analysis. M. Riemann helped to evaluate the data. Y. Jikumaru and Y. Kamiya performed hormone analysis for light experiments. M. Reichelt and A. Mithöfer performed hormone analysis for wounding experiments. K. Svyatyna, M. Takano and A. Mithöfer participated in editing the manuscript and in the design of experiments. R. Brendel, P. Nick and M. Riemann wrote the manuscript.

Conflicts of Interest

The authors declare no conflict of interest.

References

1. Briggs, W.R.; Christie, J.M. Phototropins 1 and 2: Versatile plant blue-light receptors. *Trends Plant Sci.* **2002**, *7*, 204–210.
2. Christie, J.M. Phototropin blue-light receptors. *Annu. Rev. Plant Biol.* **2007**, *58*, 21–45.
3. Lin, C.; Shalitin, D. Cryptochrome structure and signal transduction. *Annu. Rev. Plant Biol.* **2003**, *54*, 469–496.
4. Sancar, A. Structure and function of DNA photolyase and cryptochrome blue-light photoreceptors. *Chem. Rev.* **2003**, *103*, 2203–2237.
5. Furuya, M. Phytochromes—Their Molecular Species, Gene Families, and Functions. *Annu. Rev. Plant Physiol. Plant Mol. Biol.* **1993**, *44*, 617–645.
6. Nagy, F.; Schäfer, E. Phytochromes control photomorphogenesis by differentially regulated, interacting signaling pathways in higher plants. *Annu. Rev. Plant Biol.* **2002**, *53*, 329–355.
7. Quail, P.H. Phytochrome: A light-activated molecular switch that regulates plant gene expression. *Annu. Rev. Genet.* **1991**, *25*, 389–409.
8. Smith, H. Physiological and Ecological Function within the Phytochrome Family. *Annu. Rev. Plant Physiol. Plant Mol. Biol.* **1995**, *46*, 289–315.
9. Takano, M.; Inagaki, N.; Xie, X.; Kiyota, S.; Baba-Kasai, A.; Tanabata, T.; Shinomura, T. Phytochromes are the sole photoreceptors for perceiving red/far-red light in rice. *Proc. Natl. Acad. Sci. USA* **2009**, *106*, 14705–14710.
10. Takano, M.; Inagaki, N.; Xie, X.; Yuzurihara, N.; Hihara, F.; Ishizuka, T.; Yano, M.; Nishimura, M.; Miyao, A.; Hirochika, H.; *et al*. Distinct and cooperative functions of phytochromes A, B, and C in the control of deetiolation and flowering in rice. *Plant Cell* **2005**, *17*, 3311–3325.
11. Smith, H.; Holmes, M.G. The function of phytochrome in natural-environment—III. Measurement and calculation of phytochrome photo-equilibria. *Photochem. Photobiol.* **1977**, *25*, 547–550.
12. Botto, J.F.; Sanchez, R.A.; Whitelam, G.C.; Casal, J.J. Phytochrome a mediates the promotion of seed germination by very low fluences of light and canopy shade light in *Arabidopsis*. *Plant Physiol.* **1996**, *110*, 439–444.
13. Pjon, C.-J.; Furuya, M. Phytochrome action in *Oryza sativa* L.: I. Growth responses of etiolated coleoptiles to red, far-red and blue light. *Plant Cell Physiol.* **1967**, *8*, 709–718.
14. Xie, X.; Shinomura, T.; Inagaki, N.; Kiyota, S.; Takano, M. Phytochrome-mediated inhibition of coleoptile growth in rice: Age-dependency and action spectra. *Photochem. Photobiol.* **2007**, *83*, 131–138.
15. Hirose, F.; Shinomura, T.; Tanabata, T.; Shimada, H.; Takano, M. Involvement of rice cryptochromes in de-etiolation responses and flowering. *Plant Cell Physiol.* **2006**, *47*, 915–925.
16. Fujioka, S.; Li, J.; Choi, Y.H.; Seto, H.; Takatsuto, S.; Noguchi, T.; Watanabe, T.; Kuriyama, H.; Yokota, T.; Chory, J.; *et al*. The Arabidopsis *deetiolated2* mutant is blocked early in brassinosteroid biosynthesis. *Plant Cell* **1997**, *9*, 1951–1962.
17. Li, J.; Nagpal, P.; Vitart, V.; McMorris, T.C.; Chory, J. A role for brassinosteroids in light-dependent development of Arabidopsis. *Science* **1996**, *272*, 398–401.
18. Haga, K.; Iino, M. Phytochrome-mediated transcriptional up-regulation of ALLENE OXIDE SYNTHASE in rice seedlings. *Plant Cell Physiol.* **2004**, *45*, 119–128.

19. Riemann, M.; Müller, A.; Korte, A.; Furuya, M.; Weiler, E.W.; Nick, P. Impaired induction of the jasmonate pathway in the rice mutant hebiba. *Plant Physiol.* **2003**, *133*, 1820–1830.
20. Svyatyna, K.; Riemann, M. Light-dependent regulation of the jasmonate pathway. *Protoplasma* **2012**, *249*, S137–S145.
21. Wasternack, C.; Hause, B. Jasmonates: Biosynthesis, perception, signal transduction and action in plant stress response, growth and development. An update to the 2007 review in Annals of Botany. *Ann. Bot.* **2013**, *111*, 1021–1058.
22. Staswick, P.E.; Tiryaki, I. The oxylipin signal jasmonic acid is activated by an enzyme that conjugates it to isoleucine in Arabidopsis. *Plant Cell* **2004**, *16*, 2117–2127.
23. Sheard, L.B.; Tan, X.; Mao, H.; Withers, J.; Ben-Nissan, G.; Hinds, T.R.; Kobayashi, Y.; Hsu, F.F.; Sharon, M.; Browse, J.; *et al*. Jasmonate perception by inositol-phosphate-potentiated COI1-JAZ co-receptor. *Nature* **2010**, *468*, 400–405.
24. Xie, D.X.; Feys, B.F.; James, S.; Nieto-Rostro, M.; Turner, J.G. COI1: An *Arabidopsis* gene required for jasmonate-regulated defense and fertility. *Science* **1998**, *280*, 1091–1094.
25. Chini, A.; Fonseca, S.; Fernandez, G.; Adie, B.; Chico, J.M.; Lorenzo, O.; Garcia-Casado, G.; Lopez-Vidriero, I.; Lozano, F.M.; Ponce, M.R.; *et al*. The JAZ family of repressors is the missing link in jasmonate signalling. *Nature* **2007**, *448*, 666–671.
26. Thines, B.; Katsir, L.; Melotto, M.; Niu, Y.; Mandaokar, A.; Liu, G.; Nomura, K.; He, S.Y.; Howe, G.A.; Browse, J. JAZ repressor proteins are targets of the SCF(COI1) complex during jasmonate signalling. *Nature* **2007**, *448*, 661–665.
27. Riemann, M.; Haga, K.; Shimizu, T.; Okada, K.; Ando, S.; Mochizuki, S.; Nishizawa, Y.; Yamanouchi, U.; Nick, P.; Yano, M.; *et al*. Identification of rice Allene Oxide Cyclase mutants and the function of jasmonate for defence against *Magnaporthe oryzae*. *Plant. J.* **2013**, *74*, 226–238.
28. Riemann, M.; Bouyer, D.; Hisada, A.; Müller, A.; Yatou, O.; Weiler, E.W.; Takano, M.; Furuya, M.; Nick, P. Phytochrome A requires jasmonate for photodestruction. *Planta* **2009**, *229*, 1035–1045.
29. Sineshchekov, V.A.; Loskovich, A.V.; Riemann, M.; Nick, P. The jasmonate-free rice mutant *hebiba* is affected in the response of phyA'/phyA'' pools and protochlorophyllide biosynthesis to far-red light. *Photochem. Photobiol. Sci.* **2004**, *3*, 1058–1062.
30. Radhika, V.; Kost, C.; Mithöfer, A.; Boland, W. Regulation of extrafloral nectar secretion by jasmonates in lima bean is light dependent. *Proc. Natl. Acad. Sci. USA* **2010**, *107*, 17228–17233.
31. Svyatyna, K.; Jikumaru, Y.; Brendel, R.; Reichelt, M.; Mithöfer, A.; Takano, M.; Kamiya, Y.; Nick, P.; Riemann, M. Light induces jasmonate-isoleucine conjugation via OsJAR1-dependent and -independent pathways in rice. *Plant Cell Environ.* **2013**, doi:10.1111/pce.12201.
32. Riemann, M.; Takano, M. Rice JASMONATE RESISTANT 1 is involved in phytochrome and jasmonate signalling. *Plant Cell Environ.* **2008**, *31*, 783–792.
33. Yamada, S.; Kano, A.; Tamaoki, D.; Miyamoto, A.; Shishido, H.; Miyoshi, S.; Taniguchi, S.; Akimitsu, K.; Gomi, K. Involvement of OsJAZ8 in jasmonate-induced resistance to bacterial blight in rice. *Plant Cell Physiol.* **2012**, *53*, 2060–2072.
34. Terol, J.; Domingo, C.; Talon, M. The GH3 family in plants: Genome wide analysis in rice and evolutionary history based on EST analysis. *Gene* **2006**, *371*, 279–290.

35. Jain, M.; Kaur, N.; Tyagi, A.K.; Khurana, J.P. The auxin-responsive GH3 gene family in rice (*Oryza sativa*). *Funct. Integr. Genomics* **2006**, *6*, 36–46.
36. Ye, H.; Du, H.; Tang, N.; Li, X.; Xiong, L. Identification and expression profiling analysis of TIFY family genes involved in stress and phytohormone responses in rice. *Plant Mol. Biol.* **2009**, *71*, 291–305.
37. Costigan, S.E.; Warnasooriya, S.N.; Humphries, B.A.; Montgomery, B.L. Root-localized phytochrome chromophore synthesis is required for photoregulation of root elongation and impacts root sensitivity to jasmonic acid in *Arabidopsis*. *Plant Physiol.* **2011**, *157*, 1138–1150.
38. Zhai, Q.; Li, C.B.; Zheng, W.; Wu, X.; Zhao, J.; Zhou, G.; Jiang, H.; Sun, J.; Lou, Y.; Li, C. Phytochrome chromophore deficiency leads to overproduction of jasmonic acid and elevated expression of jasmonate-responsive genes in *Arabidopsis*. *Plant Cell Physiol.* **2007**, *48*, 1061–1071.
39. Chen, J.; Sonobe, K.; Ogawa, N.; Masuda, S.; Nagatani, A.; Kobayashi, Y.; Ohta, H. Inhibition of arabidopsis hypocotyl elongation by jasmonates is enhanced under red light in phytochrome B dependent manner. *J. Plant Res.* **2013**, *126*, 161–168.
40. Sun, J.; Xu, Y.; Ye, S.; Jiang, H.; Chen, Q.; Liu, F.; Zhou, W.; Chen, R.; Li, X.; Tietz, O.; *et al.* *Arabidopsis* ASA1 is important for jasmonate-mediated regulation of auxin biosynthesis and transport during lateral root formation. *Plant Cell* **2009**, *21*, 1495–1511.
41. Wakuta, S.; Suzuki, E.; Saburi, W.; Matsuura, H.; Nabeta, K.; Imai, R.; Matsui, H. OsJAR1 and OsJAR2 are jasmonyl-L-isoleucine synthases involved in wound- and pathogen-induced jasmonic acid signalling. *Biochem. Biophys. Res. Commun.* **2011**, *409*, 634–639.
42. Mithöfer, A.; Wanner, G.; Boland, W. Effects of feeding *Spodoptera littoralis* on lima bean leaves. II. Continuous mechanical wounding resembling insect feeding is sufficient to elicit herbivory-related volatile emission. *Plant Physiol.* **2005**, *137*, 1160–1168.
43. Trewavas, A.J. Growth substance sensitivity—The limiting factor in plant development. *Physiol. Plant.* **1982**, *55*, 60–72.
44. Qiao, F.; Petrášek, J.; Nick, P. Light can rescue auxin-dependent synchrony of cell division in a tobacco cell line. *J. Exp. Bot.* **2010**, *61*, 503–510.

Appendix

Figure A1. Transcript levels of *OsJAR2* in wild type and mutant seedlings in response to blue light treatment. Nipponbare (grey bar) and *phyAphyC* mutant (white bar) seedlings were irradiated with blue light (10 μmol/m^{2}s) for the times indicated. qPCR analysis was carried out using two standard genes, eEF-1α and OsUBI5, for normalisation. Data represent the average of three independent experiments. No statistically significant induction of *OsJAR2* was found.

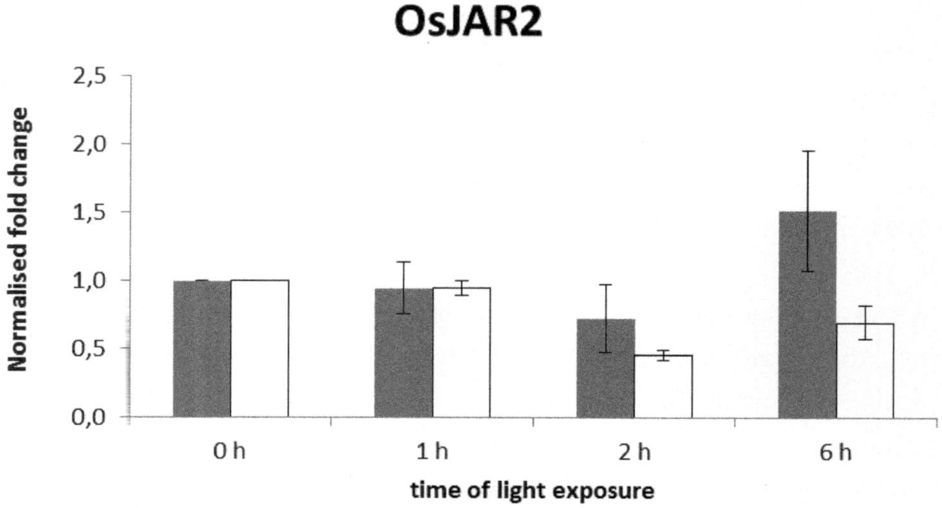

Figure A2. IAA levels of rice seedlings in response to blue light. Etiolated wild type (grey bars) and *phyAphyC* (white bars) seedlings were irradiated with blue light (10 μmol/m^{2}s) at day 5 for the times indicated. Hormone levels are shown relative to sample fresh weight and represent an average of three independent experiments. Statistically significant differences to the wild type are indicated by two asterisks (Student's *t*-test, $p < 0.01$).

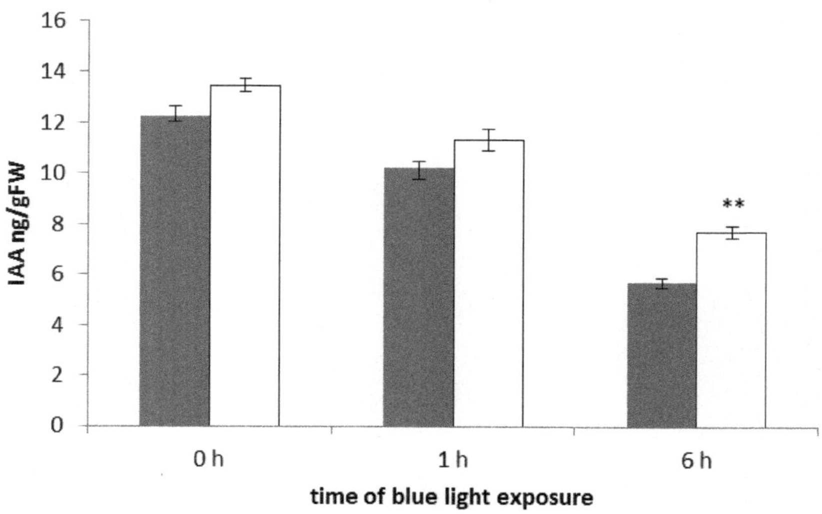

5

Cell Wall Loosening in the Fungus, *Phycomyces blakesleeanus*

Joseph K. E. Ortega *, Jason T. Truong †, Cindy M. Munoz † and David G. Ramirez †

Bioengineering Laboratory, Mechanical Engineering Department, University of Colorado Denver, Denver, CO 80217, USA; E-Mails: jcttruong@gmail.com (J.T.T.); Cindy.M.Munoz@ucdenver.edu (C.M.M.); ramiredg@gmail.com (D.G.R.)

† These authors contributed equally to this work.

* Author to whom correspondence should be addressed; E-Mail: Joseph.Ortega@ucdenver.edu

Academic Editor: Philip J. Harris

Abstract: A considerable amount of research has been conducted to determine how cell walls are loosened to produce irreversible wall deformation and expansive growth in plant and algal cells. The same cannot be said about fungal cells. Almost nothing is known about how fungal cells loosen their walls to produce irreversible wall deformation and expansive growth. In this study, anoxia is used to chemically isolate the wall from the protoplasm of the sporangiophores of *Phycomyces blakesleeanus*. The experimental results provide direct evidence of the existence of chemistry within the fungal wall that is responsible for wall loosening, irreversible wall deformation and elongation growth. In addition, constant-tension extension experiments are conducted on frozen-thawed sporangiophore walls to obtain insight into the wall chemistry and wall loosening mechanism. It is found that a decrease in pH to 4.6 produces creep extension in the frozen-thawed sporangiophore wall that is similar, but not identical, to that found in frozen-thawed higher plant cell walls. Experimental results from frozen-thawed and boiled sporangiophore walls suggest that protein activity may be involved in the creep extension.

Keywords: *Phycomyces*; expansive growth; wall loosening; chemorheology

---

1. Introduction

Cellular expansive growth is central to the development, morphogenesis and sensory responses of plants, algae and fungi. Plant, algal and fungal cells have walls. These cells regulate the mechanical behavior of their walls during expansive growth to control the growth rate, morphogenesis and growth responses to environmental stimuli [1–3]. During expansive growth, water uptake produces turgor pressure that deforms the cell wall. Walls exhibit both irreversible and reversible (elastic) deformation during expansive growth [4–6]. The wall deformation behavior is similar to that of a Maxwell–Bingham viscoelastic model, which consists of a dashpot (filled with a Bingham fluid) in series with an elastic spring [4,6]. The constitutive equation (stress-strain relationship) for a Maxwell–Bingham viscoelastic model, Equation (1), describes the strain rate as equal to the sum of the irreversible extension rate of the dashpot filled with a Bingham fluid and the reversible (elastic) extension rate of a spring [4].

$$\frac{1}{L}\frac{\mathrm{d}L}{\mathrm{d}t} = \frac{1}{\mu}(\sigma - \sigma_C) + \frac{1}{E}\frac{\mathrm{d}\sigma}{\mathrm{d}t} \tag{1}$$

Strain rate = irreversible extension rate of a dashpot + reversible extension rate of a spring

where L is the length of the viscoelastic model, t is the time, μ is the dynamic viscosity of the Bingham fluid, σ is the stress, σ_C is the critical stress of the Bingham fluid and E is the longitudinal elastic modulus of the spring. The term, $(\mathrm{d}L/\mathrm{d}t)/L$, is the strain rate, and the term, $(\sigma - \sigma_C)/\mu$, is the irreversible extension rate of the dashpot filled with a Bingham fluid, which behaves like a Newtonian fluid after a critical stress, σ_C, is exceeded. The term, $(\mathrm{d}\sigma/\mathrm{d}t)/E$, is the reversible (elastic) extension rate of the spring.

Both irreversible and reversible wall deformations are required for expansive growth. Reversible wall deformation is responsible for most of the turgor pressure production during water uptake. Irreversible wall deformation is responsible for the permanent increase in cell volume. This wall deformation behavior is implicit in the augmented growth equation, Equation (2), that was derived from the constitutive equation for a Maxwell–Bingham model, Equation (1), and which provides a quantitative mathematical description of the expansive growth rate as a function of the wall deformation rate of cells with walls [4].

$$\frac{1}{V}\frac{\mathrm{d}V}{\mathrm{d}t} = \phi\,(P - P_C) + \frac{1}{\varepsilon}\frac{\mathrm{d}P}{\mathrm{d}t} \tag{2}$$

Volume expansion rate = irreversible wall deformation rate + elastic wall deformation rate

where V is the volume of the cell wall chamber, t is the time, ϕ is the relative wall extensibility, P is the turgor pressure, P_C is the critical turgor pressure and ε is the volumetric elastic modulus. The term, $\mathrm{d}V/V\mathrm{d}t$, represents the relative rate of change in volume of the cell wall chamber; the term, $\phi\,(P - P_C)$, represents the relative irreversible wall deformation rate; and the term, $(\mathrm{d}P/\mathrm{d}t)/\varepsilon$, represents the relative reversible (elastic) wall deformation rate.

The augmented growth equation has been adapted and experimentally validated for diffuse growth, tip growth, intercalary growth, cellular growth within tissue and growth responses to environmental stimuli [2,5–7]. It is hypothesized that chemistry within the wall breaks and makes load-bearing bonds between wall polymers to loosen and harden the wall and regulate temporal and spatial irreversible and elastic wall deformations (chemorheology) during expansive growth and morphogenesis [1,2,5–7]. Furthermore, it is hypothesized that wall chemistry and chemorheological processes control the magnitude and behavior of the inclusive biophysical variables within the augmented growth equation to regulate the expansive growth rate [1,2,5–7]. Interestingly, the augmented growth equation was recently derived from fundamental thermodynamic principles as a particular case of a more general growth equation [8], and the evolution of ϕ and ε as a function of dV/Vdt was correctly predicted.

Investigations with higher plant tissue provide experimental evidence indicating acidic pH initiates wall chemistry that promotes wall loosening [9]. It is shown that low pH produces continuous deformation (creep) of frozen-thawed cell walls of higher plant cells when a constant stress is applied [10,11]. Thus, a chemorheological process promotes creep in frozen-thawed cell walls. Investigators have identified and isolated pH-dependent proteins (expansins) that mediate the wall loosening and creep in frozen-thawed plant tissue [1,11]. Expansins and their genes have been found in higher plants, fungi and bacteria [1,12]. Importantly, expansins have been shown to promote creep and irreversible wall deformation in frozen-thawed tissue from many different plants and are hypothesized to play a major role in the expansive growth of higher plants [1]. Endogenous wall enzymes are also hypothesized to play a role in the expansive growth of higher plants (e.g., xyloglucan endotransglycosylase/hydrolase and endo-β-1,4-glucanases), but they have not been shown to promote creep in frozen-thawed cell walls [1,13]. Experimental evidence indicates that another chemorheological process (calcium pectate cycle) plays a major role in the expansive growth of pollen tubes [14] and algae [15] and may play a role in the expansive growth of higher plant cells [3,8].

Importantly, almost nothing is known about the mechanism of wall loosening in fungal cell walls, although a considerable amount is known about fungal cell wall structure, synthesis and assembly [16–18]. In the current study, experiments are conducted to obtain insight into the wall loosening mechanism of the large single-celled sporangiophores of the fungus, *Phycomyces blakesleeanus* [17,18]. The cell wall mechanical properties of these sporangiophores have been studied extensively during expansive growth and exhibit wall deformation characteristics that are similar to those of higher plant cell walls [2,6,7]. It was shown that their walls exhibit deformation behavior similar to that of a Maxwell-Bingham viscoelastic model and that the augmented growth equations accurately describe their expansive growth rate [2,4,6,7].

In the present experimental investigation, anoxia is used to chemically isolate the wall from the protoplasm of the sporangiophores of *P. blakesleeanus*. Anoxia terminates the sporangiophore's metabolism [17,19] and essentially stops the export of wall polymers and materials from the protoplasm to the wall, thus chemically isolating the wall. The experimental results provide direct evidence of the existence of chemistry and chemorheology within the fungal wall that are responsible for wall loosening, irreversible wall deformation and elongation growth.

In addition, constant-tension extension experiments are conducted on frozen-thawed sporangiophore walls to obtain insight into the wall chemistry and chemorheology that produce irreversible wall deformation. It is found that a decrease in pH to 4.6 produces creep in frozen-thawed sporangiophore walls, similar to higher plant cells. Frozen-thawed sporangiophore walls that were boiled for 15 s before constant-tension extension experiments did not exhibit creep, again similar to higher plant cells. This inactivation of endogenous wall-loosening activity may suggest mediation by an expansin-like protein and/or mechanism. However, unlike higher plant cells, the creep of the sporangiophores' walls continues for minutes, not hours. This result may indicate that the chemistry and chemorheology that produce creep and wall loosening in the sporangiophore's wall may be fundamentally different from that in higher plant cells.

2. Results

2.1. Anoxia Experiments

Growing Stage IV sporangiophores of the fungus *P. blakesleeanus* (Figure 1) were subjected to an anoxic environment in order to terminate the sporangiophore's metabolism [17,19], thus eliminating the export of wall polymers and materials from the protoplasm to the wall. Experiments were conducted as follows. A Stage IV sporangiophore was selected and placed in a chamber made of transparent acrylic that was constructed specifically for these experiments. The sporangiophore was adapted in the experimental chamber to atmospheric air (~21% oxygen concentration), constant and symmetric distribution of light and other environmental conditions for 20 min. The light and environmental conditions were maintained constant throughout the adaptation period and during the experiment to prevent growth responses. After the adaptation period, the elongation growth was measured for the remainder of the experiment (Figure 2). Ten minutes afterwards, the sporangiophore was impaled with the micro-capillary tip of a pressure probe (downward pointing arrow on the time scale and labeled "Impaled"), and the turgor pressure was measured for the remainder of the experiment. At approximately 27 min on the time scale (shown by the downward point arrow labeled "Anoxia"), the chamber was filled with nitrogen gas, which decreased the oxygen concentration to less than 1%.

The results presented in Figure 2 show that both the elongation rate (slope of the elongation curve, ΔL) and turgor pressure (P) remain nearly constant between 10 min and 27 min on the time scale. The sharp and immediate decrease in turgor pressure occurs at the exact time when the oxygen concentration decreased from 21% to less than 1%. It can be seen that the turgor pressure continues to decrease for the remainder of the experiment. Interestingly, the turgor pressure decreases slowly for approximately 40 min and then suddenly decreases exponentially. This turgor pressure behavior was typical of more than five experiments conducted.

The elongation decreases to zero within a few minutes after the initiation of anoxia and continues to decrease slowly for another 40–50 min. Afterwards, the elongation decreases faster for the remainder of the experiment until the sporangium falls over onto the stalk. The decrease in elongation represents a decrease in the length of the sporangiophore stalk. The shrinking in length can be explained as "recovered" elastic wall deformation that occurs when the turgor pressure decreases in magnitude. This recovered elastic wall deformation will obscure any irreversible wall deformation (expansive growth) that may occur during anoxia. It is apparent that the turgor pressure must remain constant in order to measure irreversible wall deformation.

The pressure probe was used to measure and control the turgor pressure in the sporangiophore before and during anoxia. Figure 3 shows the results from one of these experiments. As before, a Stage IV sporangiophore was adapted to constant environmental conditions in the chamber for 20 min before elongation measurements began, and the conditions remained constant for the remainder of the experiment. The turgor pressure measurements began at approximately 12 min on the time scale, and the oxygen concentration was reduced to less than 1% at 25 min on the time scale. The turgor pressure was held constant at the value that was measured before the initiation of anoxia. Careful inspection of the elongation curve shows that elongation continues slowly for 15 min after the initiation of anoxia.

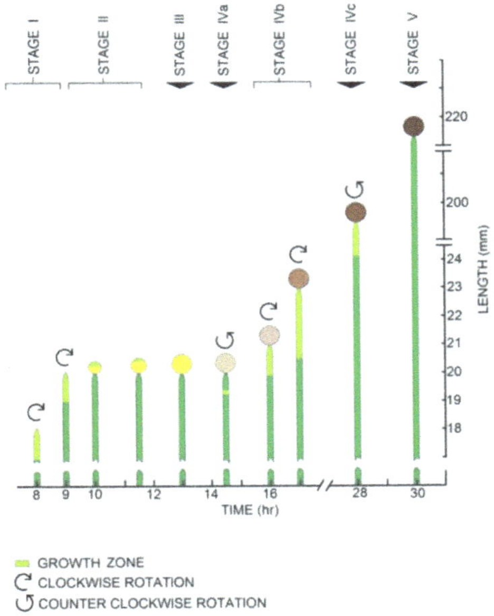

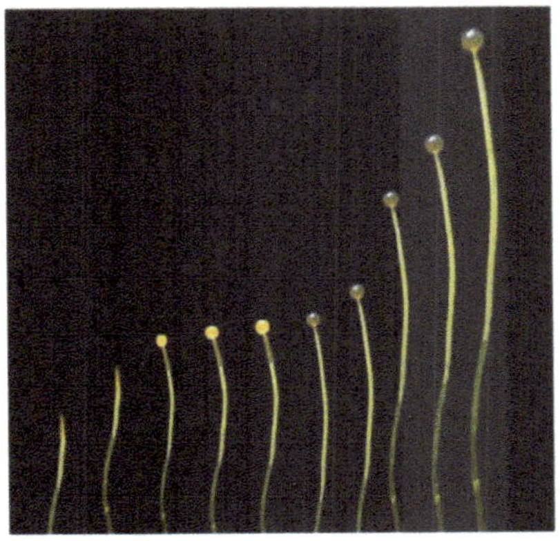

Figure 1. The schematic illustration (**left**) and corresponding photograph (**right**) show the developmental stages of the sporangiophore of *P. blakesleeanus*. Sporangiophore development is divided into five stages (Stages I, II, III, IV and V), and Stage IV is further divided into three sub-stages (IVa, IVb and IVc). The sporangiophores exhibit expansive growth and irreversible wall deformation in a region termed the "growth zone". The growth zone is located at the apical tip (tip growth) in Stage I and Stage II and adjacent to the sporangium (intercalary growth) in Stage IV. The Stage I sporangiophore is a single pointed cell that grows longitudinally at the apical tip in the growth zone, 1–2 mm in length. In the schematic, the growth zone is shown as light green and the non-growing stalk is dark green. Clockwise rotation (when viewed from above) and elongation growth occur concurrently during Stage I, producing left-handed helical growth. Stage II begins with spherical growth at the apical tip without elongation and rotation growth. The diameter of the sporangium continues to increase until Stage III, where the diameter is constant (~0.5 mm). During Stage III, there is no visible expansive growth. Stage IVa begins with elongation growth concurrent with counter-clockwise rotation growth (right-handed helical growth) in a short growth zone located approximately 0.6 mm below the sporangium. The growth zone, elongation growth rate and rotation growth rate increase in magnitude as the right-handed helical growth continues for approximately one hour. Then, the rotation rate gradually decreases to zero, and clockwise rotation begins without interruption of elongation growth. Stage IVb begins with the initiation of left-handed helical growth. Stage IVb exhibits nearly constant growth zone length (~2.5 mm), elongation growth rates (~35 $\mu m \cdot min^{-1}$) and rotation growth rates (~12 $degrees \cdot min^{-1}$) for many hours. Stage IVb sporangiophores are typically used for most biophysical experimental studies. Stage IVc is initiated by counter-clockwise rotation and right-handed helical growth. Stage V does not exhibit visible expansive growth. The large cylindrical single-celled sporangiophores are approximately 150 μm in diameter and can grow in length to ten or more centimeters. The sporangiophores can detect many environmental stimuli (e.g., gravitational acceleration, ethylene, mechanical stretch, gases, temperature, wind, light intensity, spatially asymmetric distribution of light, the presence of solid objects and turgor pressure) and respond to these stimuli with symmetric and asymmetric changes in expansive growth rate (e.g., growth responses and tropic responses) [7,17,18].

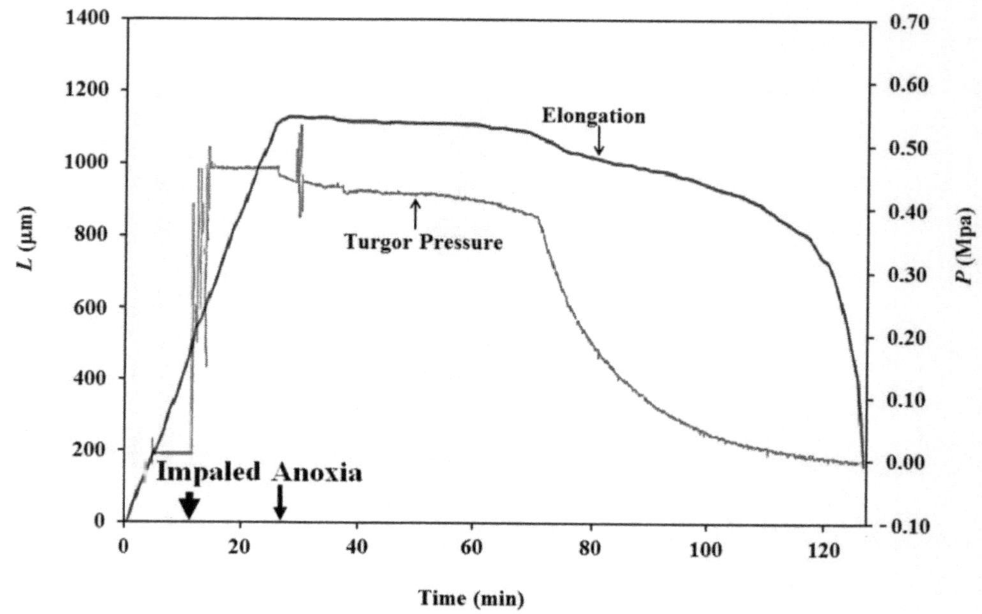

Figure 2. The change in elongation, ΔL, and turgor pressure, P, of a single Stage IV sporangiophore is plotted against the same time scale. Turgor pressure measurements begin at 10 min on the time scale and are marked by the downward pointing arrow labeled "Impaled". The turgor pressure curve (gray) is the trace from the strip-chart recorder that measures the output of the pressure transducer in the pressure probe. The oxygen concentration is decreased from 21% to less than 1% at approximately 27 min on the time scale and is marked by the downward pointing arrow labeled "Anoxia". The immediate small decrease in turgor pressure from a constant value indicates the exact time when the oxygen concentration was decreased to less than 1%. Jessica E.C. Olson and Elena L. Ortega (two graduate students) conducted this experiment while working in Joseph K.E. Ortega's laboratory.

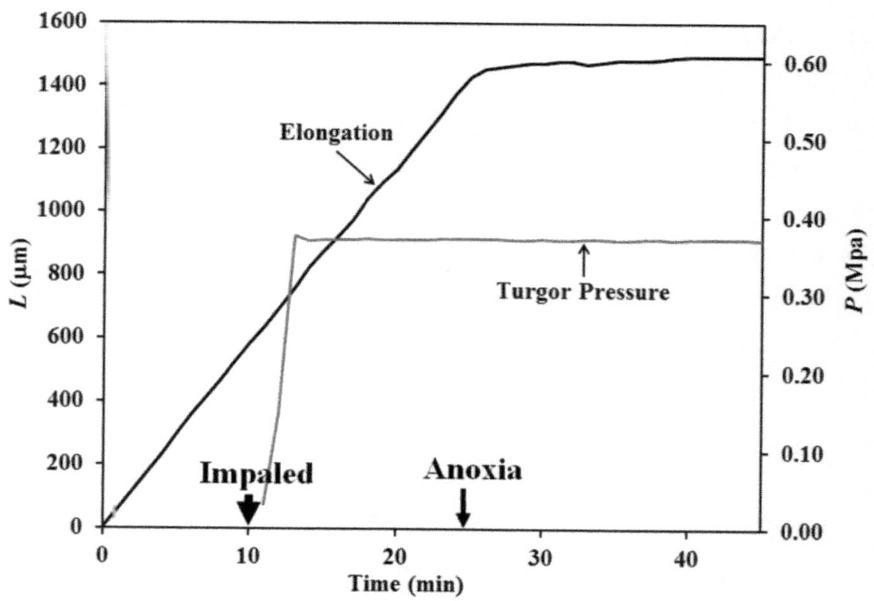

Figure 3. Both the change in elongation and turgor pressure of a Stage IV sporangiophore are plotted as a function of time. The turgor pressure was held constant with the pressure probe during anoxia.

There was some concern that the oil-cell sap interface within the micro-capillary tip of the pressure probe was unable to move while the turgor pressure was held constant during anoxia. The cell sap of the sporangiophore is very sticky and has a tendency to adhere to the inner wall of the micro-capillary tip and prevents an accurate measurement of the turgor pressure. In order to prevent the oil-cell sap from sticking in one location, small step-ups in turgor pressure were produced by injecting inert silicon oil inside the micro-capillary tip into the vacuole of the sporangiophore (Figure 4). As can be seen, turgor pressure measurements began at approximately ten minutes on the time scale, and an anoxic environment was produced at 22 min. Careful inspection of the elongation curve reveals that elongation growth occurred 2–5 min after each turgor pressure step-up (when P = constant) until approximately 45 min. The elongation that occurs one minute after the step-up in turgor pressure is predominately an elastic response and is not used to determine elongation growth and irreversible wall deformation. For this experiment, it is concluded that elongation and irreversible wall deformation continue for 23 min after anoxia.

Similar step-up experiments were conducted on Stage I sporangiophores (Figure 5). The Stage I sporangiophore was impaled to measure the turgor pressure at 10 min on the time scale, and anoxia was produced at 21 min. Inspection of the elongation curve shows that elongation growth and irreversible wall extensibility continued until approximately 27 min, or 6 min after anoxia. It was found that the duration of elongation growth and irreversible wall deformation after anoxia was shorter for Stage I sporangiophores compared to Stage IV sporangiophores.

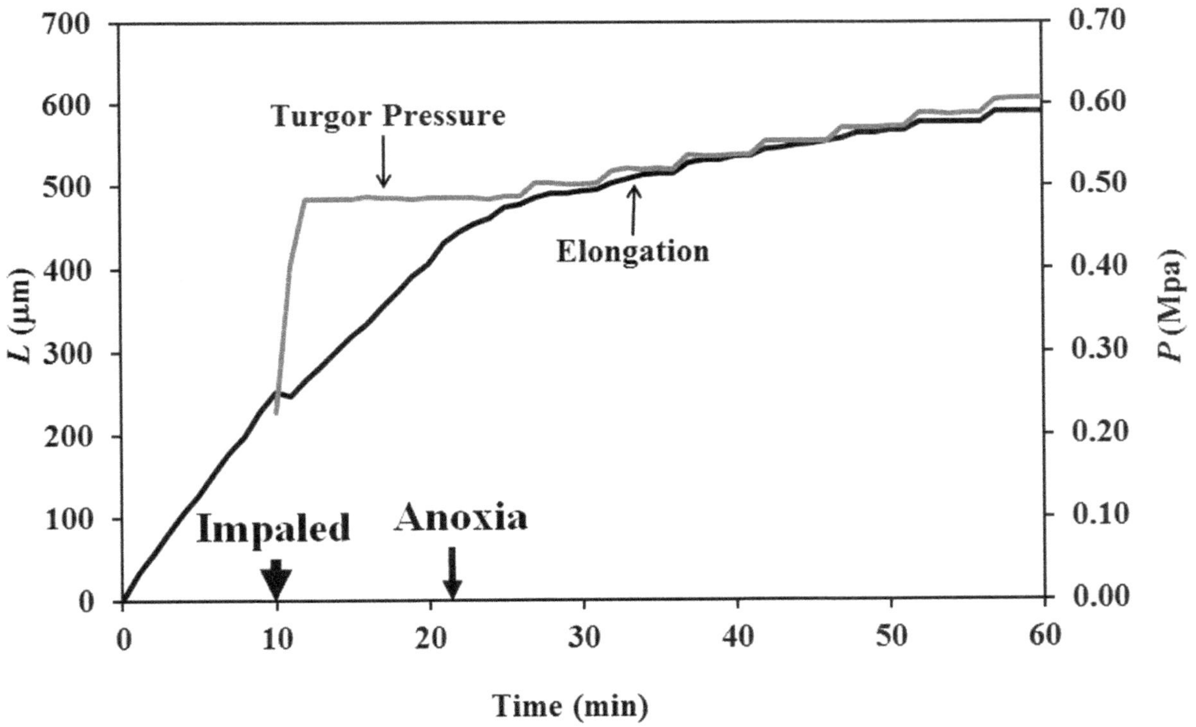

Figure 4. Both the change in elongation and turgor pressure of a Stage IV sporangiophore are plotted as a function of time. Approximately 5 min after anoxia, turgor pressure step-ups (approximately 0.021 MPa in magnitude and 5 min in duration) were produced with the pressure probe.

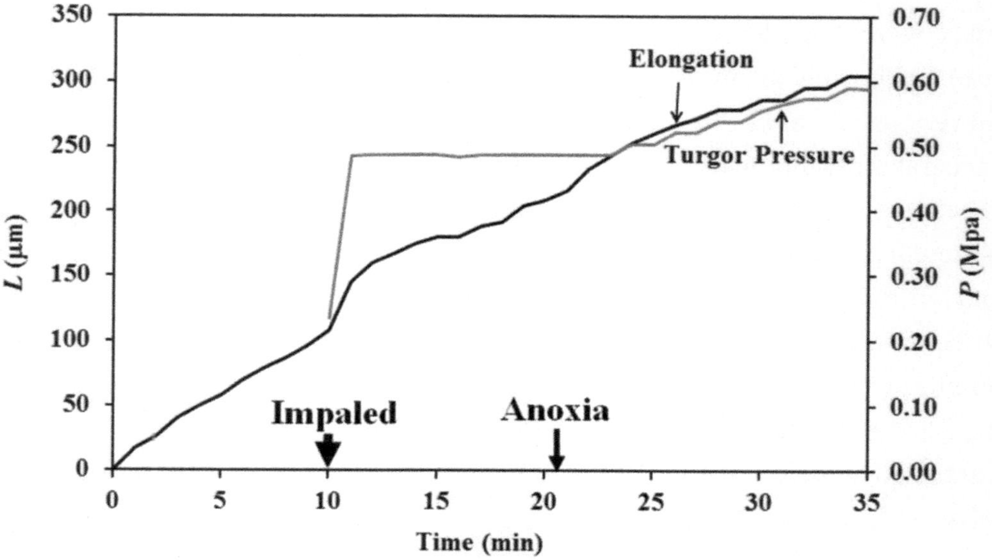

Figure 5. Change in elongation and turgor pressure for a Stage I sporangiophore are plotted as a function of time. Two minutes after anoxia, turgor pressure step-ups (approximately 0.021 MPa in magnitude and 2 min in duration) were produced with the pressure probe.

Some of the results of these experiments are summarized in Table 1, which compares the "duration of elongation during anoxia" obtained from Stage IV sporangiophores when the turgor pressure is held constant and when the turgor pressure is stepped up. Results of Student's *t*-test indicate that there is not a significant difference between the values obtained when *P* was held constant and when *P* was stepped-up. Because the values are statistically the same, they are combined. The results indicate that, on average, the Stage IV sporangiophore continues to exhibit elongation growth and irreversible wall deformation for approximately 17 min during anoxia. Because the Student's *t*-test indicates that there is no significant difference between the durations obtained from experiments when *P* was constant and when *P* was stepped up, only the step-up in *P* experiments was conducted for Stage I sporangiophores. The results presented in Table 1 indicate that the duration of elongation growth for Stage I sporangiophores is smaller than the corresponding values obtained from Stage IV sporangiophores. Student's *t*-tests confirm this observation: the duration of elongation during anoxia of Stage I sporangiophores is significantly smaller ($p < 0.05$) than that obtained from Stage IV sporangiophores when *P* was constant, when *P* was stepped up and when the values were combined.

Table 1. Duration of elongation growth during anoxia for Stage IV and Stage I sporangiophores.

Stage	Turgor Pressure during Anoxia	Elongation Duration during Anoxia (min)
IV	P = constant	14.8 ± 2.1 SE (n = 5)
IV	Step-up in P	18.7 ± 3.7 SE (n = 6)
IV	Combined	16.9 ± 2.1 SE (n = 11)
I	Step-up in P	10.0 ± 1.2 SE (n = 9)

2.2. Frozen-Thawed Sporangiophores

Constant-tension extension experiments were conducted on frozen and then thawed walls of the Stage IV sporangiophore to obtain insight into the wall chemistry and chemorheological process. A five millimeter-long section of the wall from a Stage IV sporangiophore, which includes the growth zone, was adapted in an experimental chamber to a bathing solution of pure water (70 mL, pH = 7, at 20 °C) for 20 min. Then, a tensile load of 1.24 g was applied to the wall. The tensile load produces longitudinal stress and longitudinal extension of the wall. After the extension, the length remained constant for four minutes without creep; see Figure 6 (0–4 min). At four minutes on the time scale, the pH of the solution was lowered to 4.6 by adding a predetermined amount of pH Red 4.0 buffer to the bathing water. Immediately, the wall extends (red curve with X data points and labeled "Frozen-thawed") and continues to extend for six minutes (4–10 min), exhibiting five minutes of creep (5–10 min). Also shown in Figure 6 are the results of another experiment where the same experimental protocol was conducted on another frozen-thawed wall, except that the frozen-thawed wall was immersed in boiling water for 15 s before being subjected to the experimental protocol (blue curve with • data points and labeled "Frozen-thawed-boiled"). It can be seen that the wall extends after the pH of the bathing solution was reduced to 4.6; however, the initial extension is small, and the wall does not continue to extend afterwards, *i.e.*, it does not creep.

These experiments were repeated many times. Figure 7 is a plot of the average change in extension at each minute on the time scale of 15 different frozen-thawed walls (red curve with X data points and labeled "Frozen-thawed") and eight frozen-thawed-boiled walls (blue curve with • data points and labeled "Frozen-thawed-boiled"). The results plotted in Figure 7 indicate that the frozen-thawed walls creep for eight minutes (5–13 min) after the pH of the bathing solution is reduced from 7.0 to 4.6.

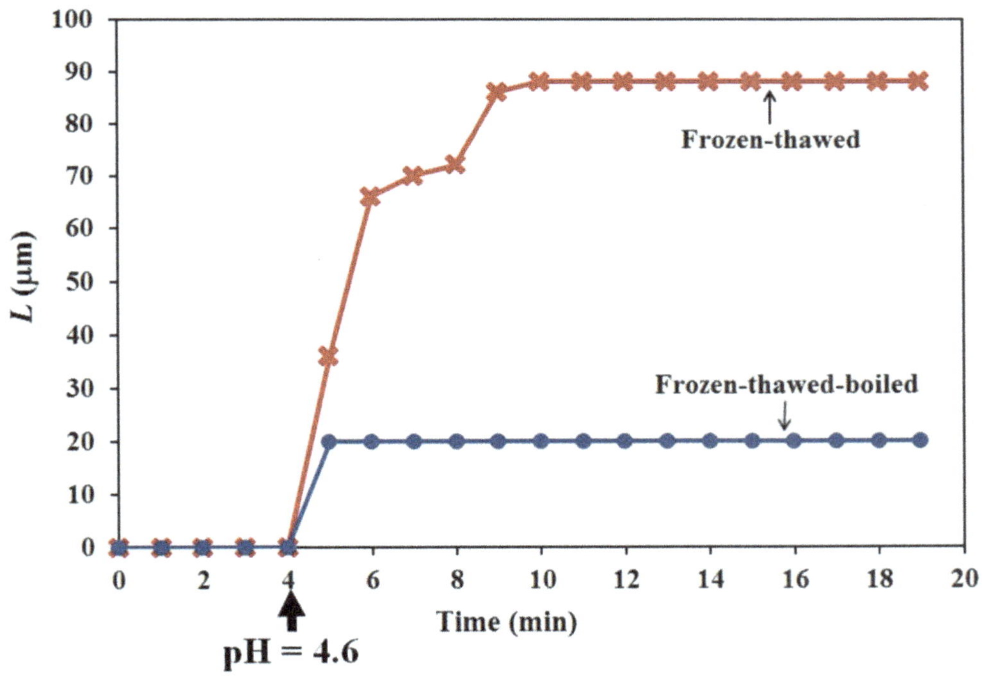

Figure 6. Extension behavior of two 5 mm-long sections from two different Stage IV sporangiophores that were frozen and then thawed is shown. Each wall section includes the growth zone of a Stage IV sporangiophore. The red curve with X data points and labeled "Frozen-thawed" shows the extension behavior of a frozen-thawed wall before and after the pH of the bathing solution was reduced from 7.0 to 4.6. The blue curve with • data points and labeled "Frozen-thawed-boiled" shows the extension behavior of a frozen-thawed wall that was boiled in water for 15 s before being subjected to the same experimental protocol.

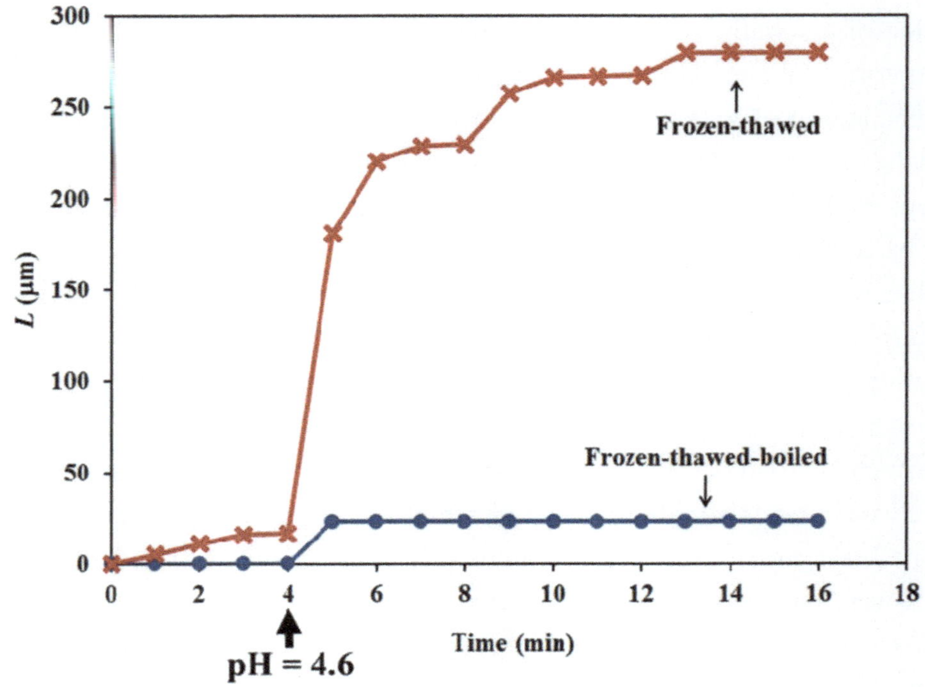

Figure 7. The average change in extension *versus* time of 15 different frozen-thawed walls is shown (red curve with X data points and labeled "Frozen-thawed"). The average change in extension *versus* time of eight frozen-thawed-boiled walls (blue curve with • data points and labeled "Frozen-thawed-boiled") is shown.

Student's *t*-tests were conducted to determine whether the extension rates obtained from frozen-thawed walls after the pH is lowered to 4.6 are significantly different from those of frozen-thawed-boiled walls. The results are shown in Table 2. The results indicate that the extension rates for the frozen-thawed walls are significantly higher compared to the frozen-thawed-boiled walls ($p < 0.05$) for 2–5 min after the pH was reduced. The interpretation is that creep occurs for four minutes (2–5 min) after the pH was reduced to 4.6. Fewer results indicate that in some experiments, there is creep until the ninth minute after the pH is lowered to 4.6.

Table 2. Extension rate for each minute and the result of the Student's *t*-tests for frozen-thawed and frozen-thawed-boiled cell walls.

Minutes after pH = 4.6	Frozen-Thawed Extension Rate (µm/min)	Frozen-Thawed-Boiled Extension Rate (µm/min)	Student's *t*-Test
1	191 ± 109.1 SE ($n = 13$)	23.3 ± 2.8 SE ($n = 8$)	$p = 0.77$
2	74.6 ± 32.7 SE ($n = 9$)	0	$p = 0.024$
3	17.6 ± 5.4 SE ($n = 7$)	0	$p = 0.005$
4	6.0 ± 2.8 SE ($n = 3$)	0	$p = 0.018$
5	59.6 ± 27.1 SE ($n = 7$)	0	$p = 0.024$
6	43.0 ± 45.9 SE ($n = 3$)	0	$p = 0.154$
7	10.0 ($n = 1$)	0	------
8	5.0 ($n = 2$)	0	------
9	181.0 ($n = 1$)	0	------

3. Discussion

3.1. Anoxia Experiments

Anoxia was used to chemically isolate the wall from the protoplasm of the sporangiophores of *P. blakesleeanus* and to reveal the wall chemistry and chemorheology that produces irreversible wall extension and elongation growth. The sporangiophores of *P. blakesleeanus* strictly require oxygen for growth and development [17,19]. The relationship between elongation growth and the metabolism of the sporangiophores was studied by using anoxia to inhibit oxidative phosphorylation [17]. Bergman *et al.* [17] report, "Many fungi can grow anaerobically for long periods of time, but *Phycomyces* sporangiophores do not have this capacity. When oxygen is removed from sporangiophores by placing them in nitrogen, streaming and growth rapidly stop." Studies conducted in our laboratory by Morgan A. Scott (Master's report, see Acknowledgments) show that streaming inside the sporangiophore stops within a minute; $t = 39.5 \pm 2.4$ (SE) s, $n = 10$. The finding that the turgor pressure continually decreases after the initiation of anoxia is consistent with the results of earlier studies that anoxia terminates the protoplast's metabolism.

The finding that elongation growth continues during anoxia for ten minutes or longer when the turgor pressure is held constant with the use of a pressure probe demonstrates that a wall chemistry continues the chemorheological process to produce irreversible wall deformation during anoxia. The wall deformation is irreversible, because the elongation growth occurs when the turgor pressure is constant (Figures 3–5). Inspection of the previously-established augmented growth equation, Equation (2), can show this. The augmented growth equation can be modified to describe the elongation growth of cells with a growth zone, such as the sporangiophores of *P. blakesleeanus*; Equation (3) [20].

$$\frac{dL}{dt} = m_g\,(P - P_C) + \frac{L_g}{E_g}\frac{dP}{dt} + \frac{L_s}{E_s}\frac{dP}{dt} \qquad (3)$$

Elongation rate = irreversible rate in growth zone + elastic rate in growth zone + elastic rate in stalk

where L is the length of the sporangiophore, dL/dt is the elongation rate, m_g is the longitudinal irreversible wall extensibility of the growth zone, P is the turgor pressure, P_C is the critical turgor pressure and L_g and L_s are the lengths of the growth zone and stalk, respectively. E_g and E_s are the longitudinal components of the volumetric elastic modulus within the growth zone and non-growing stalk, respectively. The term, $m_g\,(P - P_C)$, represents the longitudinal irreversible deformation rate of the wall in the growth zone; the term, $(L_g/E_g)\,dP/dt$, represents the longitudinal elastic deformation rate of the wall in the growth zone; and the term, $(L_s/E_s)\,dP/dt$, represents the longitudinal elastic deformation rate of the wall in the non-growing stalk.

It is noted in Equation (3) that when the turgor pressure is constant (P = constant, then $dP/dt = 0$), the elongation rate is only a function of the irreversible longitudinal deformation rate of the wall in the growth zone; Equation (4). Equation 4 is essentially the same equation previously derived by Lockhart [21]. Thus, the elongation rate that occurs when P = constant (for 15 min after the initiation of anoxia in Figure 3, between 2–5 min after the first five turgor pressure step-ups in Figure 4 and after the first two turgor pressure step-ups in Figure 5) represents the irreversible wall deformation rate.

$$\frac{dL}{dt} = m_g (P - P_C) \tag{4}$$

Elongation rate = irreversible longitudinal deformation rate in growth zone

The irreversible wall deformation and elongation growth that occur during anoxia are direct evidence of a wall-loosening chemistry or chemorheological process. In the sporangiophores, the wall chemistry and chemorheological process continue for a limited duration during anoxia, typically 10–20 min. Two explanations for the limited duration of elongation growth and associated wall-loosening chemistry during anoxia are considered. The first explanation is that irreversible wall deformation continues during the wall hardening process that begins at the initiation of anoxia. The wall hardening process converts an irreversible extensible wall to a reversible extensible wall by making bonds between microfibrils and other wall polymers. It is predicted that the irreversible wall deformation stops when the wall hardening process is complete. Second, wall building polymers and other relevant wall materials extruded into the periplasm and inner wall surface before the initiation of anoxia, are used to continue the wall-loosening chemistry and maintain the irreversible wall extensibility. After these polymers and materials are depleted, then the wall hardening process begins and continues until completion.

The first explanation predicts that the duration of elongation growth during anoxia for Stage I and Stage IV sporangiophores are approximately the same, because it is expected that the wall hardening process for both stages is similar and the rate of hardening is approximately the same. The second explanation predicts that the duration of elongation growth for Stage IV sporangiophores is larger than those of Stage I sporangiophores, because the elongation growth rates of Stage IV sporangiophores (34.1 ± 3.1 (SE) $\mu m \cdot min^{-1}$, $n = 20$) are significantly larger than those of Stage I sporangiophores (7.1 ± 0.7 (SE) $\mu m \cdot min^{-1}$, $n = 17$) [7]. The larger elongation growth rates of Stage IV sporangiophores require a faster delivery rate of wall-building polymers and relevant wall-building materials to the periplasm and inner wall surface compared to Stage I sporangiophores. It is reasonable to expect that the amount of wall-building polymers and relevant wall-building materials within the periplasm and inner wall surface at the time anoxia is initiated (and the wall is chemically isolated) is larger for Stage IV sporangiophores compared to Stage I sporangiophores. If the amount of wall building substrates for the wall chemistry is larger for Stage IV, it is predicted that the duration of the wall chemistry, wall loosening, irreversible deformation and elongation growth will be longer for Stage IV compared to Stage I. The second prediction is consistent with the results presented in Table 1.

3.2. Frozen-Thawed Sporangiophores

Extension experiments were conducted on frozen-thawed sporangiophore walls to obtain insight into the wall loosening chemistry. The constitutive equation for a Maxwell–Bingham viscoelastic model, Equation (1), can be modified to assist in the interpretation of the results. Equation (5) is derived from Equation (1) and describes the longitudinal deformation (extension) rate of the frozen-thawed wall section after a tensile load, or longitudinal stress (σ), is applied.

$$\frac{dL}{dt} = m_f (\sigma - \sigma_C) + \frac{L_o}{E_f}\frac{d\sigma}{dt} \tag{5}$$

Longitudinal deformation rate = irreversible deformation rate + elastic deformation rate where L is the length of the frozen-thawed wall section, $\mathrm{d}L/\mathrm{d}t$ is the longitudinal deformation (extension) rate, m_{f} is the longitudinal irreversible wall extensibility of the frozen-thawed growth zone wall section, σ is the longitudinal stress, σ_{C} is the critical longitudinal stress, L_{o} is the initial length and E_{f} is the longitudinal elastic modulus. The term, $m_{\mathrm{f}}\,(\sigma - \sigma_{\mathrm{C}})$, represents the longitudinal irreversible deformation rate of the wall in the growth zone, and the term, $(L_{\mathrm{o}}/E_{\mathrm{f}})\ \mathrm{d}\sigma/\mathrm{d}t$, represents the longitudinal elastic deformation rate of the wall in the growth zone.

Equation (5) is similar to an equation previously used by Takahashi *et al.* [22] to analyze the relationship between the extension behavior of frozen-thawed tissue sections of cucumber hypocotyls and expansin. Here, in Equation (5), we explicitly recognize that m_{f} (longitudinal irreversible wall extensibility that results from unidirectional stress) is not equal to ϕ or m_{g} (irreversible wall extensibility or the longitudinal component of the irreversible wall extensibility, respectively, that results from multidirectional stresses produced by turgor pressure *in vivo*), and E_{f} (the longitudinal elastic modulus that results from unidirectional stress) is not equal to ε (the volumetric elastic modulus that results from multidirectional stresses produced by turgor pressure *in vivo*).

Constant tension-extension experiments were conducted using a tensile force of 1.24 gf, or 0.012 N (F = mass × gravitational acceleration = $(1.24 \times 10^{-3}$ kg$)$ $(9.81\ \mathrm{m \cdot s^{-2}})$ = 0.012 N) that was applied to the frozen-thawed walls. The applied longitudinal wall stress can be estimated using an average sporangiophore diameter of D = 150 μm and a wall thickness of τ = 0.6 μm; $\sigma = F/\pi D\tau \cong 42$ MPa. Immediately after applying the longitudinal wall stress (tensile force of 0.012 N), initial longitudinal wall deformations (extensions) ranged from 1–2 mm. The initial wall extension in response to the application of the wall stress can be determined using Equation (5), where $\mathrm{d}\sigma/\mathrm{d}t$ is finite and relatively large (typically, the wall stress of 42 MPa is applied within a few seconds). It is apparent that the initial longitudinal extensions are predominately elastic; $\mathrm{d}L/\mathrm{d}t = (L_{\mathrm{o}}/E_{\mathrm{f}})\ \mathrm{d}\sigma/\mathrm{d}t$. After the initial longitudinal extension, the longitudinal stress and length of the frozen-thawed wall section remained constant. Because σ is constant in Equation (5) (*i.e.*, $\mathrm{d}\sigma/\mathrm{d}t = 0$), subsequent extensions that occur after decreasing the pH of the bathing solution to 4.6 represent irreversible deformations and creep, Equation (6).

$$\frac{\mathrm{d}L}{\mathrm{d}t} = m_{\mathrm{f}}\,(\sigma - \sigma_{\mathrm{C}}) \tag{6}$$

Longitudinal deformation rate = Irreversible deformation rate

The applied longitudinal wall stress, $\sigma \cong 42$ MPa, is larger than the estimated longitudinal stress produced by the turgor pressure *in vivo*. Previously, the pressure probe was used to determine the average turgor pressure (P = 0.32 ± 0.01 (SE) MPa, n = 20) and average critical turgor pressure (P_{C} = 0.26 ± 0.01 (SE) MPa, n = 20) ([7] and the references within). The longitudinal stress produced by turgor pressure can be estimated using the equation, $\sigma = PR/2\tau$, where P is the turgor pressure, R is the radius (75 μm) and τ is the wall thickness (0.6 μm). The longitudinal wall stress produced by the average turgor pressure (P = 0.32 MPa) is estimated to be, σ = 20 MPa, which exceeds the estimated critical wall stress produced by the average critical turgor pressure (P_{C} = 0.26 MPa), $\sigma_{\mathrm{C}} \cong 16.3$ MPa.

The objective of the constant-tension extension experiments conducted on frozen-thawed sporangiophore walls is to learn if lowering the pH of the bathing solution will elicit irreversible extension and creep, similar to what was found in higher plant cells. The results shown in Figures 6 and 7 (red curve with X data points and labeled "Frozen-thawed") demonstrate that lowering the pH of the bathing solution from

7.0 to 4.6 elicits a chemorheological process, irreversible wall extension (see Equation (6)) and wall creep. In contrast, if the wall sections were immersed in boiling water for 15 s before the constant tension-extension experiments, they did exhibit a nearly instantaneous irreversible deformation, but did not exhibit creep, after the pH of the bathing solution was reduced to 4.6.

Constant-tension extension experiments were previously conducted on frozen-thawed sections of growing cucumber (*Cucumis sativus* L.) hypocotyls and oat (*Avena sativa* L.) coleoptiles immersed in a bathing solution [10,11]. When the pH of the bathing solution was reduced to 4.5, the frozen-thawed walls begin to extend, and the extension continued for hours afterwards, thus exhibiting creep. The experimental results are consistent with the "acid growth" hypothesis [9], which states that plant cells excrete protons onto the inner surface of the cell wall to decrease the pH and activate an "unknown" wall loosening process. When the frozen-thawed cucumber and oat walls were boiled in water for 15 s before the extension experiment, the walls did not exhibit creep when the pH of the bathing solution was lowered to 4.5. These and other experimental results suggested that protein activity may mediate wall loosening at low pH. Subsequently, wall-loosening proteins were isolated that are not enzymes, expansins [1]. Expansins are hypothesized to catalyze wall stress relaxation and irreversible wall deformation directly by disrupting hydrogen bonds between the cellulose microfibrils and hemicellulose [23]. Expansins have been found in a variety of plants and in bacteria and fungi [1,12]. Crude extracts and purified expansins have been shown to mediate "acid-induced" extension of boiled isolated primary walls of a variety of plants [1]. Wall loosening by expansins is consistent with the acid growth hypothesis. Expansins are thought to be strong candidates for primary wall-loosening agents for cells in the organs of higher plants (e.g., stems, roots and leaves). However, endogenous wall enzymes, such as xyloglucan endotransglycosylase/hydrolase and endo-β-1,4-glucanases, are also hypothesized to play an important role in expansive growth of higher plants [13]. Importantly, a considerable amount of experimental evidence indicates that pectin plays a major role in the expansive growth of pollen tubes ([14] and the references within) and algae ([15] and the references within). Furthermore, pectin may play a role in the expansive growth of higher plant cells [3,8]. Interestingly, a small amount of pectin is found in the sporangiophore's cell wall [24], and one wonders if pectin plays a role in the expansive growth of sporangiophores and fungal cells in general.

The general structure of the sporangiophore's wall can be described as chitin and β-glucans microfibrils embedded in an amorphous matrix composed predominately of chitosan, glycoproteins and lipids [16,18], with a small amount of pectin [24]. The wall polymers are linked together by covalent bonds, hydrogen bonds, hydrophobic interaction and ionic associations [16]. The microfibrils are extruded onto the inner wall surface by chitin synthases embedded in the plasma membrane [16]. Chitin synthases and other enzymes and wall polymers are transported to the plasma membrane in vesicles (chitosomes) and delivered to the periplasm via exocytosis [16]. Importantly, fibrillogenesis and microfibril networks have been shown to occur *in vitro* by incubating purified chitin synthases with substrate (UDP-GLcNAc) and activators [25,26]. Furthermore, more recently, the genome of *P. blakesleeanus* was sequenced, and genes involved in sensory growth responses and cell wall components are being characterized (http://genome.jgi-psf.org/Phybl2/Phybl2.home.html).

Generally, the microfibrils can be cross-linked directly (by hydrogen bonds, hydrophobic interaction and ionic associations) and by matrix polymers (chitosan, glycoproteins, and pectin). We hypothesize that wall chemistry and molecular agents that effectively break and make load-bearing cross-links between microfibrils produce chemorheology that regulates wall mechanical properties and wall deformation behavior. Breaking the cross-links directly, or by disconnecting the microfibrils from the matrix network and/or each other, will allow the microfibrils to separate and/or slide passed each other when the wall is stressed by turgor pressure *in vivo* and by an applied stress in extension experiments. In turn, this chemorheological process can produce a controlled polymer creep, often referred to as "wall loosening". Load-bearing cross-links that are not broken or that were broken and reconnected (after slippage and/or separation of microfibrils) and subsequently become load-bearing produce the elastic behavior of the wall. The wall loosening will initiate wall stress relaxation and turgor pressure relaxation that, in turn, produces water uptake, irreversible wall deformation and an increase in cell volume.

The extension curves obtained from frozen-thawed and frozen-thawed-boiled sporangiophore walls after the pH is lowered to 4.6 are qualitatively similar to those obtained for cucumber (*Cucumis sativus* L.) hypocotyls and oat (*Avena sativa* L.) coleoptiles [10,11]. The experimental results from the frozen-thawed sporangiophore walls are consistent with the "acid growth" hypothesis [9]. An important difference is that the duration of the creep for the sporangiophores' walls is limited to less than ten minutes, compared to hours for cucumber hypocotyls and oat coleoptiles. The finding that boiling the sporangiophore walls for 15 s eliminates creep may suggest that protein activity mediates the creep response, but other explanations must also be considered until more experiments are conducted.

The finding that the duration of the creep for the sporangiophores' walls is limited to less than ten minutes, compared to hours for cucumber hypocotyls and oat coleoptiles, is interesting and probably significant. This finding, together with the fact that the molecular polymers, structure and endogenous enzymes in the sporangiophore's wall are different from those in plant and algal cell walls, may indicate that the molecular wall loosening mechanism is different for fungal cells. It may be significant that the duration of elongation growth during anoxia and the duration of creep in frozen-thawed walls sections are both on the order of ten minutes. Interestingly, the duration of creep produced by decreasing the pH is approximately the same as the duration of the light growth response and avoidance growth response (transient increases in elongation growth rate) exhibited by the sporangiophores [7,17,18]. A mechanism is suggested by these findings in which proton effluxes mediate sensory growth responses of the sporangiophores. One may speculate that the sporangiophore's protoplast excretes protons at an increased rate onto the inner surface of the growth zone wall to elicit a transient increase in irreversible wall extensibility [27,28] that produces a transient increase in elongation growth rate, without increasing the turgor pressure [29]; see Equation (4). The phototropic response (growing towards a light source) and avoidance response (growing away from solid objects) could be produced by spatially increasing the proton efflux on the distal side (phototropic response) and proximal side (avoidance response) of the sporangiophore's growth zone. Future research will include adding wall building substrates, activators and endogenous wall enzymes to the bathing solution during constant-tension extension experiments to learn if they will increase the duration of creep in the frozen-thawed sporangiophore wall.

4. Experimental Section

4.1. Biological Material

The wild-type strain, NRRL1555(−), of *P. blakesleeanus* was obtained from the Biology Division, California Institute of Technology, Pasadena, USA. Potato dextrose agar was prepared and sterilized in 12 × 35 mm glass vials. Vegetative spores were heat-shocked in a water bath held at approximately 45 °C for 20 min. After heat-shocking, the spores were triturated and inoculated on the potato dextrose agar in the glass vials. The sporangiophore cultures in the vials were kept at 21 °C ± 0.5 °C in a ventilated, wood chamber and illuminated from above by four fluorescent lamps. Three days after inoculation, sporangiophores begin to appear. Long sporangiophores are removed from the mycelium (plucked) daily to obtain big and robust sporangiophores for experimentation on the next day. Typically sporangiophores from 4 to 6 day-old cultures are used for experimentation.

4.2. Anoxia Experiments

4.2.1. Elongation Growth Measurements

The elongation growth is determined by measuring the change in length of the sporangiophore, ΔL, at one-minute time intervals. The change in length is measured using a long focal length horizontal microscope (Gaertner; 7011K eyepiece and 32 m/m EFL objective) mounted to a 3D micromanipulator (Line Tool Co., Allentown, PA, USA; Model H-2, with digital micrometers). An electronic timer is used to measure the time intervals.

4.2.2. Turgor Pressure Measurements

The turgor pressure of the sporangiophore is measured with a manual version of the pressure probe [20]. A gage pressure transducer is used in the pressure probe, which measured the difference between the absolute pressure and the local atmospheric pressure; the gage pressure transducer was purchased from Kulite Semiconductor Products, Ridgefteld, NJ, USA (Model XT-190-300G), and calibrated inside the pressure probe using a Heise Bourdon Tube Pressure Gauge (Dresser Industries, Newton, CT, USA; Model CMM, 0-200 PSIG Range). The transducer's output is recorded on a Houston Omniscribe Stripchart Recorder (Ametek, Berwyn, PA, USA; Model D5217-2). The pressure probe was mounted on a 3D micromanipulator so that the micro-capillary tip (typically 5–10 μm outer diameter) can be guided to impale the sporangiophore under visual observation using a horizontally-mounted EZM-2TR Trinocular Zoom Stereomicroscope (Meiji Labax Co., Tokyo, Japan). The micro-capillary of the pressure probe was filled with inert silicone oil (Dow Corning Corp., Midland, MI, USA; fluid 200, 1–2 centistoke viscosity). After the cell was impaled, the cell sap-oil interface was maintained at a fixed location within the micro-capillary tip to measure the turgor pressure of the sporangiophore. Turgor pressure step-ups were produced by advancing a manually-controlled control rod within the pressure probe to inject inert silicone oil into the vacuole.

4.2.3. Experimental Protocol

A Stage IV sporangiophore was selected and placed in an experimental chamber that was constructed for these experiments (Figure 8). The sporangiophore was adapted in the experimental chamber to atmospheric air, constant and symmetric distribution of light and other environmental conditions for 20 min. The light and environmental conditions were maintained constant throughout the adaptation period and during the experiment to prevent growth responses. After the adaptation, measurement of the elongation growth was initiated and continued for the remainder of the experiment using a long focal length horizontal microscope. After a period of time (usually 10 min), the sporangiophore stalk was impaled with the pressure probe's micro-capillary tip with the assistance of a horizontally-mounted stereomicroscope, and the turgor pressure was continuously measured and monitored with the pressure probe for the remainder of the experiment. Then, both the elongation rate and turgor pressure were monitored for a period of time to ensure that the sporangiophore was growing at a normal rate. Afterwards, nitrogen gas was injected from the bottom of the chamber to fill the apparatus chamber containing the sporangiophore and to create an anoxic environment. The oxygen concentration was measured and monitored with a meter (Model 50SD-2, 0–20 mL/min, McMillan Company, Georgetown, TX, USA) inside the chamber and located next to the sporangiophore's growth zone. Typically, the oxygen concentration decreased from approximately 21% to less than 1% within a minute after the nitrogen gas was injected into the chamber. Nitrogen gas was continuously injected into the chamber at a low rate to maintain an oxygen concentration level of less than 1%.

4.3. Frozen Thawed Experiments

Sporangiophores, which are frozen for one or two days and then thawed, were used for these experiments. A 5-mm section from the sporangiophore stalk that contains the growth zone was cut with a razor blade and used for experimentation.

4.3.1. Experimental Apparatus

A small acrylic rectangular box was constructed and used as the testing apparatus for the extension experiments (Figure 9). In the testing apparatus, the frozen-thawed wall section is attached to two glass slides, one at each end (glued with Gorilla Super Glue). One glass slide is attached to the testing apparatus and fixed. The other glass slide is attached to a small weight (1.24 g) with a string and is free to slide within the box. The weight is supported on a platform to prevent tension on the wall section before the extension protocol is initiated. A microscope assembly attached to a micrometer is used to measure the change in length of the wall section (Figure 9A). A reference point (end of the wall section) is used throughout the experiment to measure the change in length of the wall section as a function of time. Initially, the frozen-thawed wall section is aligned to the reference point in the microscope, so that changes in length can be measured during the experiment by realigning to the reference point every minute and measuring the change on the digital micrometer.

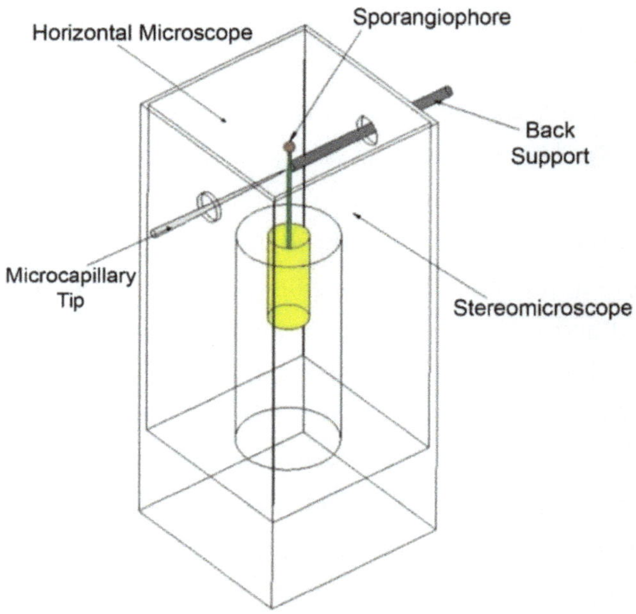

Figure 8. An illustration of the experimental chamber used for anoxia experiments. A vial containing a sporangiophore is inserted into a holder inside the chamber. Afterwards, a back support is advanced manually to the sporangiophore stalk. The back support keeps the stalk stationary while the micro-capillary tip of the pressure probe is advanced to impale the stalk below the growth zone. Two people are required to conduct a single experiment. One person uses a stereomicroscope to impale the sporangiophore stalk, monitor the cell sap-oil interface and generally operate the pressure probe. On the opposite side, another person uses a horizontal microscope, attached to a 3D micromanipulator with digital micrometers, to measure the change in elongation as a function of time.

4.3.2. Experimental Protocol

A Stage IV sporangiophore is selected and frozen for at least 24 h. A 5-mm wall section that includes the growth zone of the frozen sporangiophore is removed with a razor blade. Then, each end of the wall section is glued to a glass slide inside the testing apparatus, one of which is fixed (stationary) and the other (attached to a weight by a string) is free to slide. The frozen wall section thaws while the glue dries. After the glue is dry and the wall section is securely clamped to the two glass slides, distilled water is added to the apparatus until the top part of the apparatus is half full and left to adapt for 10 min (Figure 9A). Following this adaptation period, the change in length is measured every minute for the remainder of the experiment, and the platform supporting the weight is slowly lowered to apply tension to the wall section until the weight hangs freely. Five minutes after the wall section has been in constant tension, a predetermined amount of buffer solution was added to the water to lower the pH to approximately 4.6. A pH pen (Large Display ATC pH Pen, Model 850051, VWR International) is used to measure the pH of the bathing solution when the buffer solution is first added and at the end of the experiment. The change in length is measured continuously until no extension is observed (Figure 9B). The experimental protocol for the frozen-thawed-boiled experiments is the same, except that after removing the sporangiophore from the freezer, it is transferred into a water bath and boiled for 15 s using a microwave before the extension protocol is conducted.

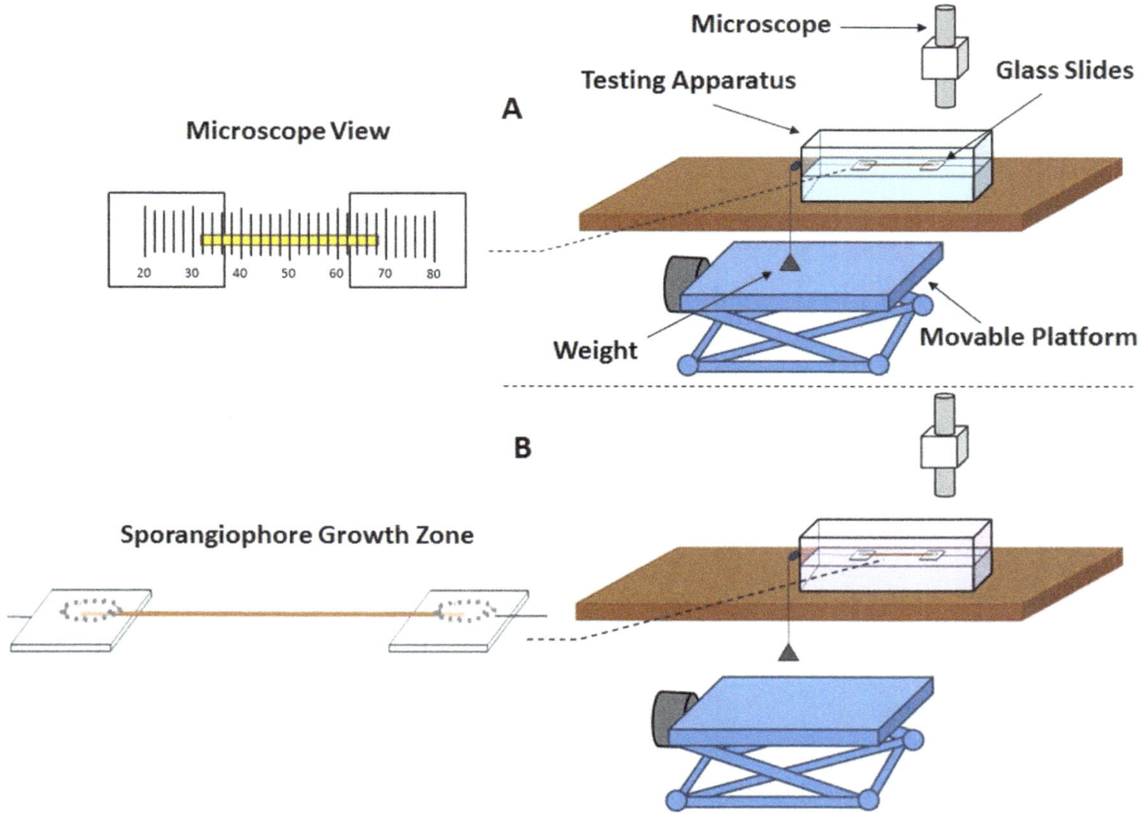

Figure 9. Schematic illustration of the experimental set-up used for the constant-tension extension experiments. (**A**) Testing arrangement for the frozen-thawed wall section during the adaptation phase. The testing apparatus is filled with water, and the platform supporting the weight is not lowered (right side). A microscope view of the aligning process (left side) of the frozen-thawed wall section to the reference point in the microscope, so that changes in length can be measured during the experiment by realigning to the reference point every minute and measuring the change in length on the digital micrometer; (**B**) Testing arrangement for the frozen-thawed wall section during the extension testing phase. The platform is lowered, and the weight hangs freely during the test (right side). Buffer solution (pink) is added to the testing apparatus to lower the pH to 4.6. A microscope view of the wall section in tension is depicted (left side).

5. Conclusions

Experiments that used anoxia to terminate the metabolism of the protoplasm of Stage I and Stage IV sporangiophores of *P. blakesleeanus* and chemically isolate their walls reveal a wall chemistry and chemorheology that continues for ten minutes or longer to produce elongation and irreversible wall extension (Figures 3–5 and Table 1). Constant-tension extension experiments were conducted on frozen-thawed wall sections of Stage IV sporangiophores to investigate the wall chemistry and chemorheology revealed in the anoxia experiments. The objective was to learn if lowering the pH of the bathing solution from 7.0 to 4.6 will elicit irreversible extension and creep. The results (Figures 6 and 7 and Table 2) demonstrate that lowering the pH produces irreversible wall extension and creep in the frozen-thawed wall sections (which include the growth zone) of the Stage IV sporangiophore. This finding is consistent with the "acid growth" hypothesis that was proposed for plant cell walls [9]. Interestingly, frozen-thawed wall sections that were subject to boiling water for 15 s do not exhibit creep after the pH is lowered to 4.6. This finding suggests that protein activity may be involved in the observed creep response, but other experimental results are needed to support this suggestion. Except for the short duration of creep, the constant tension-extension behavior of the frozen-thawed and frozen-thawed-boiled wall sections of the Stage IV sporangiophore is qualitatively similar to that obtained from cucumber (*Cucumis sativus* L.) hypocotyls and oat (*Avena sativa* L.) coleoptiles [10,11]. It is thought that the limited and short duration of creep exhibited by the frozen-thawed sporangiophore wall to acidic pH may be significant. This finding, together with the fact that the molecular polymers, structure and endogenous enzymes in the sporangiophore walls are different from those in plant and algal cell walls, may suggest that the molecular wall loosening mechanism is different from those of higher plant cells and algal cells.

Acknowledgments

This research was supported by an NSF Grant, MCB-0948921, awarded to Joseph K.E. Ortega. Cindy M. Munoz acknowledges the support from an NSF Graduate STEM fellowship in K-12 Grant, DGE-0742434. Cindy M. Munoz and David G. Ramirez acknowledge the support from an NSF Bridge to the Doctorate Grant, HRD-1301885 (sub-award, G-8960-1). The authors acknowledge the early research conducted by Jessica E.C. Olson, Elena L. Ortega and Morgan A. Scott, who studied the elongation and turgor pressure behavior of the sporangiophores of *P. blakesleeanus* during anoxia. Furthermore, the authors acknowledge that Jessica Olson and Elena Ortega conducted the experiment whose results are shown in Figure 2 and that Morgan Scott and Elena Ortega conducted early turgor pressure step-up experiments whose results are similar to those shown in Figure 4, but not the results used in Figure 4. Other interesting results are reported in a master thesis by Jessica Olson and a master report by Morgan Scott. Olson, Jessica E.C. *Phycomyces*: Helical growth and turgor pressure behavior during death by asphyxiation. M.S. thesis, University of Colorado, Denver, 2000, 1–164. Scott, Morgan A. *Phycomyces blakesleeanus*: An investigation of elongation growth and turgor pressure of wild-type sporangiophores during anoxia, and the growth and physiological characteristics of two mutant sporangiophores. M.E. Report, University of Colorado Denver, 2008, 1–77.

Author Contributions

Joseph K.E. Ortega initiated, supervised and provided financial support for the all of the research conducted and reported in this paper. In addition, he was predominately responsible for the development of the experimental protocols, directing the analysis of the data, interpretation of the results and writing the paper. Jason T. Truong, Cindy M. Munoz and David G. Ramirez conducted all of the experiments whose results are reported in this paper (except for Figure 2) and contributed equally to this work. In addition, these authors conducted most of the data analysis, assisted in the development of the experimental protocols and assisted in the writing of the text in the Materials and Methods.

Conflicts of Interest

The authors declare no conflict of interest.

References

1. Cosgrove, D.J. Growth of the plant cell wall. *Nat. Rev. Mol. Cell Biol.* **2005**, *6*, 850–861.
2. Geitmann, A.; Ortega, J.K.E. Mechanics and modeling of plant cell growth. *Trends Plant Sci.* **2009**, *14*, 467–478.
3. Peaucelle, A.; Braybrook, S.; Hofte, H. Cell wall mechanics and growth control in plants: The role of pectin revisited. *Front. Plant Sci.* **2012**, *3*, 121.
4. Ortega, J.K.E. Augmented growth equation for cell wall expansion. *Plant Physiol.* **1985**, *79*, 318–320.
5. Ortega, J.K.E. Plant cell growth in tissue. *Plant Physiol.* **2010**, *154*, 1244–1253.
6. Ortega, J.K.E.; Welch, S.W.J. Mathematical models for expansive growth of cells with walls. *Math. Model. Nat. Phenom.* **2013**, *8*, 35–61.
7. Ortega, J.K.E. Growth rate regulation of cells with walls: The sporangiophores of *Phycomyces blakesleeanus* used as a model system. *Recent Res. Dev. Plant Physiol.* **2012**, *5*, 1–19.
8. Barbacci, A.; Lahaye, M.; Magnenet, V. Another Brick in the cell wall: Biosynthesis dependent growth model. *PLOS ONE* **2013**, *8*, e74400.
9. Rayle, D.L.; Cleland, R.E. The acid growth theory of auxin-induced cell elongation is alive and well. *Plant Physiol.* **1992**, *99*, 1271–1274.
10. Cosgrove, D.J. Characterization of long-term extension of isolated cell wall from growing cucumber hypocotyls. *Planta* **1989**, *177*, 121–130.
11. McQueen-Mason, S.; Durachko, D.M.; Cosgrove, D.J. Two endogenous proteins that induce cell wall extension in plants. *Plant Cell* **1992**, *4*, 1425–1433.
12. Nikolaidis, N.; Doran, N.; Cosgrove, D.J. Plant Expansins in Bacteria and Fungi: Evolution by Horizontal Gene Transfer and Independent Domain Fusion. *Mol. Biol. Evol.* **2014**, *31*, 376–386.
13. Park, Y.B.; Cosgrove, D.J. A revised architecture of primary cell walls based on biomechanical changes induced by substrate-specific endoglucanases. *Plant Phyciol.* **2012**, *158*, 1933–1943.
14. Parre, E.; Geitmann, A. Pectin and the role of the physical properties of the cell wall in pollen tube growth of *Solanum chacoense*. *Planta* **2005**, *220*, 582–592.
15. Proseus, T.E.; Boyer, J.S. Calcium deprivation disrupts enlargement of *Chara corallina* cells: Further evidence for the calcium pectate cycle. *J. Exp. Bot.* **2012**, *63*, 3953–3958.
16. Ruiz-Herrera, J. *Fungal Cell Wall: Structure, Synthesis, and Assembly*; CRC Press, Taylor & Francis Group: New York, NY, USA, 2012.
17. Bergman, K.; Burke, P.V.; Cerda-Olmedo, E.; David, C.N.; Delbruck, M.; Foster, K.W.; Goodell, E.W.; Heinsenberg, M.; Meissner, G.; Zalokar, M.; *et al*. Phycomyces. *Bacteriol. Rev.* **1969**, *33*, 99–157.
18. Cerda-Olmedo, E.; Lipson, E.D. *Phycomyces*; Cold Spring Harbor Laboratory: Cold Spring Harbor, NY, USA, 1987.
19. De Boer, S.R. Respiration of *Phycomyces*. *Rec. Trav. Bot. Neerl.* **1928**, *25*, 117–239.
20. Ortega, J.K.E.; Munoz, C.M.; Blakley, S.E.; Truong, J.T.; Ortega, E.L. Stiff mutant genes of *Phycomyces* affect turgor pressure and wall mechanical properties to regulate elongation growth rate. *Front. Plant Sci.* **2012**, *3*, 99.
21. Lockhart, J.A. An analysis of irreversible plant cell elongation. *J. Theor. Biol.* **1965**, *8*, 264–275.
22. Takahashi, K.; Hirate, S.; Kido, N.; Katou, K. Wall-yielding properties of cell walls from elongating cucumber hypocotyls in relation to the action of expansin. *Plant Cell Physiol.* **2006**, *47*, 1520–1529.

23. Durachko, D.M.; Cosgrove, D.J. Disruption of hydrogen bonding between plant cell wall polymers by proteins that induce wall extension. *Proc. Natl. Acad. Sci. USA* **1994**, *91*, 6574–6578.
24. Van Laere, A.J.; Carlier, A.R.; van Asscje, J.A. Cell wall carbohydrates in *Phycomyces blakesleeanus* Burgeff. *Arch. Microbiol.* **1977**, *112*, 203–206.
25. Ruiz-Herrera, J.; Sing, V.O.; van der Woude, W.J.; Bartnicki-Garcia, S. Microfibril assembly by granules of chitin sythetase. *Proc. Natl. Acad. Sci. USA* **1975**, *72*, 2706–2710.
26. Bracker, C.E.; Ruiz-Herrera, J.; Bartnicki-Garcia, S. Structure and transformation of chitin synthetase particles (chitiosomes) during microfibril synthesis *in vitro*. *Proc. Natl. Acad. Sci. USA* **1976**, *73*, 4570–4574.
27. Ortega, J.K.E.; Gamow, R.I. An increase in mechanical extensibility during the period of light-stimulated growth. *Plant Physiol.* **1976**, *57*, 456–457.
28. Ortega, J.K.E.; Gamow, R.I. An increase in mechanical extensibility during the avoidance growth response. *Plant Physiol.* **1977**, *60*, 805–806.
29. Ortega, J.K.E.; Manica, K.J.; Keanini, R.G. *Phycomyces*: Turgor pressure behavior during the light and avoidance growth responses. *Photobiol. Photochem.* **1988**, *48*, 697–703.

6

Plasmolysis: Loss of Turgor and Beyond

Ingeborg Lang [1,*], Stefan Sassmann [1], Brigitte Schmidt [1] and George Komis [2]

[1] Cell Imaging and Ultrastructure Research, University of Vienna, Althanstrasse 14, A-1090 Vienna, Austria; E-Mails: stefan.sassmann@univie.ac.at (S.S.); brigitte.schmidt@univie.ac.at (B.S.)

[2] CR-Hana, Palacký University Olomouc, Šlechtitelů 586/11, 783 71 Olomouc-Holice, Czech Republic; E-Mail: georgios.komis@upol.cz

* Author to whom correspondence should be addressed; E-Mail: ingeborg.lang@univie.ac.at

---

Abstract: Plasmolysis is a typical response of plant cells exposed to hyperosmotic stress. The loss of turgor causes the violent detachment of the living protoplast from the cell wall. The plasmolytic process is mainly driven by the vacuole. Plasmolysis is reversible (deplasmolysis) and characteristic to living plant cells. Obviously, dramatic structural changes are required to fulfill a plasmolytic cycle. In the present paper, the fate of cortical microtubules and actin microfilaments is documented throughout a plasmolytic cycle in living cells of green fluorescent protein (GFP) tagged *Arabidopsis* lines. While the microtubules became wavy and highly bundled during plasmolysis, cortical filamentous actin remained in close vicinity to the plasma membrane lining the sites of concave plasmolysis and adjusting readily to the diminished size of the protoplast. During deplasmolysis, cortical microtubule re-organization progressed slowly and required up to 24 h to complete the restoration of the original pre-plasmolytic pattern. Actin microfilaments, again, recovered faster and organelle movement remained intact throughout the whole process. In summary, the hydrostatic skeleton resulting from the osmotic state of the plant vacuole “overrules” the stabilization by cortical cytoskeletal elements.

Keywords: *Arabidopsis* hypocotyl; cytoskeleton; microtubules; actin microfilaments; plasmolysis; deplasmolysis; GFP-MAP4; GFP-TUA6; GFP-ABD

---

1. Introduction

The process of plasmolysis is probably still known to many from their student days. In hyperosmotic solutions such as sucrose, mannitol or sorbitol, water is extruded from the vacuole causing a loss of turgor pressure. If this state persists, the protoplast retracts further, causing the detachment of the plasma membrane from the rigid cell wall. Two major types of plasmolysis are known, depending on: The cell type, the viscosity of the cytoplasm, and the osmoticum used [1]. In convex plasmolysis, the protoplast is rounded up exhibiting symmetrical convex ends (Figure 1a). In concave plasmolysis, the plasma membrane separates from the cell wall by the formation of several concave pockets (Figure 1b). Plasmolysis is reversible and the addition of hypotonic solutions or plain water will lead to the re-expansion of the protoplast and the reinstatement of the original turgor pressure [1]. The central vacuole is the major compartment of osmotic water flow during plasmolysis but obviously, the abrupt change in protoplast size and shape impacts the subcellular architecture as a whole. In this research, we followed the organization of plant cytoskeletal elements namely: Cortical microtubules and actin microfilaments, during a plasmolytic cycle and documented the entire process in living cells.

Figure 1. Schematic of the two major plasmolysis forms; (**a**) convex plasmolysis; (**b**) concave plasmolysis.

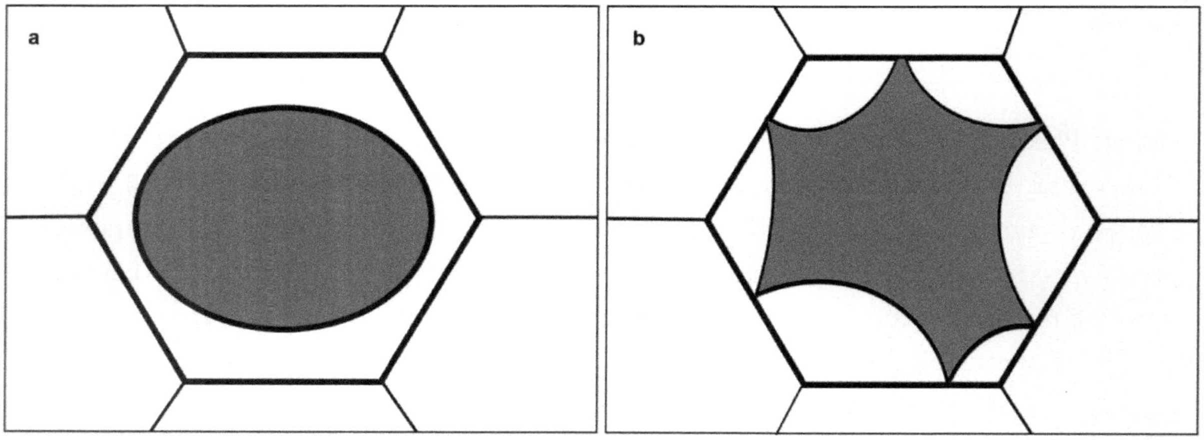

Aside from the central vacuole, a rigid cell wall is required for plasmolysis. This structure forms a solid shell encasing the osmo-sensitive and membrane-covered protoplast. Consequently, plasmolysis occurs in walled cell types ranging from bacteria, to fungi and finally plants [1–4]. When large molecules with a size above the cell wall exclusion limit are used as osmotica (e.g., polyethylene glycols with a MW above 20 kDa) [5], then hyperosmolarity induces the collapse of the entire cell wall—plasma membrane continuum in a phenomenon known as cytorrhysis [6].

In basic biology courses, plasmolysis is used to demonstrate plant cell turgor and its relation to the mechanical rigidity of plant organs. During plasmolysis, the plasma membrane is separated from the cell wall, and this process is easily demonstrated. Specific chemicals like potassium salts lead to swelling of the cytoplasm thereby allowing a distinction between the tonoplast and the plasma membrane in so called cap plasmolysis [7,8]. Furthermore, plasmolysis is an active process characteristic of viable cells; therefore, it is used to test the cellular viability against treatments including heavy metals and other stress factors [9–11]. The connection of the plasma membrane and the cell wall is still widely discussed subject [4,12–16] and plasmolysis is essentially used to analyze the space and link between these two structures.

The term plasmolysis was defined by de Vries [17] upon the invention of a method to determine a plant's turgor pressure using hypertonic solutions. Later, Hecht [18] intensively studied plasmolysed onion epidermal cells. He observed a network-like structure and fine strands (Hechtian strands) in the periplasmic space (the space between the cell wall and the retracted protoplast) that link the protoplast to the inner side of the cell wall of plasmolysed cells. Plasmodesmata have been discussed as candidates for Hechtian attachment sites; these minute channels between adjacent cells have been magnificently observed by transmission electron microscopy in plasmolysed plant tissue [19,20]. However, Hechtian strands can also be formed in the periplasmic space despite cell walls lacking plasmodesmata (e.g., trichomes or outer epidermis walls), suggesting additional structural connections like RGD-containing peptides [13,14], arabinogalactan proteins [21] or growing cellulose microfibrils [22]. A comprehensive review on the process of plasmolysis is given by Oparka [1]. Since then, cytoskeletal elements have been analyzed during a plasmolytic cycle in various plant species and cell types, [23–25] but mainly in fixed plant material which was used to visualize microtubules and/or actin microfilaments by means of immunolocalization. In this research, the plasmolytic fate of cortical microtubules and actin microfilaments was followed in epidermal hypocotyl cells of appropriate *Arabidopsis* lines stably transformed with GFP-tagged cytoskeletal markers allowing the documentation of the whole process *in vivo*.

2. Results and Discussion

2.1. A Plasmolytic Cycle

During a plasmolytic cycle, the semipermeable membranes, plasma membrane and tonoplast, were forced to adjust to the loss of water from the vacuole in hypertonic solutions (plasmolysis), or to the water uptake until full turgor is reinstated (deplasmolysis). Plasmolysis started immediately after contact with the plasmolytic solution and in *Arabidopsis* hypocotyl cells, it was complete after 30 min following exposure to 0.8 M mannitol solution. Cells survived in the plasmolysed state for longer than 24 h, depending on the experimental design and cell type used, [26–28] while in exceptional cases they recovered after prolonged exposure to hyperosmoticum reinstating turgor and cortical microtubule organization, suggesting the function of volume regulatory increase mechanisms [26] and references therein, [29]. Fine Hechtian strands and a network like structure (Hechtian reticulum) provided the contact of the plasma membrane to the cell wall during plasmolysis, while preserving the plasma membrane surplus resulting from protoplast reduction [22,26,30,31]. The process of plasmolysis was easily observed with good bright field optics. However, the investigation of subcellular changes in living cells required the use of GFP-tagged *Arabidopsis* lines, as is recognized in the present study using microtubule associated protein 4 (MAP4; [32]) and tubulin alpha 6 (TUA6; [33]) lines for labeling microtubules, and a GFP-tagged actin binding domain of fimbrin 1 (ABD; [34]) for visualization of actin microfilaments in living cells.

2.2. Microtubules

In interphase cells, plasmolysis (which is the disruption of the cell wall—plasma membrane—cortical cytoskeleton continuum) is expected to exert the strongest impact on cortical microtubules since they are closely linked to the plasma membrane, exerting a role in oriented cellulose microfibril deposition [15,16,28,35–37]. In fully turgid *Arabidopsis* hypocotyl cells, we observed the biased organization of cortical microtubules arranged in parallel order, and in oblique to transverse orientations in the cortical cytoplasm (Figures 2a and 3f). At the onset of plasmolysis, cortical microtubules became wavy in order to accommodate the decreased shape of the protoplast (Figure 2b). Concurrently, the microtubules come together to form bundles (Figure 3). In the bundled state, cortical microtubules were preserved for up to 24 h while still in the plasmolytic medium. Interestingly, the cellulase KORRIGAN (KOR) showed similar microtubule arrays [38] but a role of microtubules in the transport of cellulose synthase (CESA) proteins from their origin in the Golgi towards the plasma membrane, e.g., by microtubule-associated CESA compartments (MASCs) or small CESA compartments (SmaCCs) is still under discussion (for a review, see [39]). Cortical microtubules labeled with GFP-MAP4 and GFP-TUA6 exhibit the same behavior under plasmolytic conditions but the background in GFP-TUA6 cells was higher (Supplemental Figure S1). Therefore, only data for GFP-MAP4 are submitted hereafter. Microtubules or microtubule bundles were also present in Hechtian strands (Figure 3j). In deplasmolysis, microtubules remained organized in thick bundles, but they gradually separated from each other during the course of the protoplast swelling to its original size (Supplemental Movie S1). In some samples, fluorescently labeled spots are observed (Figure 3h–j); these are located along microtubule bundles. It took up to 24 h until fine cortical microtubules were reinstated but even then, some thick tubular structures—probably corresponding to residual microtubule bundles—persisted (data not shown).

Figure 2. Green fluorescent protein (GFP)-tagged microtubules (GFP-MAP4) in *Arabidopsis* hypocotyl cells; (**a**) Interphase cells before plasmolysis; (**b**) Bundles and wavy microtubules in plasmolysed cells, treatment with 0.8 M mannitol for 30 min; bar: 10 µm.

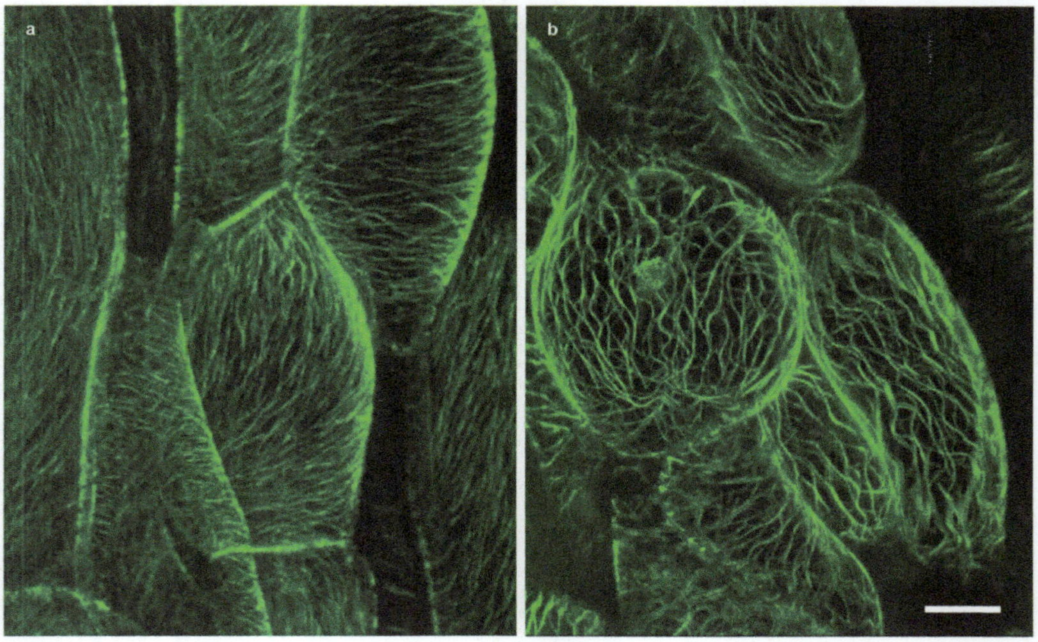

Figure 3. Cortical microtubules (GFP-MAP4) in *Arabidopsis* hypocotyl cells during plasmolysis (**a**–**c**; **f**–**j**) and deplasmolysis (**d**,**e**); plasmolysis/deplasmolysis times marked on the micrographs. (**a**) Bright field image showing cell walls, occasional stomata and chloroplasts (arrowheads); (**b**) Bundling of microtubules (arrows) at the onset of plasmolysis in 0.8 M mannitol; (**c**) After 40 min, cortical microtubules are showing waves and bundles; they remain aligned within the cortical cytoplasm; (**d**) Bundles of microtubules persist in deplasmolysis when the protoplasts become realigned at the cell wall; (**e**) Overlay of fluorescence and bright field image at the end of the plasmolytic cycle; chloroplasts appear as dark grey dots; (**f**–**j**) Higher magnification of plasmolysing hypocotyl cells; inlay in (**j**) shows microtubules in Hechtian strands (arrows). Bar: 10 μm.

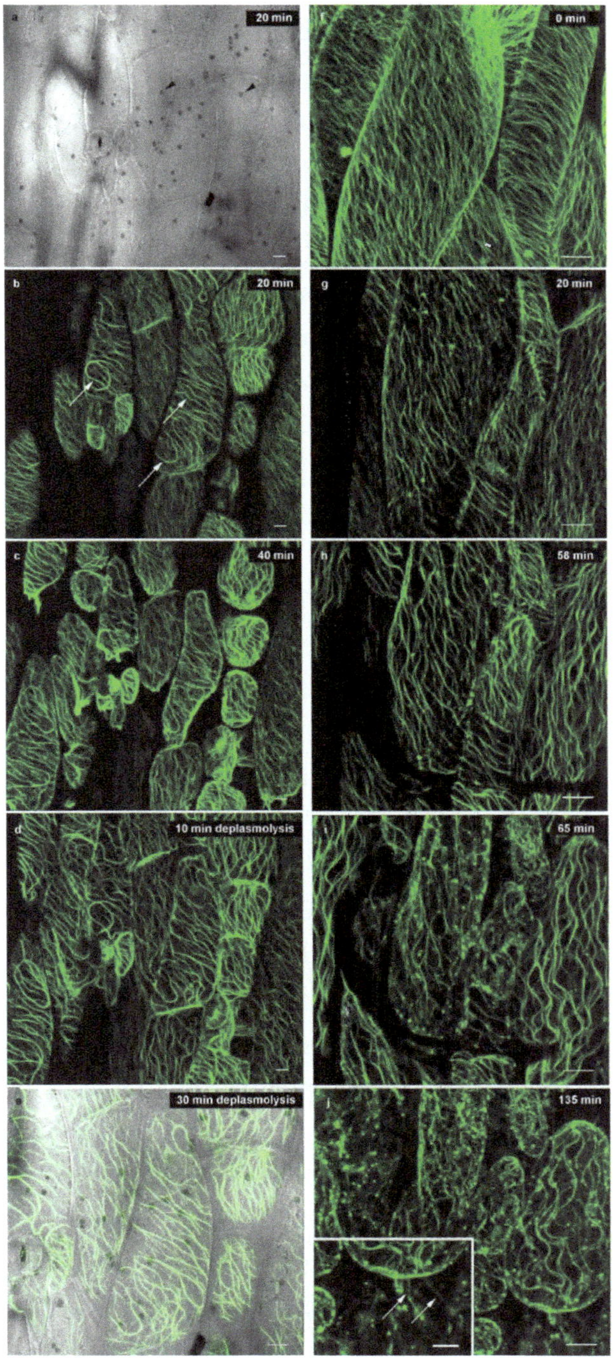

2.3. Actin Microfilaments

Filamentous actin, as visualized by GFP-ABD in *Arabidopsis* hypocotyl cells, presented a fine network (Figure 4a). It consisted of radiating microfilaments and microfilament bundles which extended from the nuclear envelope towards the cortical parts of the cell (Figure 4b). Endoplasmic actin was present in transvacuolar cytoplasmic strands that traversed the vacuole thereby connecting the nuclear cytoplasm to the thin cortical cytoplasmic layer immediately below the plasma membrane. Since the subcortical cytoplasm was highly dynamic and organelles therein are constantly streaming in interphase cells, actin microfilaments needed to adjust accordingly. Therefore, it is easily understandable that actin microfilaments were able to reorganize rapidly in order to accommodate the shrinking protoplast during plasmolysis. The actin filaments radiating from the nuclear surface were still visible in plasmolysed cells (Figure 4b,c), while cortical actin lined the plasma membrane (arrowheads; Figure 4c,d, inserts) and stretched out towards the cell wall in Hechtian strands (Figure 4f,g, arrows). The actin filaments did not become wavy as observed for microtubules. The plant cytoplasm contained a pool of globular actin in order to constantly assemble filaments; these in turn can rapidly disassemble if F-actin remodeling is necessary. Actin microfilament reorganization after hyperosmotic treatment was linked to Ca^{2+} signaling and was associated with membrane mechanical integrity ([26] and references therein). However, it remains to be clarified if this Ca^{2+} mobilization triggered actin filament formation directly by the increased polymerization of globular actin [40], or indirectly through a Ca^{2+} signaling cascade [41].

At the onset of deplasmolysis, filamentous actin was clearly visible at the sites of big cytoplasmic aggregations: Around the nucleus and lining of the concave membrane-wall detachments. A weak fluorescent signal of GFP-ABD protein was observed along the plasma membrane of the expanding protoplast, even when there was only a very thin cortical layer. Hechtian strands became incorporated into the expanding protoplast. The cytoarchitecture at the start of deplasmolysis was preserved throughout the process (Figure 4d,e/Supplemental Movie S2) until the protoplast was reconnected with the cell wall. Shortly after the plasmolytic cycle was completed, actin microfilaments in *Arabidopsis* hypocotyl cells were organized similarly to non-plasmolyzed control cells. Consequently, most filamentous actin was found to be present at sites of high cytoplasmic density as well as in the thin cortical layer below the plasma membrane. Organelle movement denoting cytoplasmic streaming was maintained throughout the whole plasmolytic cycle (Supplemental Movie S3). In leaf cells of *Chlorophytum comosum*, the behavior of cortical actin filaments during plasmolysis has been described in terms of plasma membrane integrity and protoplast volume regulation [26]. In this study, actin microfilaments were shown to line the plasma membrane at areas of intense mechanical strain as in the case of concave plasmolysis. The authors report that many actin filament bundles formed a network, lining the areas of detached plasma membrane during concave plasmolysis. In addition, these fibers were compared to stress fibers found in animal cells. It is suggested that the formation of numerous cortical, subcortical and endoplasmic actin filaments was necessary to regulate shape and volume in plasmolysis (see also [42,43]). Furthermore, Komis and co-workers [26] describe the disappearance of thin cortical actin filaments during plasmolysis. This, however, could be attributed to the fact that cells were fixed and most of the actin in the cortical layer was lost. The GFP-tagged actin *Arabidopsis* line in the present study allowed for the use of living material. A faint fluorescence signal was observed along the protoplast, even at sites that contain hardly any cytoplasm.

Figure 4. Actin microfilaments (GFP-ABD) in *Arabidopsis* hypocotyl cells during a plasmolytic cycle; plasmolysis/deplasmolysis times marked on the micrographs. (**a**,**b**) Before plasmolysis, actin microfilaments were located in the cortical cytoplasm (**a**) and radiated from the nucleus (**b**); (**c**) In the plasmolysed state, actin microfilaments conserved this structure, but also lined the detached plasmalemma regions (arrows); (**d**) The protoplast rounded up at the onset of deplasmolysis, incorporating Hechtian strands (arrowheads) and (**e**) a "stabilizing layer" of bundled actin microfilaments followed the expanding protoplast; (**f**,**g**) High magnification of Hechtian strands (arrows) with actin microfilaments (**f**) and the corresponding bright field image (**g**), 30 min of plasmolysis in 0.8 M manntiol; arrowheads indicate the plasma membrane. Inserts (**c**,**d**): Single focal planes to show the cortical array. Bar: 10 μm.

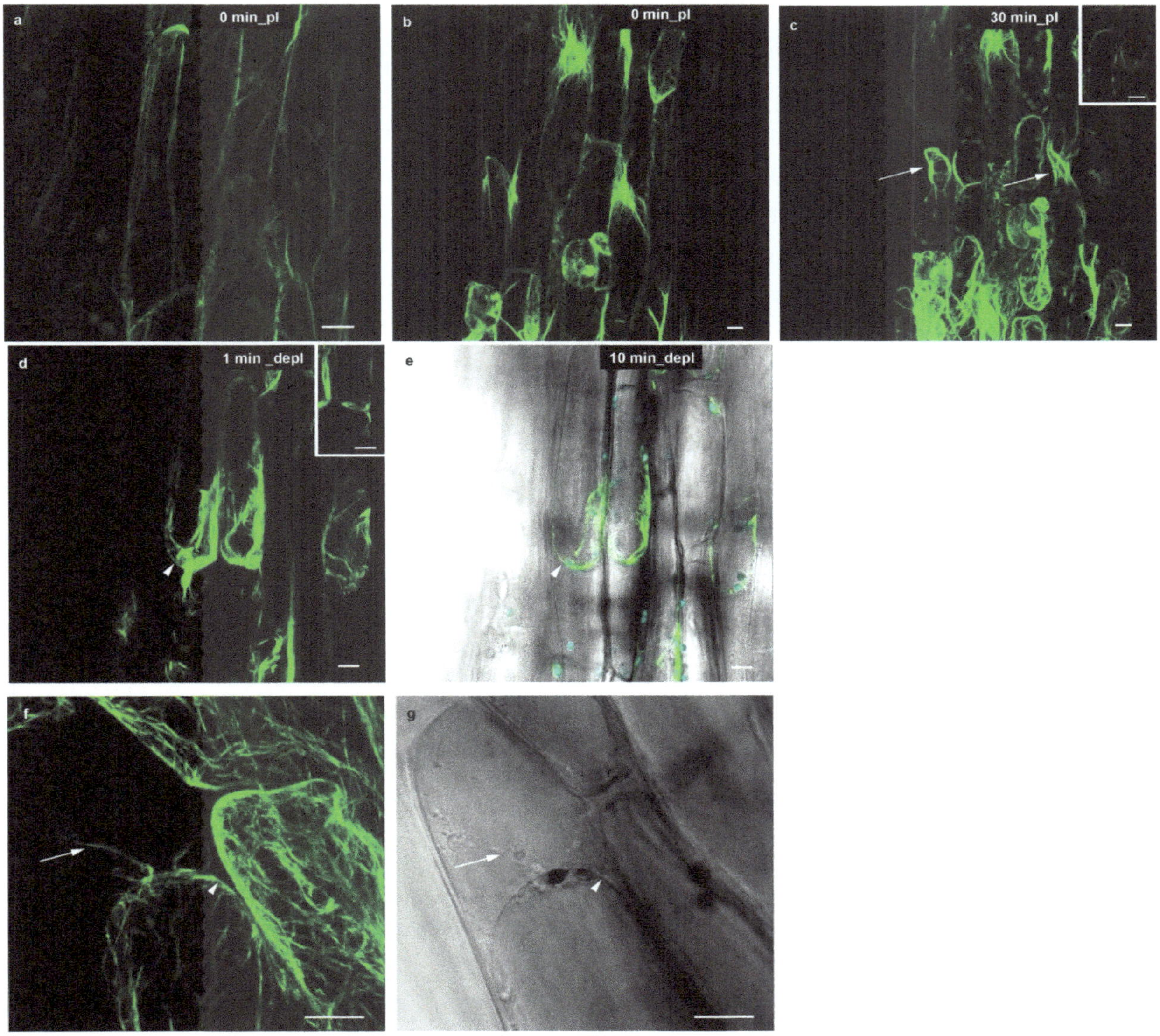

2.4. A Broader View

Plasmolysis is not only used in laboratory experiments, it has been reported to occur naturally due to extracellular water withdrawal in freezing conditions [44]. A lot of current knowledge on the water balance of plants is based on extensive studies by Stadelmann [5] and his co-workers. In this study, although the focus was maintained on the cytoskeleton, keeping in consideration that it is part of a complex ER-cytoskeleton-plasma membrane-cell wall continuum, and plays an essential role in signaling and mechanosensing [21,45,46].

3. Experimental Section

Hypocotyl cells of *Arabidopsis thaliana* plants were used in this study. The plants were grown on ½ MS medium [47] under sterile culture conditions for five days. Apart from wild-type plants (Col 0), cytoskeletal elements were followed in green fluorescent protein (GFP)-tagged *Arabidopsis* lines. In order to visualize microtubules, *Arabidopsis* lines expressing a microtubule associated protein coupled to GFP (GFP-MAP4; [32]) were utilized; seeds were generously gifted by Professor Jozef Šamaj as well as an alpha tubulin encoding line (GFP-TUA6; [33]); kind gift from Dr. Sidney Shaw. Actin microfilaments were detected in *Arabidopsis* plants expressing a GFP-tagged actin binding domain of fimbrin 1 (GFP-ABD; [34]); another kind gift from Professor Jozef Šamaj.

Seedlings were secured between a microscope slide and a coverslip spaced by Parafilm stripes, and sealed using liquid petroleum jelly. The immobilization in the application of petroleum jelly prevented dislocation of the plantlets during liquid exchange and allowed for the observation of the same cells during a whole plasmolytic cycle. To induce plasmolysis, a 0.8 M mannitol solution was applied on one side of the coverslip and carefully removed from the opposite side using filter paper. Plasmolysis was completed after 30 to 40 min. Deplasmolysis was initiated by perfusing a 0.4 M mannitol solution for 15 min followed by the perfusion of distilled water.

Both plasmolysis and deplasmolysis, were observed under a confocal laser scanning microscope (Leica CTR SP5 linked to a DM6000CS stand) using a 63× water immersion lens. Z-stacks were taken at specific time frames (e.g., onset of plasmolysis, after 10 min, 30 min, 60 min, 12 h, and finally after 24 h). Time lapse video clips were also produced to follow the dynamic processes. For the videos shown as supplemental data, single images were taken at 2 min time intervals. The single image series were exported as .avi files by the Leica LASAF software at 12 fps. Editing was done with Adobe Premiere Pro version CS4 and CS6 (Adobe, San Jose, CA, USA).

4. Conclusions

Although plasmolysis is used in many cell biology experiments and student courses, the process itself and the incurred cytoarchitectural rearrangements remain to be fully understood. In this research paper, we describe cortical microtubule and actin microfilament organization during a plasmolytic cycle. Both structures are forced to adjust accordingly in the diminishing/expanding protoplast driven by vacuolar water efflux/influx. Further functional studies using stabilizing or disrupting agents on the cytoskeleton will allow for a more in-depth view on the role of cytoskeletal elements in plasmolysis.

Acknowledgments

This work was supported by a travel and accommodation grant (project INTERHANA, CZ.1.07/2.3.00/20.0165) to G.K. We gratefully acknowledge the gift of GFP-MAP4 and GFP-ABD lines by

Jozef Šamaj (CR-Hana, Palacky University, Olomouc, Czech Republic) and of GFP-TUA6 by Sidney Shaw (Biology Department, Indiana University, Bloomington, IN, USA).

This article was supported by the Open Access Publishing Fund of the University of Vienna.

Author Contributions

I.L. initiated this contribution to the special issue on "Plant Vacuole" and wrote the manuscript with editorial assistance from S.S. and G.K. Plasmolysis experiments were performed by I.L., S.S. and G.K. B.S. edited the videos for supplemental data. All authors discussed and interpreted the results.

Conflicts of Interest

The authors declare no conflict of interest.

References

1. Oparka, K.J. Plasmolysis: New insights into an old process. *New Phytol.* **1994**, *126*, 571–591.
2. Rojas, E.; Theriot, J.A.; Huang, K.C. Response of *Escherichia coli* growth rate to osmotic shock. *Proc. Natl. Acad. Sci. USA* **2014**, *111*, 7807–7812.
3. Bitsikas, V.; Karachaliou, M.; Gournas, C.; Diallinas, G. Hypertonic conditions trigger transient plasmolysis, growth arrest and blockage of transporter endocytosis in *Aspergillus nidulans* and *Saccharomyces cerevisiae*. *Mol. Membr. Biol.* **2011**, *28*, 54–68.
4. Chitcholtan, K.; Harris, E.; Yu, Y.; Harland, C.; Garrill, A. An investigation into plasmolysis in the oomycete *Achlya bisexualis* reveals that membrane-wall attachment points are sensitive to peptides containing the sequence RGD and that cell wall deposition can occur despite retraction of the protoplast. *Can. J. Microbiol.* **2012**, *58*, 1212–1220.
5. Stadelmann, E. Plasmolyse und Deplasmolyse. In *Encyclopedia of Plant Physiology*; Ruhland, W., Ed.; Springer Verlag: Berlin/Göttingen/Heidelberg, Germany, 1956; Volume 2, pp. 71–115.
6. Lenné, T.; Bryant, G.; Hocart, C.H.; Huang, C.X.; Ball, M.C. Freeze avoidance: A dehydrating moss gathers no ice. *Plant Cell Environ.* **2010**, *33*, 1731–1741.
7. Küster, E. Über Inhaltsverlagerungen in plasmolysierten Zellen. *Flora* **1910**, *100*, 267–287.
8. Stadelmann, E. Zu Plasmolyse und Deplasmolyse von *Allium*-Epidermen. *Protoplasma* **1964**, *59*, 14–68.
9. Fischer, J.M.C.; Peterson, C.A.; Bols, N.C. A new fluorescent test for cell vitality using calcofuor white M2R. *Stain Technol.* **1985**, *60*, 69–79.
10. Luyet, B.J.; Gehenio, P.M. The survival of moss vitrified in liquid air and its relation to water conten. *Biodynamica* **1938**, *42*, 1–7.
11. Strugger, S. *Praktikum der Zell- und Gewebephysiologie der Pflanze*, 2nd ed.; Springer Verlag: Berlin/Göttingen/Heidelberg, Germany, 1949; Volume 1, pp. 89–93.
12. Ruoslahti, E.; Pierschbacher, M.D. New perspectives in cell adhesion: RGD and integrins. *Science* **1987**, *238*, 491–497.
13. Hostetter, M.K. RGD-mediated adhesion in fungal pathogens of humans, plants and insects. *Curr. Opin. Microbiol.* **2000**, *3*, 344–348.

14. Lu, B.; Chen, F.; Gong, Z.H.; Xie, H.; Zhang, J.H.; Liang, J.S. Intracellular localization of integrin-like protein and its roles in osmotic stress-induced abscisic acid biosynthesis in *Zea mays*. *Protoplasma* **2007**, *232*, 35–43.
15. Ambrose, C.; Allard, J.F.; Cytrynbaum, E.N.; Wasteneys, G.O. A CLASP-modulated cell edge barrier mechanism drives cell-wide cortical microtubule organization in *Arabidopsis*. *Nat. Commun.* **2011**, *2*, e430.
16. Lei, L.; Li, S.; Bashline, L.; Gu, Y. Dissecting the molecular mechanism underlying the intimate relationship between cellulose microfibrils and cortical microtubules. *Front. Plant Sci.* **2014**, *5*, e90.
17. De Vries, H. Eine Methode zur Analyse der Turgorkraft. *Jahrb. Wiss. Bot.* **1877**, *14*, 427–601.
18. Hecht, K. Studien über den Vorgang der Plasmolyse. *Beitr. Biol. Pflanz.* **1912**, *11*, 133–189.
19. Sitte, P. Zellfeinbau bei plasmolyse. *Protoplasma* **1963**, *57*, 304–333.
20. Tilney, L.G.; Cooke, T.J.; Connelly, P.S.; Tilney, M.S. The structure of plasmodesmata as revealed by plasmolysis, detergent extraction, and protease digestion. *J. Cell Biol.* **1991**, *112*, 739–747.
21. Pont-Lezica, R.F.; McNally, J.G.; Pickard, B.G. Wall-to-membrane linkers in onion epidermis: Some hypotheses. *Plant Cell Environ.* **1993**, *16*, 111–123.
22. Lang, I.; Barton, D.A.; Overall, R.L. Membrane-wall attachements in plasmolysed plant cells. *Protoplasma* **2004**, *224*, 231–243.
23. Cleary, A.L. Plasma membrane-cell wall connections: Roles in mitosis and cytokinesis revealed by plasmolysis of *Tradescantia virginiana* leaf epidermal cells. *Protoplasma* **2001**, *215*, 21–34.
24. Lang-Pauluzzi, I.; Gunning, B.E.S. A plasmolytic cycle: The fate of cytoskeletal elements. *Protoplasma* **2000**, *212*, 174–185.
25. Komis, G.; Quader, H.; Galatis, B.; Apostolakos, P. Macrotubule-dependent protoplast volume regulation in plasmolysed root-tip cells of *Triticum turgidum*: Involvement of phospholipase D. *New Phytol.* **2006**, *171*, 737–750.
26. Komis, G.; Apostolakos, P.; Galatis, B. Hyperosmotic stress-induced actin filament reorganization in leaf cells of *Chlorophyton comosum*. *J. Exp. Bot.* **2002**, *53*, 1699–1710.
27. Komis, G.; Apostolakos, P.; Galatis, B. Altered patterns of tubulin polymerization in dividing leaf cells of *Chlorophyton comosum* after a hyperosmotic treatment. *New Phytol.* **2001**, *149*, 193–207.
28. Ambrose, C.; Wasteneys, G.O. Nanoscale and geometric influences on the microtubule cytoskeleton in plants: Thinking inside and outside the box. *Protoplasma* **2012**, *249*, S69–S76.
29. Volgger, M.; Lang, I.; Ovečka, M.; Lichtscheidl, I. Plasmolysis and cell wall deposition in wheat root hairs under osmotic stress. *Protoplasma* **2010**, *243*, 51–62.
30. Oparka, K.J.; Prior, D.A.M.; Crawford, J.W. Behaviour of plasma membrane, cortical ER and plasmoldesmata during plasmolysis of onion epidermal cells. *Plant Cell Environ.* **1994**, *17*, 163–171.
31. Oparka, K.J.; Prior, D.A.M.; Crawford, J.W. Membrane conservation during plasmolysis. In *Membranes: Specialized Functions in Plants*; Smallwood, M., Knox, J.P., Bowles, D.J., Eds.; BIOS Scientific Publishers Ltd.: Milton, UK, 1996; pp. 39–56.
32. Marc, J.; Granger, C.L.; Brincat, J.; Fisher, D.D.; Kao, T.H.; McCubbin, A.G.; Cyr, R.J. A GFP-MAP4 reporter gene for visualizing cortical microtubule rearrangements in living epidermal cells. *Plant Cell* **1998**, *10*, 1927–1939.
33. Shaw, S.L.; Kamyar, R.; Ehrhardt, D.W. Sustained microtubule treadmilling in *Arabidopsis* cortical arrays. *Science* **2003**, *300*, 1715–1718.

34. Voigt, B.; Timmers, A.C.J.; Samaj, J.; Muller, J.; Baluska, F.; Menzel, D. GFP-FABD2 fusion construct allows *in vivo* visualization of the dynamic actin cytoskeleton in all cells of *Arabidopsis* seedlings. *Eur. J. Cell Biol.* **2005**, *84*, 595–608.
35. Baskin, T.I. On the alignment of cellulose microfibrils by cortical microtubules: A review and a model. *Protoplasma* **2001**, *215*, 150–171.
36. Gutierrez, R.; Lindeboom, J.J.; Paredez, A.R.; Emons, A.M.C.; Ehrhardt, D.W. *Arabidopsis* cortical microtubules position cellulose synthase delivery to the plasma membrane and interact with cellulose synthase trafficking compartments. *Nat. Cell Biol.* **2009**, *11*, 797–806.
37. Somerville, C. Cellulose synthesis in higher plants. *Annu. Rev. Cell Dev. Biol.* **2006**, *22*, 53–78.
38. Vain, T.; Crowell, E.F.; Timpano, H.; Biot, E.; Desprez, T.; Mansoori, N.; Trindade, L.M.; Pagant, S.; Robert, S.; Hoefte, H.; *et al*. The cellulase KORRIGAN is part of the cellulose synthase complex. *Plant Physiol.* **2014**, *165*, 1521–1532.
39. Bashline, L.; Du, J.; Gu, Y. The trafficking and behavior of cellulose synthase and a glimpse of potential cellulose synthesis regulators. *Front. Biol.* **2011**, *6*, 377–383.
40. Strzelecka-Golaszewska, H. Divalent cations, nucleotides, and actin structure. *Results Probl. Cell Differ.* **2001**, *32*, 23–41.
41. Wen, Q.; Janmey, P.A. Polymer physics of the cytoskeleton. *Curr. Opin. Solid State Mater. Sci.* **2011**, *15*, 177–182.
42. Liu, Q.; Qiao, F.; Ismail, A.; Chang, X.; Nick, P. The plant cytoskeleton controls regulatory volume increase. *BBA Biomembr.* **2013**, *1828*, 2111–2120.
43. Wojtaszek, P.; Baluska, F.; Kasprowicz, A.; Luczak, M.; Volkmann, D. Domain-specific mechanosensory transmission of osmotic and enzymatic cell wall disturbances to the actin cytoskeleton. *Protoplasma* **2007**, *230*, 217–230.
44. Singh, J. Ultrastructural alterations in cells of hardened and non-hardened winter rye during hyperosmotic and extracellular freezing stresses. *Protoplasma* **1979**, *98*, 329–341.
45. Lichtscheidl, I.K.; Url, W.G. Organization and dynamics of cortical endoplasmic reticulum in inner epidermal cells of onion bulb scales. *Protoplasma* **1990**, *157*, 203–215.
46. Pickard, B.G.; Ding, J.P. The mechanosensory calcium-selective ion-channel: Key component of a plasmalemmal control center. *Aust. J. Plant Physiol.* **1993**, *20*, 439–459.
47. Murashige, T.; Skoog, F. A revised medium for rapid growth and bio assays with tobacce tissue cultures. *Physiol. Plant.* **1962**, *15*, 473–497.

7

Penium margaritaceum: A Unicellular Model Organism for Studying Plant Cell Wall Architecture and Dynamics

David S. Domozych

Department of Biology and Skidmore Microscopy Imaging Center, Skidmore College, Saratoga Springs, NY 12866, USA; E-Mail: ddomoz@skidmore.edu

External Editor: Philip Harris

Abstract: *Penium margaritaceum* is a new and valuable unicellular model organism for studying plant cell wall structure and developmental dynamics. This charophyte has a cell wall composition remarkably similar to the primary cell wall of many higher plants and clearly-defined inclusive zones containing specific polymers. *Penium* has a simple cylindrical phenotype with a distinct region of focused wall synthesis. Specific polymers, particularly pectins, can be identified using monoclonal antibodies raised against polymers of higher plant cell walls. Immunofluorescence-based labeling is easily performed using live cells that subsequently can be returned to culture and monitored. This feature allows for rapid assessment of wall expansion rates and identification of multiple polymer types in the wall microarchitecture during the cell cycle. Cryofixation by means of spray freezing provides excellent transmission electron microscopy imaging of the cell, including its elaborate endomembrane and cytoskeletal systems, both integral to cell wall development. *Penium*'s fast growth rate allows for convenient microarray screening of various agents that alter wall biosynthesis and metabolism. Finally, recent successful development of transformed cell lines has allowed for non-invasive imaging of proteins in cells and for RNAi reverse genetics that can be used for cell wall biosynthesis studies.

Keywords: *Penium*; cell wall; pectin; live cell labeling; monoclonal antibody; protoplast; transformed cell lines

1. Introduction

The utilization of new research methods and strategies emerging from technologies in molecular genetics, immuno-binding/cytochemical labeling, high resolution microscopy and spectroscopy have recently provided significant insight into deciphering the complexities of the plant cell wall [1–6]. The current, yet rapidly evolving, working model of the cell wall is one of a fibrous composite of polymers, primarily polysaccharides and glycoproteins, which are incorporated into specific loci of the wall and are subsequently modified at various intervals during the life of the plant cell [7–10]. Microfibrillar cellulose forms the load-bearing framework of the wall and is tethered by hemicelluloses into a stable network and embedded in a matrix of pectins. Structural proteins (e.g., extensin and arabinogalactan proteins), enzymes (e.g., pectin methyl esterase or PME, wall-associated kinases or WAKs), cations, polyphenolics, like lignins, and water also contribute to wall infrastructure. The components of the cell wall form specific structural domains around the plant cell that, in turn, are vital for such functions as controlled expansion and division, rigidity, defense and cell-cell adhesion, to name just a few. A significant portion of the plant cell's genome is devoted to the biosynthesis and metabolism of its cell wall [11] that, in turn, is regulated by complex signal transduction cascades responding both to internal development prompts and environmental stresses [12–14]. Further elucidation of these cell wall-based phenomena will be critically important in many realms of basic and applied biology. This includes enhancing our knowledge of the living dynamics of a plant cell, the evolution of plants, especially over the last 500 million years of land plant evolution, and the ever increasing use of cell walls in the food, paper, building, textile, pharmaceutical and biofuels industries.

Much of what we have learned about the plant cell wall has been derived from studies of multicellular plants and, in particular, outstanding model organisms, like *Arabidopsis thaliana*. Over the past few decades, the emergence of highly detailed molecular data and protocols and their synthesis with experimental data derived from an easily-manipulatable organism, like *Arabidopsis*, has provided an efficacious system for resolving specific questions about the cell wall dynamics [15–17]. Likewise, the use of tissue culture-based systems, like tobacco cell suspension cultures, has aided in resolving the basics of the cell walls of individual cells [18]. However, these systems have some inherent limitations. For example, it is exceptionally difficult to resolve the cell wall structure and dynamics of a single cell situated in a plant organ/tissue composed of thousands of cells. Likewise, tissue culture-based cells are maintained under notably artificial conditions, and results derived from studies with them must be interpreted very carefully. Ideally, the utilization of a model plant that is naturally single celled, whose cell cycle can be synchronized (*i.e.*, cell wall developmental stages carefully controlled) and one that is easy to manipulate for experimental analyses would be ideal for dissecting the structure and function of the cell wall. The Charophycean green algae or charophytes (Streptophyta), the group of green algae most closely related and ancestral to land plants, may provide us with these models.

Recently, biochemical and immunobinding-based screening (including, comprehensive polymer profiling, CoMPP, and immunocytochemistry) has revealed that the charophytes have cell wall polymer composition profiles that are remarkably similar to those of land plants [19–21]. This includes a cellulose microfibril infrastructure with associated hemicelluloses (mannans, xylans, xyloglucans) embedded in a pectin matrix. Glycoproteins, such as extensins and arabinogalactan proteins (AGPs), as well as lignin have also been found in charophyte walls. Most charophytes also exhibit simple phenotypes when compared with those found in land plants. This simplicity has been exploited in many cell wall studies focused on fundamental phenomena, such as pectin dynamics and cell expansion (*Chara*) [22], as well as wall development during cellular morphogenesis (*Micrasterias*) [23,24]. Recently, my laboratory isolated a clone of the charophyte, *Penium margaritaceum,* and it has become increasingly valuable in cell wall studies. This alga is *unicellular*, produces only a primary cell wall that has clearly defined domains of cellulose, pectin and other wall polymers and can be easily manipulated for both microscopy-based analyses and experimental screening [25,26]. Recently, *Penium* was successfully transformed [27], and current efforts to sequence the genome are nearly complete. All of this contributes significantly to *Penium*'s present and future role as a model organism for cell wall studies. In this paper, the attributes of, and protocols for, using *Penium* in multiple areas of cell wall research are presented.

2. General Biology and Growth Characteristics

Penium margaritaceum is a unicellular desmid whose shape is that of a simple elongate cylinder with rounded poles (Figure 1A). Each cell measures approximately 17 μm in diameter at the cell center or isthmus and attains lengths of 120 to 240 μm. A nucleus is positioned for most of the cell cycle at the isthmus and is surrounded by two elongated chloroplasts. Several different types of vacuoles are found throughout the cytoplasm, as are numerous crystals, most likely barium or silica salts, a feature that is common in desmids [28]. The peripheral cytoplasm is also the site of rapid cytoplasmic streaming and consists of narrow channels appressed to the plasma membrane that run approximately parallel to the long axis of the cell. The cell wall is approximately 1 μm thick throughout most of the cell surface [25]. The outermost cell wall layer is characterized by a distinct collection of projections or "warts" (Figure 1B) that forms a regular pattern. This layer has been designated as the pectin "lattice". However, the wall is notably thinner and is devoid of the projections in a 1-μm band that encircles the isthmus [26]. Likewise, several narrow peripheral bands lined by different-sized projections are also found on the cell wall. The isthmus is the major site of the cell wall expansion mechanism, which begins prior to cell division [29]. Here, new wall components are deposited in a narrow band found at the center of the isthmus that, in turn, displaces older wall regions outward toward both poles (*i.e.*, "bipolar" expansion). Additionally, other areas of wall deposition have been noted in development [29]. During cell division, post-mitotic daughter nuclei are rapidly translocated from the isthmus, each to the center of a soon-to-be daughter cell. The exact mechanism for this event is not known for *Penium*, but is most likely microtubule-based, as noted for other desmids [30]. Cytokinesis occurs at the isthmus shortly after mitosis and entails both a cell plate and a small furrow. After cytokinetic separation, the poles of each daughter cell may expand for a short time thereafter. *Penium* also secretes a gel-like extracellular polymeric substance (EPS) beyond the cell wall [31]. The EPS is a xylose-, fucose- and glucuronic acid-rich polysaccharide(s) that is often produced in very large amounts, especially as the cells age or if the cells are exposed to high light, high temperature or desiccation conditions. The EPS may be imaged using immunofluorescence labeling of live cells (Figure 1C).

Figure 1. General features of *Penium* and its cell wall. (**A**) *Penium* is an elongated cylindrical uni-cell with rounded edges. The central isthmus ("isthmus") contains the nucleus and is also the focal point of cell wall expansion. New wall polymers are deposited here in a narrow band that, in turn, displaces older cell wall outward toward the poles ("mode of expansion", arrow); (**B**) variable pressure scanning electron microscopy (VPSEM) image of the cell wall surface showing the numerous projections that form the outer wall lattice ("lattice projections", arrows); (**C**) *Penium* also produces a gel-like EPS that attaches to the wall and extends outward (arrow). The cell was labeled with an anti-EPS antibody and viewed with a confocal laser scanning light microscope (CLSM); (**D**) For culturing, cells are grown in flasks of sterile liquid medium; (**E**) cells may also be successfullygrown on agarose-solidified medium. In this image, the cells are growing on cellophane on top of Woods Hole medium (WHM) medium solidified with 1.5% agarose. EPS, extracellular polymeric substance.

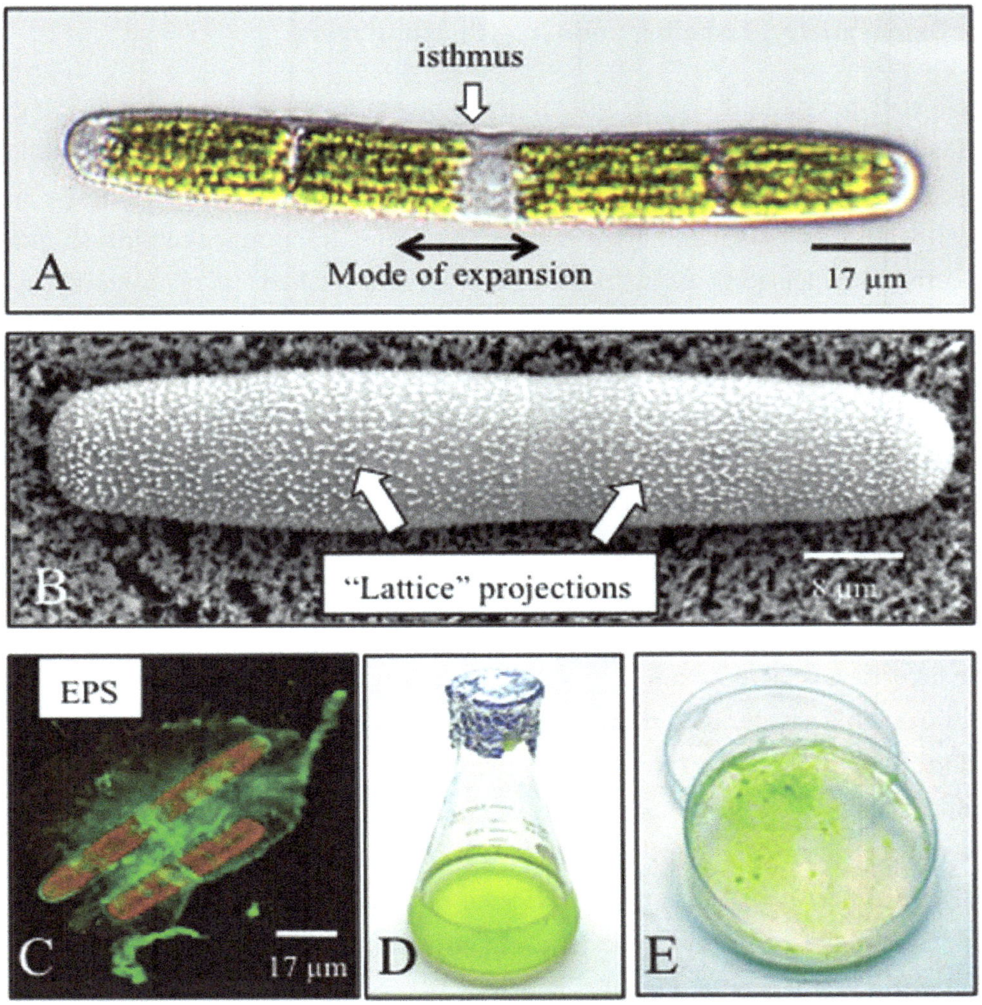

Penium cultures may be maintained in many types of algal freshwater media. My laboratory routinely uses liquid Woods Hole MBL (Marine Biology Laboratory) medium (WHM) [32], either alone or supplemented with 5% soil extract. Culture volumes of 75 to 150 mL in sterile 125- or 250-mL flasks are convenient and yield log phase cultures within seven days (Figure 1D). Subcultures are typically made every 10–14 days by transferring 5 mL of algal suspension to a new flask with fresh medium. Cells are also successfully grown on agarose-solidified WHM (1%–2% agarose), either directly on agarose-solidified medium or on sheets of sterile dialysis membrane laid on the surface of agarose-solidified medium (Figure 1E).

For harvesting cultures, cells grown in liquid culture are collected by gentle centrifugation at 500–1000 × *g* for 1 min in 15-mL or 50-mL centrifuge tubes. For washing and to remove EPS gel from the cells (EPS gel on the cell surface may alter subsequent labeling efficacy), post-centrifugation cell pellets are re-suspended in fresh growth medium, shaken vigorously for 10–15 s and then re-centrifuged. The supernatant containing EPS is discarded. This is repeated several times, and centrifugation results in tight green cell pellets free of EPS. These cells can then be used for cytochemical labeling, experimentation or cell wall isolation. For the production of large volumes of cells, cultures may be upscaled to 1-L volumes in 2-L flasks. If cultures become contaminated, washed cells may be treated with an antibiotic/antimycotic solution treatment (e.g., Sigma A5955; Sigma Chem., St. Louis, MO, USA; 1/100 dilution in WHM, overnight, followed by extensive washing in WHM).

3. The Cell Wall: Biochemistry, Labeling and Growth Monitoring

The *Penium* cell wall consists of two dominant components, a homogalacturonan (HG)-rich outer layer that is complexed with calcium (Ca^{2+}) and that is manifested by the lattice that is positioned on the cell wall surface [26]. The HG-rich outer layer is attached to an inner layer of cellulose microfibrils. Pure fractions of cell walls can be isolated using simple physical extraction techniques (Figure 2A) [25]. Log phase culture cells are collected by centrifugation, extensively washed and resuspended in an ice-cold buffered detergent solution (0.01% Triton-X-100 in 0.05 M Tris buffer, pH 7.2). The suspension is then sonicated on ice to cavitation for ten, five- to ten-s bursts. This procedure ruptures the cells and leaves the cell walls. The cell walls are collected by centrifugation, repeatedly washed with buffer and recentrifuged. This results in a pure white pellet. The purity of the pellet can be judged by light microscopy examination. If the pellet still contains whole cells, further sonication/washings are needed, and if starch grains are present, the pellet may be treated with amylase for a few hours and then washed. From pure cell wall pellets, various wall polymers can be extracted from the isolated walls using conventional cell wall extraction protocols (e.g., with chelators, base, acids or cadoxen). For example, the outer HG-rich layer is extracted from the isolated walls with 4-h incubations in 50 mM CDTA (1,2-Diaminocyclohexanetetraacetic acid monohydrate). The extract is then extensively dialyzed against deionized water and freeze dried for storage. Subsequently, when solutions of this pectin extract (e.g., 50–250 μg/mL in deionized water) are mixed with 10–25 mM solutions of exogenous Ca^{2+}-salts or other cation-salts, transparent gels form rapidly, typical of HGs. The post-CDTA wall remnants may then be extracted with other agents, and the extracts are analyzed via biochemical/immunobinding protocols. It is important to note that prior to extraction, large amounts of EPS may be isolated from the medium of cell cultures or from the washes of cell pellets (see above). The EPS in these solutions may be concentrated by centrifugation at 25,000 × *g* for 10 min, where it forms a gel-like pellet. The EPS may be dialyzed against deionized water and freeze dried for storage prior to analysis.

A valuable feature of *Penium*'s cell wall is that it can be easily labeled via immunofluorescence in live cells or after removal from cells (Figure 2B; see [25] and [29] for detailed descriptions) using monoclonal antibodies (mAbs) currently available for the study of higher plant cell walls. For example, the HG-rich outer layer readily binds to mAbs specific for HGs derived from higher plants (e.g., JIM5, LM18, LM19,

PAM1, 2F4, JIM7, LM20; [33–37]). Using a double labeling protocol, live cells may be labeled with a particular mAb and then labeled with the appropriate secondary antibody conjugated with a fluorophore (Figure 2C). The cells may be washed and then placed back into culture. Cells are not harmed by this process and continue to expand and divide at the same rate as unlabeled cells. The fluorescent label is very stable and typically maintains the same intensity level even after seven days. After a period of growth, new cell wall HG will appear as a dark zone situated between the mAb-labeled zones (Figure 2D,E). The surface areas of the dark zones may be measured to determine wall expansion rates during the cell cycle and under various experimental treatments. Likewise, after a designated period of time, the labeled cells may be labeled with the same mAb followed by labeling with a different secondary antibody-fluorophore conjugate. Imaging of the two fluorescent signals with different channels and/or filter sets of a confocal laser scanning light microscope (CLSM) allows for identification of old and new wall areas and growth patterns (Figure 2F). Similarly, another mAb with specificity for other wall polymer epitopes may be used in subsequent labeling to help reveal the precise location of different polymers in the wall and putative interpolymeric connections. For example, cells first labeled with JIM5, LM18, 2F4 (specific for relatively low methyl-esterified HG or Ca^{2+}-complexed HG), re-cultured and then labeled with JIM7 or LM20 (specific for "relatively" high methyl-esterified HG) show that "high" methylesterified HG is secreted in a narrow band at the isthmus and is surrounded by wall loci with HG that has been de-esterified and complexed with Ca^{2+}. Presumably, the enzyme, pectin methylesterase (PME), de-esterifies the HG at this band that, in turn, allows Ca^{2+} to cross-link to the HG and form the projections of the outer wall lattice. This live labeling technology is also valuable for studying the effects of specific experimental agents on the wall microarchitecture. For example, cells incubated in medium containing the enzyme, PME, for 24 h (Figure 2G) or β-galactosidase for 48 h (Figure 2H) and then labeled with JIM5 reveal significant alteration to the HG lattice at the isthmus (see also [26]). *Penium*'s "versatility" in immunocytochemical studies make it an excellent organism for screening new wall mAbs, as well as other probes, including carbohydrate binding modules (CBMs) [3], particularly those applicable for the analysis of cellulose and hemicellulosic polymers.

4. High Resolution Electron Microscopy Analyses

Penium's unicellular habit also makes it a convenient model organism for electron microscopy-based studies of the wall dynamics, especially the secretory apparatus and cytoskeletal system. Earlier studies with other unicellular desmids, like *Micrasterias* [22,23], provided important insight into the development of protocols for cell preparation for *Penium*. For transmission electron microscopy (TEM), cells may be fixed with conventional glutaraldehyde/osmium tetroxide-based protocols, dehydrated, embedded in plastic and sectioned in order to provide good quality preservation of the cell wall. However, considerable shrinkage of the protoplast occurs via this method. However, all cell ultrastructure may be preserved satisfactorily using cryo-fixation and freeze substitution methodologies. For rapid cryo-fixation, a suspension of cells may be sprayed with a commercial artist's spray brush and air compressor into liquid propane cooled in liquid nitrogen. This spray freezing is followed by freeze substitution in glutaraldehyde, osmium tetroxide and/or uranyl acetate in acetone or ethanol cooled to −80 °C for various periods of time. Cells are then warmed to either −20 °C or room temperature, washed with acetone or ethanol, infiltrated and embedded in plastic. We often employ osmium-free freeze substitution and London Resin embedding at −20 °C for cells to be subsequently sectioned and immunolabeled. For general ultrastructural examination, we use freeze substitution cocktails containing osmium and subsequently embed in Spurr's low viscosity plastic at 60 °C. For both protocols, we find it convenient to place cells in a thin layer of pre-polymerized plastic sandwiched between two sheets of the Aclar plastic. After polymerization, the Aclar sheets are removed, individual cells of interest are selected with a light microscope and are excised from the sheet. The cells are positioned/glued onto a plastic block for sectioning at various orientations. It is also important to note that if sections of osmicated cells are first treated for 5 min in 5% H_2O_2, they can be successfully used for immunogold labeling with the wall polysaccharide-specific mAbs.

Figure 2. The cell wall and labeling with mAbs. (**A**) VPSEM image of isolated cell walls obtained through sonication (see text); (**B**) JIM5 immunofluorescence of the homogalacturonan (HG) of the outer wall layer of isolated cell walls. This mAb recognizes HG with relatively low levels of methyl-esterification. Note that the projections are clearly distinguished on the outer wall surface; (**C**) JIM5 immunofluorescence of a live cell. Most of the cell wall surface is labeled except for the isthmus (small arrow) and a lateral band (arrowhead); (**D**) After live labeling with JIM5, cells are placed back into culture. Over time, newly-deposited HG/cell wall appears as a dark zone centered at the isthmus (double arrow); (**E**) New growth zones and expansion rates in live labeled cells may be assessed using fluorescence microscopy. The surface areas of the dark zones (arrows) are measured and compared to overall cell surface area; (**F**) Live labeled cells that were allowed to grow in culture for 24 h and subsequently labeled with JIM5 that is, in turn, labeled with a different fluorophore reveals old (red) and new (green) wall regions. In this cell, the position and size of old wall regions ("old wall", arrows) reveals an asymmetry in wall expansion [25]; (**G**) Live cell labeling with JIM5 is also used to assess alteration of the outer wall lattice. When treated with pectin methyl esterase (PME) for 24 h, the lattice at the isthmus is replaced by dense punctate labeling (arrow; see also [27]); (**H**) When cells are treated for 24 h with β-galactosidase, the lattice is disrupted (arrow).

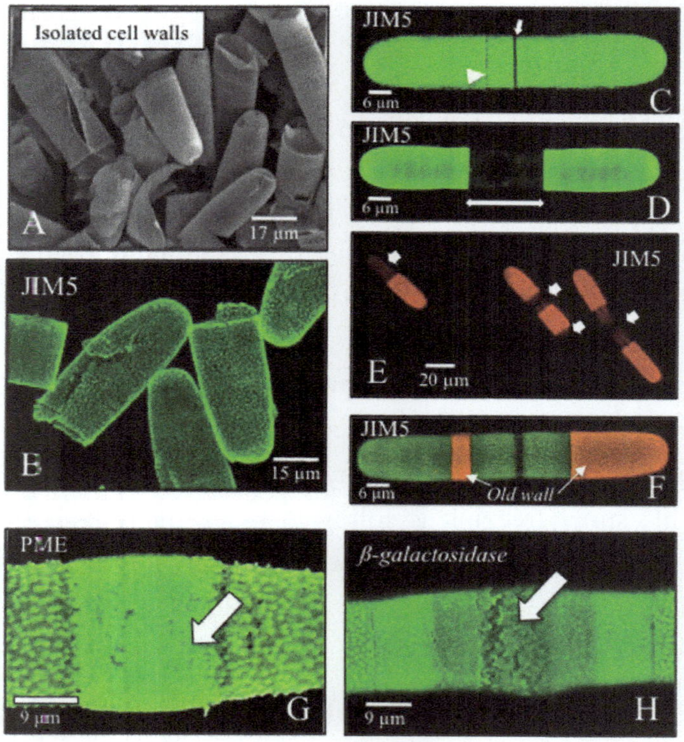

TEM analyses reveal that the cell wall is approximately 1 μm wide throughout most of the cell, with the outer HG-lattice being 550 nm and the inner cellulose-based layer being 450 nm thick (Figure 3A). EPS fibrils attach to the outer portion of the inner wall layer and extend through the HG lattice and away from the cell surface for several μm (Figure 3B). The HG-rich lattice consists of projections made of struts that extend outward and fuse to form a single terminus. Mild EDTA extraction of cells or isolated cell walls in order to remove Ca^{2+} from the outer lattice reveals that the lattice struts consist of parallel arrays of hundreds of thin fibrils (Figure 3C). Fibrils from the inner portion of the lattice penetrate and embed in the inner cellulose microfibril-rich layer. This zone constitutes the medial wall layer (Figure 3D). Immunogold labeling shows that both the lattice and medial layer contain HG, while the medial layer alone contains HG and rhamnogalacturonan-I (RG-I; Figure 3E). This suggests that RG-I may serve as a linker between the HG and cellulose [27]. At the primary site of wall expansion, the isthmus, the cellulosic inner layer is the only wall layer observed and is devoid of the lattice. The medial layer and emerging outer lattice (Figure 3F) are found on either side of the isthmus region.

The ultrastructure of the *Penium* wall may also be analyzed using scanning electron microscopy (SEM). Field emission scanning electron microscopy (FESEM) of both frozen and isolated cell walls clearly shows the lack of the lattice at the isthmus (Figure 4A), but also reveals narrow lateral bands found at more polar regions of the cell (Figure 4B). Close examination shows that the lattice projections are taller on the pole-facing side of these lateral bands than on the isthmus-facing side. The role of these bands is unknown, but may very well represent areas of growth oscillations. For example, wall synthesis does not occur during the dark period of the photocycle. It is may be that the lattice formation stops abruptly when the cell is placed in the dark, resulting in normal size projections. However, when returned to the light, the lattice formation mechanism is more gradual, and those projections are notably smaller. FESEM also has been important in revealing that the arrangement of cellulose microfibrils on the innermost surface of the cell's isthmus runs perpendicular to long axis of the cell at the isthmus [27]. FESEM is also used to identify changes to the HG lattice when cells are treated with agents that affect HG secretion and incorporation into the wall. For example, in cells grown in cultures containing 10 times the normal concentration of $CaCl_2$, large elongated ridges replace the individual projections of the typical lattice (Figure 4C).

Though not as powerful a high resolution imaging tool as FESEM, variable pressure scanning electron microscopy (VPSEM) allows for rapid assessment of wall structure. Here, cells treated in various ways can be collected and placed on nitrocellulose sheets. The sheets are plunge frozen in liquid nitrogen and secured upon on a cooled stage that is placed in the VPSEM. The vacuum pressure of the SEM column is brought to 20–30 Pascal, and within minutes, sublimation removes the ice from the cell surface, to reveal structural changes (Figure 4D). This technology has proven to be exceptionally helpful in quickly screening cells treated with a variety of experimental agents.

Figure 3. TEM imaging of the cell wall (see [25,27] for details). (**A**) Cross-section of the cell wall highlighting the distinctive outer layer (OL) that forms the lattice, the inner cellulose-based layer (IL) and the interface medial layer (ML). The medial layer is composed of fibrils from the outer layer that embed in the inner cellulose layer; (**B**) EPS fibrils (EPS) attach to the outer surface of the inner/medial wall (white arrows) and extend outward beyond the outer layer (OL); (**C**) Dark field image of the HG fibrils (arrow) that form dense aggregates that make the outer wall layer lattice. This cell was treated with a weak chelator to remove Ca^{2+} (see text); (**D**) JIM5 immunogold labeling (white arrows) of the HG-containing outer layer (OL) and medial layer (ML). The inner layer (IL) does not label with this mAb; (**E**) RG-I immunogold labeling (white arrows) of the medial layer (ML), but not the outer layer (OL); (**F**) The isthmus zone labeled with JIM5. Note the absence of the outer lattice in the isthmus ("isthmus", arrows), but the presence of the lattice on the edges of the isthmus (white arrows).

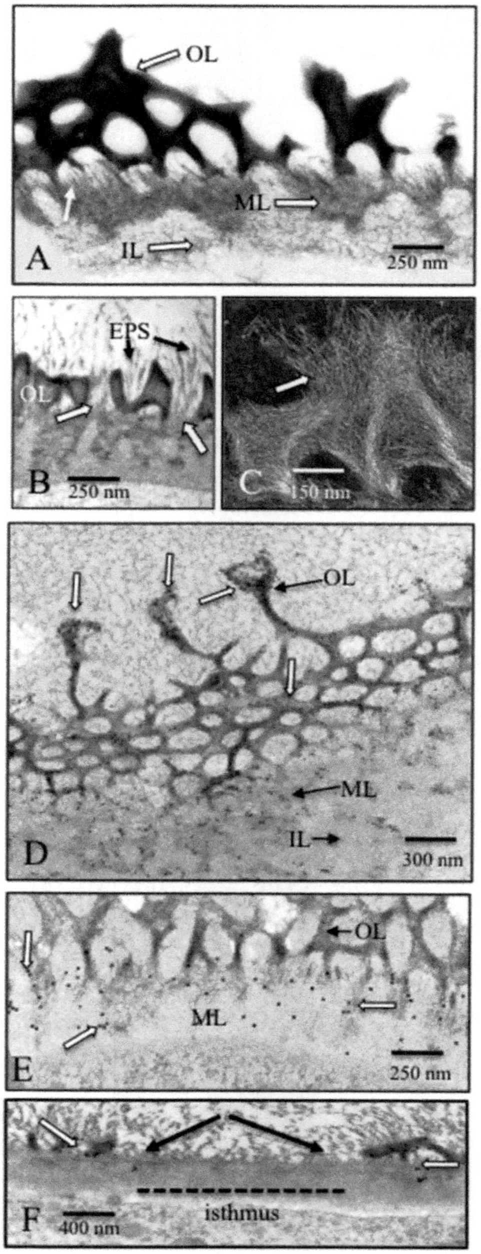

Figure 4. SEM imaging of the cell wall (see [25,27] for details). (**A**) FESEM image of the isthmus zone ("isthmus", arrows). Note that the outer wall lattice projections are not present in the isthmus; (**B**) FESEM image of a lateral band. On the outer edge of the band (*i.e.*, facing a pole of the cell), the projections are thick (large arrow). On the inner edge, the lattice is smaller (small arrow). No projections are found in the middle of the band; (**C**) FESEM image of a cell incubated for 48 h in medium supplemented with 10 × $CaCl_2$. The older wall (small arrow) possesses the typical outer lattice. However, in the newer wall zone formed when under high Ca^{2+} conditions, the lattice forms a series of elongated ridges (large arrow; see also [27]); (**D**) VPSEM image of a cell treated for 48 h with PME. Note that at the isthmus, *i.e.*, the focal point of wall expansion, the outer lattice is missing (arrow).

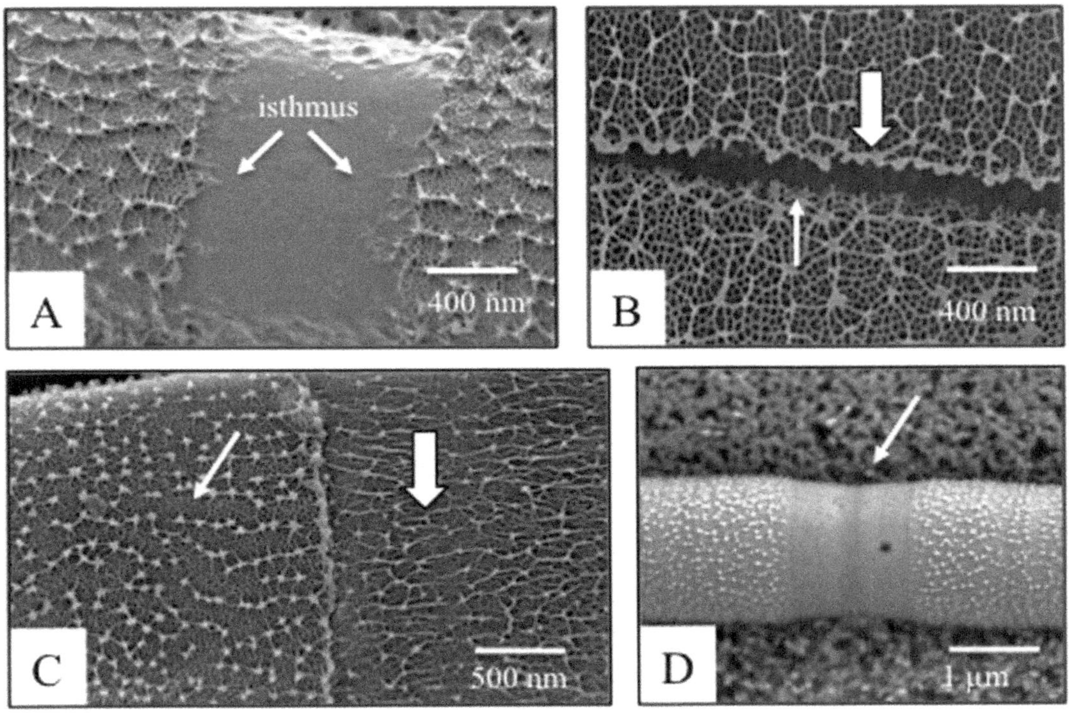

5. Experimental Analyses: Microarray or Single-Cell Approach

Penium's unicellular cylindrical shape and distinct cell wall along with a fast growth rate provide an outstanding model system for the screening of large numbers of experimental agents and gauging their effects on cell and wall structure and development. Cells incubated in medium containing various agents and concentrations thereof may be grown in 12-, 24- or 48-welled petri dishes. Typically, concentrations of 1000 cells/mL in various aliquots are used per well. The cells may also be first labeled with a wall polymer-specific mAb-fluorophore (see above) before treatment. The effects of the agent may then be conveniently viewed using an inverted light or CLSM. For cells initially labeled with a wall-specific mAb, growth rates, as identified by the surface area of non-labeled zones *vs.* whole cell surface area, may be measured. Aliquots of cells from individual wells may also be harvested and labeled for immunofluorescence or processed for electron microscopy. Likewise, aliquots of treated cells from each well may be harvested, washed and placed in fresh growth medium in order to monitor recovery dynamics after an experimental treatment. Finally, individual cells of *Penium* may be isolated using sterile tips of a 0.5–20 μL micropipettor and subsequently grown in medium contained in wells of a 96-welled petri dish. This allows for imaging individual cell histories over time and the monitoring of individual cell dynamics under specific experimental treatments. These protocols have been especially valuable in studies of wall development in cells treated with cytoskeletal poisons, various cations, enzymes and exogenous pectins [26,38,39].

6. Intracellular Studies

Penium contains an elaborate endomembrane system consisting of 100–125 Golgi bodies per cell, numerous vesicle types involved in secretory activities and a highly dynamic cytoplasmic streaming network that transports the vesicles. These subcellular components are involved in the biosynthesis and secretion of the cell wall and EPS, but many of the specific details have not yet been resolved. The Golgi bodies of *Penium* do not move notable distances in the cell, but remain stationary in the bases of valleys of cytoplasm lined by lobes of the chloroplasts. This is unlike what has been described for higher plants, whereby Golgi bodies are mobile [40,41]. Each Golgi body of *Penium* contains 12–16 tightly stacked cisternae and displays distinct *cis* and *trans* faces (Figure 5A) [42]. Large EPS-containing vesicles emerge from the peripheries of *trans* face cisternae, while wall polymers, like pectin, appear to be packaged in small vesicles arising from a more medial region of the Golgi body. All vesicles move to the cell periphery where they become part of the active cytoplasmic streaming network. This streaming mechanism is associated with bundles of actin cables that may be imaged using rhodamine phalloidin-based labeling protocols (Figure 5B) [43] or TEM. *Penium*'s microtubule network may also be imaged using freeze shattering and labeling with anti-tubulin [40]. The isthmus zone is almost always lined with a band of microtubules aligned perpendicular to the long axis of the cell. These microtubules are likely associated with wall synthesis in this area of active wall expansion, most likely as agents participating in cellulose microfibril biosynthesis and/or as directional targets for vesicle-plasma membrane fusion. This band also marks the future site of mitosis and cytokinesis, *i.e.*, very much like a preprophase band found in embryophytes [44,45]. Interestingly, when the cell starts to expand prior to cell division, satellite bands arise on either side of the isthmus band and migrate toward the poles of the cell during expansion. These bands represent the sites of future cell division and wall expansion, in effect pre-prophase bands for the next division phase [40]. Further elucidation of these processes using transformed cell lines will certainly aid in understanding the dynamic activities associated with these subcellular systems.

7. Selective Dissolution of the Cell Wall and Protoplast Formation

Protoplasts of plant cells are important tools for experimentation (e.g., transformation, biochemical syntheses) and helping to resolve key questions in plant cell biology and biochemistry [46–48]. They also provide a unique system for monitoring recovery dynamics of the cell wall after it has been removed from the cell. Protoplasts may be isolated from *Penium* when log phase cells are incubated in a mannitol-based osmoticum containing 1% Driselase (Figure 5C,D). Typically, within 20 min, high yields of protoplasts (80% or more of the cells yield protoplasts) are released from the altered cell wall and may be cultured (work to be reported elsewhere). Though viable for several days in osmoticum, the recovery dynamics of the cell wall have yet to be described. As important, live *Penium* may also be incubated in medium containing specific wall-degrading enzymes. Specific polymers of the wall are selectively removed, and the architecture of the altered wall are monitored by immunofluorescence or electron microscopy analyses. Recovery dynamics may also be observed in washed cells that are returned to culture. Here, the isthmus zone serves as the initiation site of recovery [26]. These enzymatic dissections of the cell wall in live cells represent a powerful means of assessing the role of particular polymers in cell wall architecture, cell shape maintenance and cell expansion.

Figure 5. The cell and protoplasts (see [39,40,43] for details). (**A**) A TEM image of two of the 100–125 Golgi bodies of the cell. The Golgi bodies sit at the base of lobes of cytoplasm lined by lobes of the chloroplast. Clearly defined cis (*cis*) and trans (*trans*) faces are apparent. Vesicles emerge from the trans of the medial loci of the Golgi body (arrow) and travel to the peripheral cytoplasm, where they enter the peripheral cytoplasmic streaming network; (**B**) Rhodamine phalloidin labeling of the actin bundles in the peripheral cytoplasm (arrows); (**C**) Protoplast release through the cell wall at the isthmus zone (arrow). The cell was treated with Driselase in a 12% mannitol solution (work to be reported elsewhere); (**D**) Fluorescence image of protoplast (large arrow) emerging from the cell wall (small arrow) labeled with the mAb, JIM5.

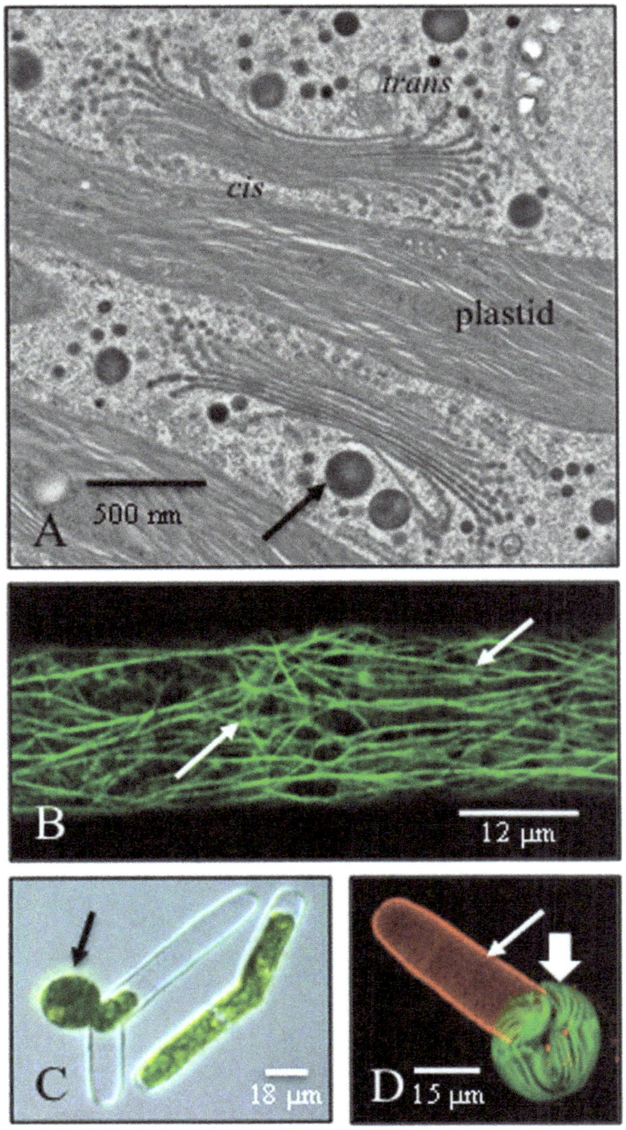

8. Molecular Biology: Transformation Technology

Surprisingly, there is a paucity of both detailed genomic sequencing and annotation in the charophytes today [49,50]. This has hindered our understanding of evolutionary trends and biosynthetic/metabolic pathways associated with the cell wall in these organisms. Recently, though, a major breakthrough occurred when *Penium* was successfully transformed [27]. *Agrobacterium*-based transformation was accomplished with cells initially incubated in nitrogen-depleted medium. This allowed for the establishment of cell lines displaying stable RFP-KDEL expression and provided novel insight into the network of endoplasmic reticulum in the cell. Likewise, RNAi-mediated gene suppression was used for the first time to elucidate PME and cellulose synthase (CESA). This exciting discovery and new technology offers a powerful and non-invasive means of studying specific proteins in *Penium*. When coupled with future genomic sequencing information, elucidation of *Penium*'s cell wall dynamics will be enhanced significantly.

9. The Future

The first decade of detailed studies of *Penium* have yielded results that demonstrate the importance of this unicellular organism as a model organism for studying plant cell walls. Its versatility in analysis using multiple and diverse technologies gives plant cell wall biologists and chemists a convenient and easily manipulable organism for obtaining insight into the fundamentals of cell wall structure and function. Some of the next major steps that need to be undertaken with *Penium* include: (1) analyzing the elaborate endomembrane system in order to identify the locations of key biosynthetic and secretory events associated with wall development; (2) identifying the pathways of endocytosis and its role in regulating expansion and wall polymer retrieval; (3) performing detailed biochemical dissections of the cell wall in order to determine the role of RG-I in HG-cellulose interactions and the location/functions of wall glycoproteins, also components of the wall; and (4) elucidating the chemistry and secretory dynamics of EPS, as it represents a major source of extracellular polysaccharide associated with the cell wall. Clearly, the use of unicellular charophytes, like *Penium*, provide new opportunities to resolve many questions about the cell wall that cannot be addressed in multicellular organisms.

Acknowledgements

This work was supported by grants from the U.S. National Science Foundation (NSF-MCB 0919925, NSF-DBI 0922805 and NSF-MRI 1337280).

Conflicts of Interest

The author declares no conflicts of interest.

References

1. Harris, D.; Bulone, V.; Ding, S.-Y.; DeBolt, S. Tools for cellulose analysis in plant cell walls. *Plant Physiol.* **2010**, *153*, 420–426.
2. Popper, Z. The plant cell wall. Methods and protocols. *Ann. Bot.* **2012**, *109*, vii–viii.

3. Gilbert, H.J.; Knox, J.P.; Boraston, A.B. Advances in understanding the molecular basis of plant cell wall polysaccharide recognition by carbohydrate-binding modules. *Curr. Opin. Struct. Biol.* **2013**, *23*, 669–677.
4. Pedersen, H.L.; Fangel, J.U.; McCleary, B.; Ruzanski, C.; Rydahl, M.G.; Ralet, M.-C.; Farkas, V.; von Schantz, L.; Marcus, S.E.; Andersen, M.C.F.; *et al.* Versatile high-resolution oligosaccharide microarrays for plant glycobiology and cell wall research. *J. Biol. Chem.* **2012**, *287*, 39429–39438.
5. Pattathil, S.; Avci, U.; Baldwin, D.; Swennes, A.G.; McGill, J.A.; Popper, Z.; Bootten, T.; Albert, A.; Davis, R.H.; Chennareddy, C.; *et al.* A comprehensive toolkit of plant cell wall glycan-directed monoclonal antibodies. *Plant Physiol.* **2010**, *153*, 514–525.
6. Lee, K.J.; Marcus, S.E.; Knox, J.P. Cell wall biology: Perspectives from cell wall imaging. *Mol. Plant* **2011**, *4*, 212–219.
7. Cosgrove, D.J.; Jarvis, M.C. Comparative structure and biomechanics of plant primary and secondary cell walls. *Front. Plant Sci.* **2012**, *3*, e204.
8. Doblin, M.S.; Pettolino, F.; Bacic, A. Plant cell walls: The skeleton of the plant world. *Funct. Plant. Biol.* **2010**, *37*, 357–381.
9. Caffall, K.H.; Mohnen, D. The structure, function, and biosynthesis of plant cell wall pectic polysaccharides. *Carbohydr. Res.* **2009**, *344*, 1879–1900.
10. Keegstra, K. Plant cell walls. *Plant Physiol.* **2010**, *154*, 483–486.
11. Popper, Z.A.; Gurvan, M.; Hervé, C.; Domozych, D.S.; Willats, W.G.T.; Tuohy, M.; Kloareg, B.; Stengel, D.B. Evolution and diversity of plant cell walls: From algae to flowering plants. *Ann. Rev. Plant Biol.* **2011**, *62*, 567–590.
12. Dhugga, K.S. Biosynthesis of non-cellulosic polysaccharides of plant cell walls. *Phytochemistry* **2012**, *74*, 8–19.
13. Ruprecht, C.; Persson, S. Co-expression of cell-wall related genes: New tools and insights. *Front. Plant. Sci.* **2012**, *3*, e83.
14. Wang, H.Z.; Dixon, R.A. On-off switches for secondary cell wall biosynthesis. *Mol. Plant* **2012**, *5*, 297–303.
15. Liepman, A.H.; Wightman, R.; Geshi, N.; Turner, S.R.; Scheller, H.V. *Arabidopsis*—A powerful model system for plant cell wall research. *Plant J.* **2010**, *61*, 1107–1121.
16. Anderson, C.T.; Wallace, I.S.; Somerville, C.R. Metabolic click-labeling with a fucose analog reveals pectin delivery, architecture, and dynamics in Arabidopsis cell walls. *Proc. Natl. Acad. Sci. USA* **2012**, *109*, 1329–1334.
17. Tan, L.; Eberhard, S.; Pattathil, S.; Warder, C.; Glushka, J.; Yuan, C.; Hao, Z.; Zhu, X.; Avci, U.; Miller, J.S.; *et al.* An *Arabidopsis* cell wall proteoglycan consists of pectin and arabinoxylan covalently linked to an arabinogalactan protein. *Plant Cell* **2013**, *25*, 270–287.
18. Yokoyama, R.; Daisuke, T.; Fujino, T.; Itoh, T.; Nishitani, K. Cell wall dynamics in tobacco BY-2 cells. In *Biotechnology in Agriculture and Technology*, 2004 ed.; Nagata, T., Hasezawa, S., Inze, D., Eds.; Springer-Verlag: Berlin, Germany, 2004; Volume 53, pp. 217–230.
19. Sørensen, I.; Pettolino, F.A.; Bacic, A.; Ralph, J.; Lu, F.; O'Neill, M.A.; Fei, Z.; Rose, J.K.C.; Domozych, D.S.; Willats, W.G.T. The charophycean green algae provide insights into the early origins of plant cell walls. *Plant J.* **2011**, *68*, 201–211.

20. Sørensen, I.; Domozych, D.S.; Willats, W.G.T. How have plant cell walls evolved? *Plant Physiol.* **2010**, *153*, 366–372.
21. Domozych, D.S.; Ciancia, M.; Fangel, J.U.; Mikkelsen, M.D.; Ulvskov, P.; Willats, W.G.T. The cell walls of green algae: A journey through evolution and diversity. *Front. Plant Sci.* **2012**, *3*, e82.
22. Proseus, T.E.; Boyer, J.S. Calcium pectate chemistry causes growth to be stored in *Chara corallina*: A test of the pectate cycle. *Plant Cell Environ.* **2008**, *31*, 1147–1155.
23. Eder, M.; Lütz-Meindl, U. Pectin-like carbohydrates in the green alga *Micrasterias* characterized by cytochemical analysis and energy filtering TEM. *J. Microsc.* **2008**, *231*, 201–214.
24. Lütz-Meindl, U.; Brosch-Salomon, S. Cell wall secretion in the green alga *Micrasterias*. *J. Microsc.* **2000**, *198*, 208–217.
25. Domozych, D.S.; Serfis, A.; Kiemle, S.N.; Gretz, M.R. The structure and biochemistry of charophycean cell walls: I. Pectins of *Penium margaritaceum*. *Protoplasma* **2007**, *230*, 99–115.
26. Domozych, D.S.; Sørensen, I.; Popper, Z.A.; Ochs, J.; Andreas, A.; Fangel, J.U.; Pielach, A.; Sacks, C.; Brechka, H.; Willats, W.G.T.; *et al.* Pectin metabolism and assembly in the cell gall of the Charophyte Green alga *Penium margaritaceum. Plant Physiol.* **2014**, *165*, 105–118.
27. Sørensen, I.; Fei, Z.; Andreas, A.; Willats, W.G.T.; Domozych, D.S.; Rose, J.K.C. Stable transformation and reverse genetic analysis of *Penium margaritaceum*: A platform for studies of Charophycean Green Algae, the immediate ancestors of land plants. *Plant J.* **2014**, *77*, 339–351.
28. Krejci, M.R.; Wasserman, B.; Finney, L.; McNulty, I.; Legnini, D. Selectivity in biomineralization of barium and strontium. *J. Struc. Biol.* **2011**, *176*, 192–202.
29. Domozych, D.S.; Lambiasse, L.; Kiemle, S.N.; Gretz, M.R. Cell-wall development and bipolar growth in the desmid *Penium margaritaceum* (Zygnematophyceae, Streptophyta): Asymmetry in a symmetric world. *J. Phycol.* **2009**, *45*, 879–893.
30. Url, T.; Höftberger, M.; Meindl, U. Microtubule-microfilament-controlled nuclear migration in the desmid *Euastrum oblongum*. *J. Phycol.* **2004**, *28*, 537–544.
31. Domozych, D.S.; Kort, S.; Benton, S.; Yu, T. The extracellular polymeric substance of the green alga *Penium margaritaceum* and its role in biofilm formation. *Biofilms* **2005**, *2*, 129–144.
32. Nichols, H.W. Growth media-freshwater. In *Handbook of Phycological Methods: Culture Methods and Growth Measurements*, 1st ed.; Stein, J.R., Ed.; Cambridge University Press: New York, NY, USA, 1973; Volume 1, pp. 39–78.
33. Clausen, M.H.; Willats, W.G.T.; Knox, J.P. Synthetic methyl hexagalacturonate hapten inhibitors of anti-homogalacturonan monoclonal antibodies LM7, JIM5 and JIM7. *Carbohydr. Res.* **2003**, *338*, 1797–1800.
34. Liners, F.; Letesson, J.J.; Didembourg, C.; van Cutsem, P. Monoclonal antibodies against pectin: Recognition of a conformation induced by calcium. *Plant Physiol.* **1989**, *91*, 1419–1424.
35. Liners, F.; Thibault, J.F.; van Cutsem, P. Influence of the degree of polymerization of oligogalacturonates and of esterification pattern of pectin on their recognition by monoclonal antibodies. *Plant Physiol.* **1992**, *99*, 1099–1104.
36. Verhertbruggen, Y.; Marcus, S.E.; Haeger, A.; Ordaz-Ortiz, J.J.; Knox, J.P. An extended set of monoclonal antibodies to pectic homogalacturonan. *Carbohydr. Res.* **2009**, *344*, 1858–1862.

37. Willats, W.G.T.; Limberg, G.; Buchholt, H.C.; van Alebeek, G.J.; Benen, J.; Christensen, T.M.; Visser, J.; Voragen, A.; Mikkelsen, J.D.; Knox, J.P. Analysis of pectic epitopes recognised by hybridoma and phage display monoclonal antibodies using defined oligosaccharides, polysaccharides, and enzymatic degradation. *Carbohydr. Res.* **2000**, *327*, 309–320.
38. Domozych, D.S.; Brechka, H.; Britton, A.; Toso, M. Cell wall growth and modulation dynamics in a model unicellular green alga—*Penium margaritaceum*: Live cell labeling with monoclonal antibodies. *J. Bot.* **2011**, *2011*, e632165.
39. Domozych, D.S.; Sørensen, I.; Sacks, C.; Brechka, H.; Andreas, A.; Fangel, J.U.; Rose, J.K.C.; Willats, W.G.T.; Popper, Z.A. Disruption of the microtubule network alters cellulose deposition and causes major changes in pectin distribution in the cell wall of the green alga, *Penium margaritaceum*. *J. Exp. Bot.* **2014**, *65*, 465–479.
40. Nebenführ, A.; Staehelin, L.A. Mobile factories: Golgi dynamics in plant cells. *Trends Plant Sci.* **2001**, *6*, 160–167.
41. DaSilva, L.L.; Snapp, E.L.; Denecke, J.; Lippincott-Schwartz, J.; Hawes, C.; Brandizzi, F. Endoplasmic reticulum export sites and Golgi bodies behave as single mobile secretory units in plant cells. *Plant Cell* **2004**, *16*, 1753–1771.
42. Domozych, D.S. Exopolymer production by the green alga *Penium margaritaceum*: Implications for biofilm residency. *Int. J. Plant Sci.* **2007**, *168*, 763–774.
43. Ochs, J.; LaRue, T.; Tinaz, B.; Yongue, C.; Domozych, D.S. The cortical cytoskeletal network and cell-wall dynamics in the unicellular charophycean green alga *Penium margaritaceum. Ann. Bot.* **2014**, doi:10.1093/aob/mcu013.
44. Ambrose, J.C.; Cyr, R.J. Mitotic spindle organization by the preprophase band. *Mol. Plant* **2008**, *6*, 950–960.
45. Rasmussen, C.G.; Wright, A.J.; Müller, S. The role of the cytoskeleton and associated proteins in determination of the plant cell division plane. *Plant J.* **2013**, *75*, 258–269.
46. Jopling, C.; Boue, S.; Belmonte, J.C.I. Dedifferentiation, transdifferentiation and reprogramming: Three routes to regeneration. *Nat. Rev. Mol. Cell Biol.* **2011**, *12*, 79–89.
47. Eeckhaut, T.; Lakshmanan, P.S.; Deryckere, D.; van Bockstaele, E.; van Huylenbroek, J. Progress in plant protoplast research. *Planta* **2013**, *238*, 991–1003.
48. Davey, M.R.; Anthony, P.; Patel, D.; Power, J.B. Plant protoplasts: Isolation, culture and plant regeneration. In *Plant Cell Culture: Essential Methods*, 1st ed.; Davey, M.R., Anthony, P., Eds.; Wiley and Sons: New York, NY, USA, 2010; pp. 153–173.
49. Hori, K.; Maruyama, F.; Fujisawa, T.; Togashi, T.; Yamamoto, N.; Seo, M.; Sato, S.; Yamada, T.; Mori, H.; Tajima, N.; *et al. Klebsormidium flaccidum* genome reveals primary factors for plant terrestrial adaptation. *Nat. Commun.* **2014**, *5*, e3978.
50. Timme, R.E.; Bachvaroff, T.S.; Delwiche, C.F. Broad phylogenomic sampling and the sister lineage of land plants. *PLoS One* **2012**, *7*, e29696.

8

Calmodulin Gene Expression in Response to Mechanical Wounding and *Botrytis cinerea* Infection in Tomato Fruit

Hui Peng [1,2], Tianbao Yang [1,*] and Wayne M. Jurick II [1]

[1] Food Quality Laboratory, Beltsville Agricultural Research Center, Agricultural Research Service, United States Department of Agriculture, Beltsville, MD 20705, USA;
E-Mails: hui.peng@ars.usda.gov (H.P.); wayne.jurick@ars.usda.gov (W.M.J.)

[2] College of Life Sciences, Guangxi Normal University, Guilin 541004, China

* Author to whom correspondence should be addressed; E-Mail: tianbao.yang@ars.usda.gov

Abstract: Calmodulin, a ubiquitous calcium sensor, plays an important role in decoding stress-triggered intracellular calcium changes and regulates the functions of numerous target proteins involved in various plant physiological responses. To determine the functions of calmodulin in fleshy fruit, expression studies were performed on a family of six calmodulin genes (*SlCaMs*) in mature-green stage tomato fruit in response to mechanical injury and *Botrytis cinerea* infection. Both wounding and pathogen inoculation triggered expression of all those genes, with *SlCaM2* being the most responsive one to both treatments. Furthermore, all calmodulin genes were upregulated by salicylic acid and methyl jasmonate, two signaling molecules involved in plant immunity. In addition to *SlCaM2*, *SlCaM1* was highly responsive to salicylic acid and methyl jasmonate. However, *SlCaM2* exhibited a more rapid and stronger response than *SlCaM1*. Overexpression of *SlCaM2* in tomato fruit enhanced resistance to *Botrytis*-induced decay, whereas reducing its expression resulted in increased lesion development. These results indicate that calmodulin is a positive regulator of plant defense in fruit by activating defense pathways including salicylate- and jasmonate-signaling pathways, and *SlCaM2* is the major calmodulin gene responsible for this event.

Keywords: calcium signaling; plant defense; salicylic acid; jasmonic acid; postharvest decay

Calcium is a universal second messenger involved in growth, development and mediating responses to a variety of abiotic and biotic stresses in plants [1–3]. Cellular changes in calcium are captured by calcium sensors containing the EF-hand motif. Calmodulin (CaM) is a ubiquitous calcium sensor in plants, and plays an important role in almost all aspects of cell activity [4–6]. In contrast to animals which have one or a few *CaM* genes encoding identical isoforms, plants have multiple genes encoding more diversified isoforms. In all plants examined, *CaM* genes, even genes encoding the same isoform, are differentially expressed in response to numerous external stimuli such as touch, heat shock, cold, light, pathogens, and to phytohormones. In *Arabidopsis*, seven *CaM* genes encode four highly conserved isoforms. A loss-of-function mutant in *Arabidopsis AtCaM2* affects pollen germination [7]. The *atcam3* knockout mutant exhibits reduced thermo-tolerance after heat treatment, whereas overexpression of *AtCaM3* significantly increases thermo-tolerance [8]. In soybean, specific CaM isoforms *SCaM-4* and *SCaM-5*, are highly induced either by a fungal elicitor or pathogen attack, whereas three other *SCaMs* show no response to these stimuli [9]. Transgenic tobacco plants overexpressing *SCaM-4* and *SCaM-5* display spontaneous lesions and constitutive expression of systemic acquired resistance-associated genes. Thus, the level of individual CaM proteins is differentially regulated in plants upon exposure to various stimuli.

Tomato (*Solanum lycopersicum* L.) is the second most important vegetable crop, and the total worldwide production in 2012 was 161.8 million tons with a farm gate value of $80 billion. However, over 25% of fresh produce including tomato fruit are lost after harvest due to the mechanical damage during handling, transportation, and decay [10]. Previous studies on the function of tomato *CaMs* have focused on their role in vegetative tissues. Bergey and Ryan (1999) reported an accumulation of *CaM* mRNA and CaM protein in tomato leaves after wounding or systemin treatment, suggesting that it plays a role in plant defense [11]. Zhao *et al.* (2013) reported that there are six *CaM* genes in tomato encoding for four isoforms [12]. *SlCaMs* in leaves were highly responsive to a variety of biotic and abiotic stimuli. Silencing *SlCaM2* and *SlCaM6* altered expression of defense-related genes and reduced resistance to pathogens. However, there is a gap in the literature concerning the importance of CaM in the ripening of fleshy fruit, as well as in response to stresses that can be encountered during postharvest handling and storage. In this study, we report an expression analysis of *CaM* gene family in response to postharvest stresses, and characterization of the functional significance of a specific *CaM* gene responsible for disease resistance in tomato fruit.

2. Results

2.1. SlCaMs Are Responsive to Mechanical Wounding

To study the effect of wounding on *SlCaMs* expression, mature green stage fruit were selected because this is the specific stage that is routinely harvested by the tomato industry, Without wounding, the most abundantly expressed genes were *SlCaM1* and *SlCaM5* (Figure 1). After wounding, the expression of all *SlCaM*s were stimulated within one hour after treatment and peaked following 2–4 h. Among them, *SlCaM2* showed the most profound stimulation. Its expression increased more than 10-fold in one hour, and peaked at 32-fold within two hours. Wounding also increased the expression of

SlCaM3, *SlCaM4* and *SlCaM6* 2–4 fold, albeit at lower levels. However, wounding had little effect on the expression of *SlCaM1* and *SlCaM5*. *SlPR2b* and *SlLAP-A1*, two genes known to be induced by wounding [13,14] were used as a positive control, and their expression was triggered by wounding as expected. These results collectively suggest that *SlCaMs* in tomato fruit are wound-responsive.

Figure 1. Expression of *SlCaMs* in wounded tomato fruit. Transcription levels of *SlCaMs* were measured by RT-qPCR. Relative gene expression levels are shown following normalization with actin transcript values. Error bars represent standard error of the mean. For each gene, different letters indicate statistically significant differences among mean values ($p < 0.05$). The results are based on at least three repeats in three independent experiments.

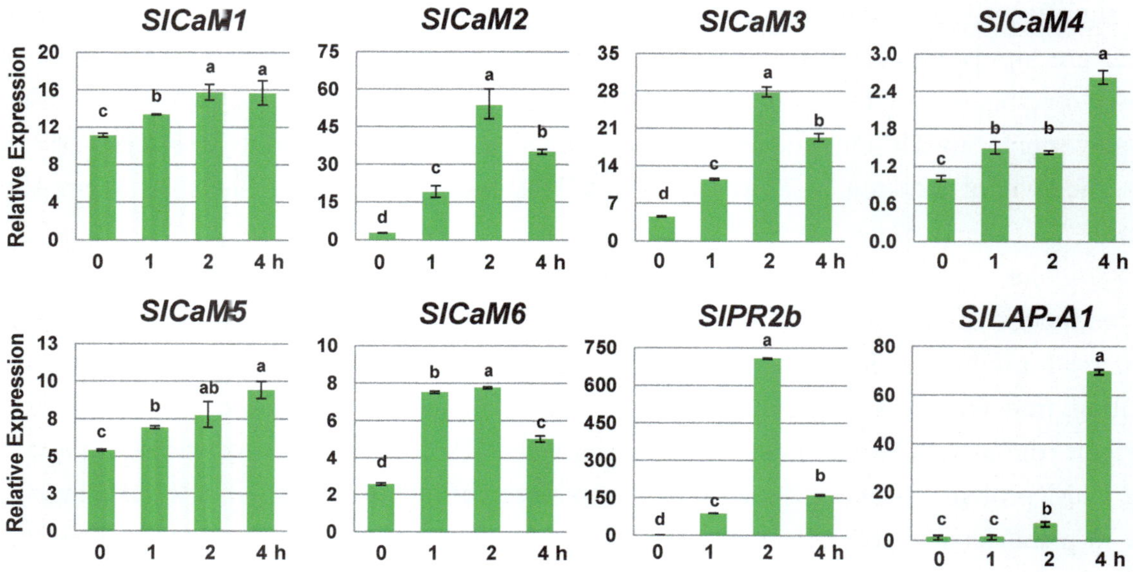

2.2. Pathogen Infection Triggers Calmodulin Gene Expression

Wounding is a prerequisite for most fungal postharvest pathogens that cause decay during storage. To investigate the effects of pathogen infection on *SlCaMs* expression, wounded tomato fruit were inoculated with the necrotrophic fungal pathogen, *Botrytis cinerea.* The growth of the fungus was observed two days postinoculation. The inoculated wounds displayed extensive water soaking, mycelial growth, and necrosis. In contrast, the control treatments (mock inoculation), wounds containing water only, were asymptomatic. Additionally, the expression levels of *SlCaMs* in response to pathogen infection were investigated. Since wounding triggered *SlCaMs* gene expression, much of the increased *SlCaMs* in early time points after both mock and pathogen-inoculation may have resulted from wounding (Figure 2). However, the pathogen-treatment had a more profound and prolonged stimulation of all *SlCaMs* than the mock inoculation. After 24 h of treatment, expression differences between pathogen- and mock inoculation were more distinguishable. For instance, *SlCaM2* exhibited the most dramatic stimulation in response to pathogen infection and mock inoculation immediately after treatment, suggesting that this stimulatory effect resulted mainly from physical wounding. However, following 24 h inoculation, the wound effect on *SlCaM2* expression declined, and the pathogen effect was evident. Similarly, pathogen treatment enhanced the expression of *SlCaM1*, *SlCaM3*, *SlCaM4* and *SlCaM5* 24 h post inoculation. *SlPR2b* was stimulated by 87–106 fold after 1 h of treatments mainly due to the effect of wounding. However, it was increased over 13,600 fold at 24 h post inoculation. Interestingly, the expression of *SlPR1* was remarkably high after 24 h post inoculation. These results suggest that *SlCaMs* could be involved in signaling events during fungal infection by *B. cinerea*.

Figure 2. Expression of *SlCaMs* in response to fungal infection. Tomato fruit at the mature-green stage were mechanically wounded and immediately treated with water (mock inoculation) or inoculated with *Botrytis cinerea* conidia. The wounded and wounded-plus-inoculated areas were excised after 0 to 48 h of incubation at 20 °C. Total RNA samples for RT-PCR were isolated from pericarp tissue and transcript levels of *SlCaMs* genes were determined by RT-qPCR. Relative gene expression levels are shown following normalization with actin transcript values. For each gene, different letters indicate statistically significant differences among mean values ($p < 0.05$). The results are based on at least three repeats in three independent experiments.

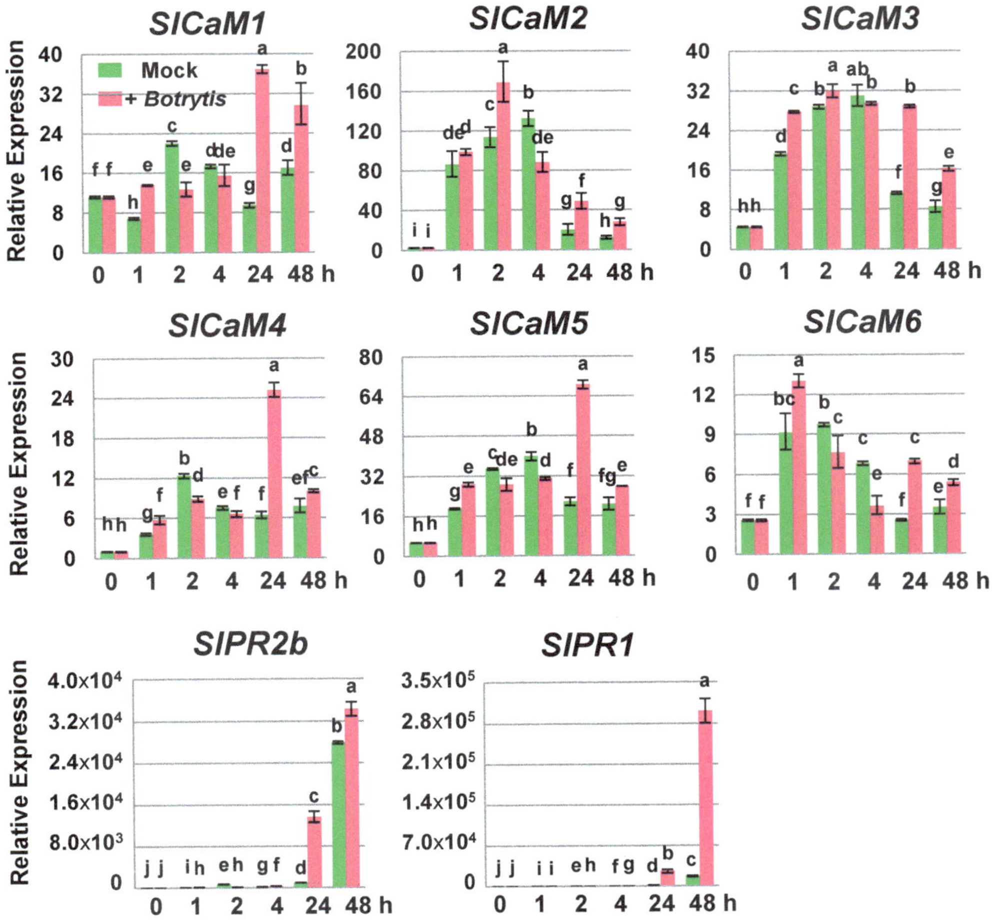

2.3. Calmodulin Genes Are Salicylic Acid-Responsive

Salicylic acid (SA) is a key signaling molecule for the activation of genes involved in systemic acquired resistance to both biotic and abiotic stresses [13,15]. Treating harvested fruits with SA can help reduce decay incidence by activating defense genes such as *PR-1* and *PR-2* [16–20]. To determine whether SA affects the expression of *SlCaMs*, fruits were treated with salicylic acid for different time periods ranging from 0 to 48 h. Previously, it was observed that *SlPR1*, a marker of SA-regulated gene expression, showed the highest induction at 4 mM [21], and thus this concentration of SA was utilized in the following experiments. All *SlCaMs* were upregulated by SA (Figure 3). There were two peaks for the expression of *SlCaM3*, *SlCaM4* and *SlCaM5*. The first peak (minor) occurred one hour after treatment, and second peak (major) appeared at or after 24 h. This pattern was similar to that of the SA-responsive gene *SlPR1*. In comparison, only one peak was observed for *SlCaM1* and *SlCaM6* after 8 or 24 h of treatment. However, the highest induction was observed after 24 h of treatment for all CaMs.

Figure 3. *SlCaMs* expression levels in response to salicylic acid treatment. Transcription levels of *SlCaMs* genes were investigated by quantitative real time RT-qPCR. Mature green stage fruit were treated with 4 mM salicylic acid for different time periods as indicated. Relative gene expression levels are shown following normalization with actin transcript values. Error bars represent the standard error of the mean. For each gene, different letters indicate statistically significant differences among the means ($p < 0.05$). The results are based on at least three repeats in three independent experiments.

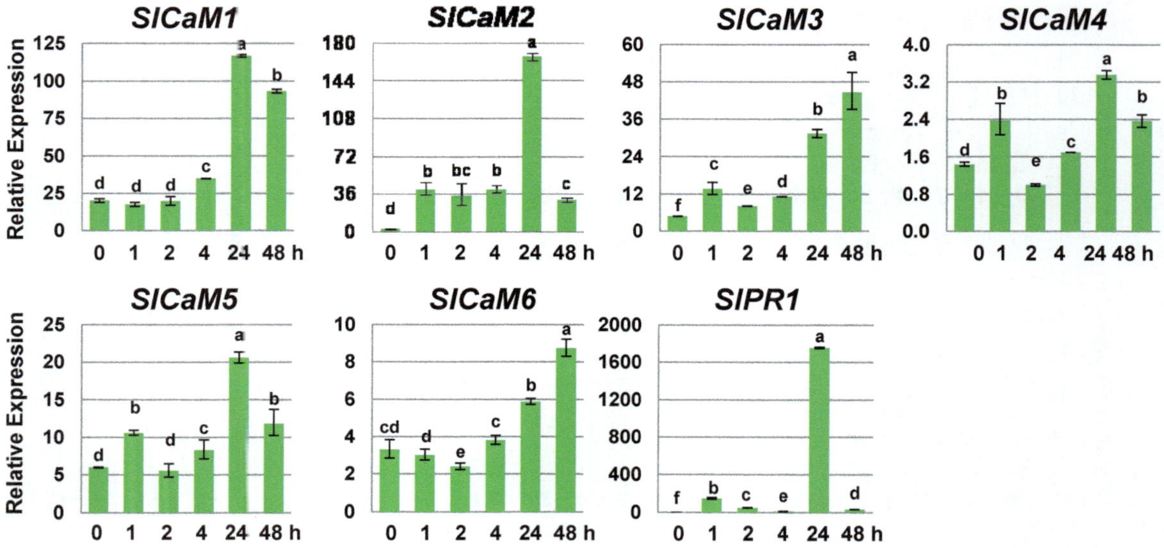

2.4. Methyl Jasmonate Stimulates Calmodulin Gene Expression

Jasmonic acid (JA) is another signal molecule that regulates plant responses to wounding/insect stress and necrotrophic pathogen attack [13,22]. To investigate the expression patterns of *SlCaMs* in response to JA, fruits were treated with 20 μM methyl jasmonate (MeJA), the methyl ester of JA. The expression of all *SlCaMs* increased after applying MeJA (Figure 4). The stimulatory response for *SlCaM2*, *SlCaM3* and *SlCaM6* was detected one hour after treatment. However, significant changes were evident 4 h post treatment for *SlCaM1*, and at 8 h post treatment for *SlCaM3*, *SlCaM4* and *SlCaM5*. The most JA-responsive genes were *SlCaM1* and *SlCaM2*. It is interesting to note that the expression of *SlCaM2* exhibited a wave-like pattern in response to MeJA. The first peak appeared at one hour, the second at 8 h, and the third at 48 h, which was similar to the expression pattern of the JA-responsive gene *SlPR2b*. These results suggest that *SlCaM2* is both a JA early responsive and late responsive gene. Taken together, *SlCaMs* positively respond to JA in tomato fruit, which is opposite to JA's effect on those genes in leaf tissue. Interestingly, the expression of all six *SlCaMs* in tomato leaves was inhibited by MeJA treatment, suggesting that there is tissue differential expression for *SlCaMs*' response to JA [12].

Figure 4. *SlCaMs* gene expression in response to methyl jasmonate treatment. Transcription levels of *SlCaMs* genes were investigated by quantitative real time RT-qPCR. Mature green stage fruit were treated with 20 μM methyl jasmonate for different periods of time. Relative gene expression levels are shown following normalization with actin transcript values. Error bars represent standard error of the mean. For each gene, different letters indicate statistically significant differences among mean values ($p < 0.05$). The results are based on at least three repeats in three independent experiments.

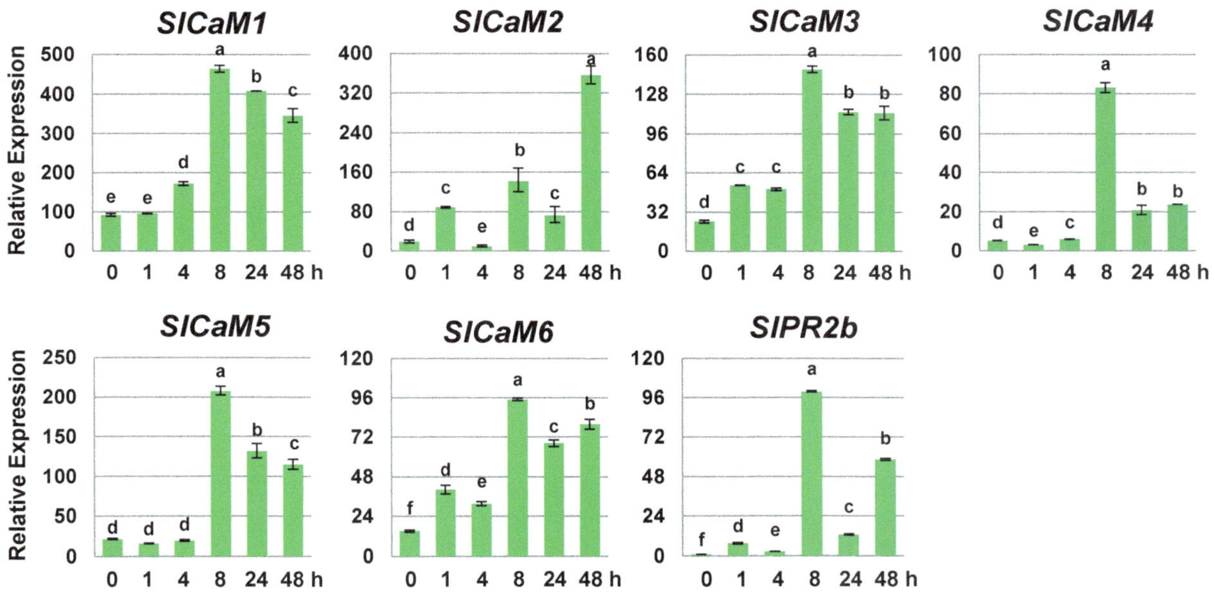

2.5. Overexpression of the SlCaM2 Gene Reduces Symptoms Incited by Botrytis Cinerea

Since *SlCaM2* was the only gene highly responsive to wounding and pathogen infection, as well as SA and MeJA, it was selected for further functional analysis *in planta*. To quickly assess transgene expression in fruit, we selected the agroinjection method to transiently express *SlCaM2* in the early mature-green stage tomato fruit. We introduced *SlCaM2*-sense and antisense constructs into tomato fruit via *Agrobacterium tumefaciens*. Forty-eight hours after injection, the expression level of *SlCaM2* in control (vector alone) was ~2–3 fold higher than non-transgenic fruit (WT) (Figure 5A), suggesting that *Agrobacterium* itself could stimulate the expression of the endogenous *SlCaM2* expression. However, as compared to empty vector alone, the expression level of *SlCaM2* in the sense fruit was over five fold higher 48 h after injection. In contrast, the expression in the antisense fruit was reduced 4–5 times as compared to the vector only control. These results demonstrate that the transient *Agrobacterium*-mediated transformation methodology was a viable choice to study candidate gene function in tomato fruits.

Figure 5. Expression level of *SlCaM2* in tomato fruit in relation to *B. cinerea* infection. Mature green stage fruits were agroinjected with *A. tumefaciens* carrying different constructs. *pDF28F*, empty vector (control); *35S:SlCaM2-S*, pDL28F carrying *SlCaM2* in the sense orientation; *35S:SlCaM2-A*, pDL28F carrying *SlCaM2* in the antisense orientation. (**A**) Examination of expression levels of *SlCaM2* in different transgenic fruits. 48 h after agroinjection, a piece of pericarp tissue from each fruit was used for RT-qPCR. The results are based on at least three repeats in three independent experiments. For each gene, different letters indicate statistically significant differences among mean values ($p < 0.05$). (**B**) Pathogen resistance and susceptibility test shows that overexpression of *SlCaM2* enhanced resistance to *B. cinerea*. The fruits agroinjected with different constructs were inoculated into a wound with water (mock) or *B. cinerea* conidia, and put in a moisture saturated container to observe decay development. The photos were taken 72 h post inoculation.

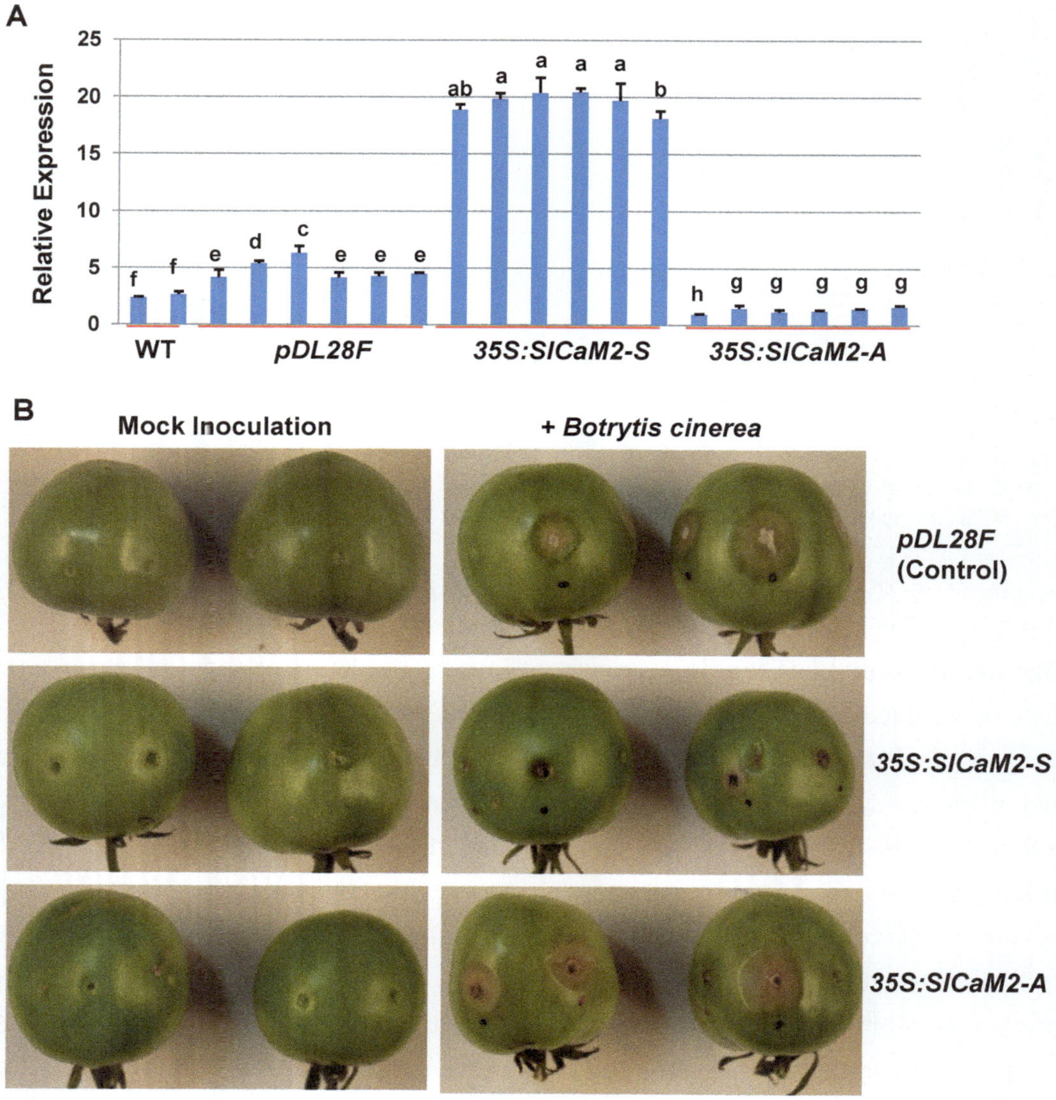

At 48 h post agroinjection, *B. cinerea* conidial suspensions or water were pipetted into wounds (via nail punctures) on transgenic fruit, and monitored for decay development. In the first 24 h, no obvious symptoms were observed. At 2 days postinoculation, control and antisense fruit started to display water soaking and necrosis, while the pathogen inoculated wounds in sense fruit showed some slight necrosis. Three days postinoculation, the control fruit exhibited severe decay, as lesion development was evident at the inoculation site which included mycelial proliferation, accompanied by soft watery circular lesions that were delineated by healthy tissue. The antisense fruit had similar symptoms of decay compared to the control (Figure 5B). However, sense fruit showed limited necrosis and limited water-soaking. All the mock inoculated fruit were asymptomatic at the inoculation sites. These results indicate that increased expression of *SlCaM2* enhanced resistance to *B. cinerea*, and ultimately limited fungal growth, lesion development, and colonization in tomato fruit.

3. Discussion

Fleshy fruits are an important part of the human diet, providing fiber, minerals, nutrients and various other substances beneficial to human health. Nearly a quarter of all fresh fruits and vegetables in the U.S. market are lost after harvest due to damage caused by abiotic and biotic stresses such as wounding and pathogen infection [10]. The importance of calcium in fruit ripening and postharvest handling has been recognized for many years. Calcium is most frequently associated with delayed ripening maintaining fruit quality (particularly firmness), and preventing decay through the strengthening of cell walls by cross-linking de-esterified pectic acid residues [23–30]. However, it is well-recognized that calcium mediated signaling is important for plant responses to a wide variety of stresses including mechanical touch and pathogen attack [1–3,31]. Intracellular calcium changes in response to those stimuli are perceived and decoded by calcium sensors including CaM [4–6]. Further, *CaM* genes themselves respond to the biotic and abiotic signals. In tomato, wounding or systemin increased the accumulation of *CaM* in leaves [11]. Zhao *et al.* (2013) reported that *SlCaMs* in tomato leaves were highly responsive to a variety of biotic and abiotic stimuli [12]. Silencing of *SlCaM2* and *SlCaM6* led to the reduced resistance to tobacco rattle virus and *Pythium aphanidermatum*. However, it did not affect the resistance to *Pseudomonas syringae* and *Xanthomonas oryzae*. In this study, we found that six *SlCaMs*, especially *SlCaM2*, in tomato fruits were upregulated by mechanical wounding, necrotrophic fungal infection, SA and JA. It is important to note that *SlCaMs* genes respond to JA in leaves and fruits very differently. Thus, there may be tissue specificities for *SlCaMs* expression in response to stresses that accompany functional changes in different tissues.

SA and JA are two major signal molecules that mediate plant defense responses. In general, the SA-dependent pathway is activated by biotrophic pathogens, while the JA pathway is triggered by necrotrophs and herbivore attack/wounding [13,22,32,33]. The interactions between those two pathways can be antagonistic or synergistic [34,35]. It has been reported that the necrotrophic pathogens such as *Botrytis cinerea* can produce an exopolysaccharide, which acts as an elicitor of SA pathway in tomato [36,37]. We observed that SA-responsive gene *SlPR1*was remarkably stimulated by *Botrytis* infection which supported their findings (Figure 2). It has been well documented that the synthesis of SA and SA signaling are under extensive regulation by calcium signaling [38–42]. Cross-talk between the JA signaling pathway and calcium signaling also occurs [43,44]. A calcium/CaM-regulated transcription factor AtSR1/CAMTA3 has been shown to be involved in both SA- and JA-dependent pathways [45,46]. Analysis of the AtSR1/CAMTA3 gene knockout mutant showed enhanced resistance to both biotrophic and necrotrophic pathogens [41,47]. Previously, the expression patterns of *SlSRs*, the orthologs of *AtSRs/CaMTAs* in tomato, were analyzed. All of them displayed a positive response to both SA and MeJA in tomato fruit [21]. Therefore *SlSRs* could be candidate targets for *SlCaMs* in controlling disease resistance. In addition, there are quite a few other CaM-target proteins involved in SA-signaling that respond both positively and negatively [38–42]. Considering six *SlCaM* genes encoding four isoforms, the regulation of wounding and disease resistance in tomato fruit by calcium/CaM is hypothesized to be complex. Nevertheless, *SlCaM2* will be a first choice for further expression and functional analysis studies in tomato fruits. Introducing *SlCaM2* into tomato, by classical breeding or through molecular means, might enhance the tolerance/resistance to mechanical injury, pathogens, and other stresses leading to the availability of high quality fresh fruit for consumption.

4. Experimental Section

4.1. Plant Materials

Tomato plants (*S. lycopersicum* cv. Moneymaker) were grown in a greenhouse at 28 °C with a 16 h/8 h (light/dark) cycle. Fruit were harvested at the mature green stage (MG), as defined by USDA-ARS criteria [48], when the fruits were physiologically mature but not yet ripe. In the industry, tomatoes are often harvested at this stage for packing and shipment, and subsequently treated with ethylene to promote ripening prior to sale. The greenhouse-grown MG fruit were held under ambient conditions overnight to reduce harvest shock prior to treatment.

4.2. Mechanical Injury, Methyl Jasmonate and Salicylic Acid Treatments

Wounding was executed by manually cutting fruit pericarp into one inch pieces with a sharp knife at room temperature. The fruit pieces were put in a plastic box with the wet towel to maintain high humidity for the indicated time period. For SA treatment, fruits were immersed in 4 mM SA solution for 0 to 48 h. For MeJA treatment, fruits were sealed in a jar with 20 μM MeJA. After each treatment, pericarp samples were immediately frozen in liquid nitrogen and stored at −80 °C.

4.3. Pathogen Infection and Decay Assay

Fruits were mechanically injured and infected by inoculation with a conidial suspension of the fungal pathogen *B. cinerea* strain 22B which was isolated from naturally infected apple fruit [49]. The fungus was propagated via single spore isolate and maintained on Potato Dextrose Agar. The inoculation was done essentially as described previously [21]. Briefly, fruit were punctured (3 mm depth, 2 mm diameter) at six sites around the equator of each fruit; 3 sites with 10 μL of conidial suspension (1×10^4 conidia/mL), and the other three with 10 μL of sterile Tween20-treated water for mock inoculation. After inoculation, the fruit were stored in plastic sealed containers with moist towels to maintain high humidity and kept at 20 °C. Pericarp tissue samples were obtained from inoculated fruits by using a cork borer to isolate the tissue immediately surrounding the inoculated area at different intervals after treatment. The pericarp tissue was collected from fruit at the different time points, frozen in liquid nitrogen and stored at −80 °C.

4.4. RNA Extraction and RT-qPCR

Total RNA was isolated from frozen tissue using RNeasy Plant Mini Kit following the manufacturer's instruction (Qiagen, Valencia, CA, USA). Reverse transcription and qPCR were performed as described [21]. Briefly, one µg of total RNA was used to synthesize cDNA with iScript™ kit (Bio-RAD, Hercules, CA, USA), RT-qPCR analysis of cDNA was performed on a CFX96 Real-time System (Bio-RAD). Gene specific primers listed in Table 1 were designed with the Primer3 software [50]. The efficiency coefficient E was calculated for all primer pairs individually by plotting the relationship between Cq value (threshold cycle) and log[cDNA]. All reactions were performed in triplicate from three independent samples. Cq was used for relative quantification of the input target number. Relative fold difference (*N*) was the number of the treated target gene copies relative to the untreated control gene copies and is calculated as follows: $N = 2^{\Delta Cq}$. ΔCq was the difference in threshold cycles for *SlCaMs* targets and the *actin* internal reference. Relative gene expression (fold changes) was calculated based on *N* with the lowest value as 1. Student's *t* test ($p < 0.05$) was used to determine the significant difference of relative expression of individual genes among different treatments and controls (Microsoft Excel 2007).

Table 1. Primers used for qPCR and cloning.

Primer Name	Oligonucleotides	Gene ID
SlCaM1-a	CCAGAGTTCCTTAACCTGATGG	Solyc01g008950
SlCaM1-b	CTTTTCGCCTAGGTTTGTCATC	
SlCaM2-a	TCTGAGGAGGAGTTGAAAGAGG	Solyc10g081170
SlCaM2-b	TCAACATCAGCTTCCCTAATCA	
SlCaM3-a	GATGGTAATGGAACCATCGACT	Solyc10g077010
SlCaM3-b	CATCAGTGAGCTTCTCACCAAG	
SlCaM4-a	TCAGATCTCGGAGTTCAAAGAAG	Solyc11g072240
SlCaM4-b	CAGGTTAAGGAACTCAGGGAAGT	
SlCaM5-a	TTAACTTGATGGCTCGGAAGAT	Solyc12g099990
SlCaM5-b	ACGAATCATCTCGTCAACCTCT	
SlCaM6-a	ATCACTTGGTCAGAATCCCACT	Solyc03g098050
SlCaM6-b	AGCTGCAGAAATAAAGCCATTC	
SlPR1-a	CTGTGAAGATGTGGGTTGATGAG	NM-001247429
SlPR1-b	TCTCCAGTTACCTGGTGGATCAT	
SlPR2b-a	TCTTGCCCCATTTCAAGTTC	M80608
SlPR2b-b	TGCACGTGTATCCCTCAAAA	
SlLAPA1-a	TGTCGCAGCATGTGAAAATATG	Solyc12g010020
SlLAPA1-b	AGCACCAGTTAATGTTGCCAGA	
SlActin-a	GAAATAGCATAAGATGGCAGACG	X55749
SlActin-b	ATACCCACCATCACACCAGTAT	
SlCaM2-S1 *	gg<u>tggtacc</u>ATGGCGGATCAGCTGACGG	
SlCaM2-S2 *	gg<u>aggatcc</u>CTTGGCCATCATGACCTTAAC	
SlCaM2-A1 *	gg<u>tggtacc</u>CTTGGCCATCATGACCTTAAC	
SlCaM2-A2 *	gg<u>aggatcc</u>ATGGCGGATCAGCTGACGG	

* Primers used for cloning full length *SlCaM2*. The underlined portion indicates the restriction enzyme site.

4.5. Construction of Ti Plasmids Carrying Sense- and Antisense-SlCaM2 Gene

The full length tomato *SlCaM2* was amplified from a mixture of fruit tissues by Pfx DNA polymerase, and subcloned to TA cloning Kit (Life Technology, Grand Island, NY, USA) using gene specific primers (Table 1). The nucleotide sequences of the positive clones were confirmed by sequencing. The full length of *SlCaM2* was subcloned into pDL28F, a derivative of pCambia1300 [41] in the sense- and antisense- orientations in the sites of *Kpn* I and *Bam*H I downstream of 35S promoter, and introduced into *Agrobacterium tumefaciens* strain GV3101. The positive clones were verified by PCR using gene-specific primers.

4.6. Fruit Agroinjection

Agroinjection of tomato fruit was carried as described [51]. Briefly, agrobacterium cultures were grown overnight from individual colonies at 28 °C in YEB medium plus selective antibiotics, transferred to induction medium (0.5% beef extract, 0.1% yeast extract, 0.5% peptone, 0.5% sucrose, 2 mM $MgSO_4$, 20 μM acetosyringone, 10 mM MES, pH 5.6) plus antibiotics, and grown again overnight. The next day, cultures were resuspended with infiltration medium (10 mM $MgCl_2$, 10 mM MES, 200 μM acetosyringone, pH 5.6 with OD_{600} of 1.0), and incubated at room temperature with gentle agitation (20 rpm) for a minimum of 2 h. Cultures were injected into the early mature green fruit using a syringe with a 0.5- × 16-mm needle. The needle was introduced into the fruit tissue through the stylar apex and blossom end. Because of differences in fruit size, the injection was terminated when the solution started running off the injection site.

5. Conclusions

Calmodulin, a ubiquitous calcium receptor in the eukaryotic cell, plays an important role in almost all aspects of cell activity in plants [4–6]. Previous studies have suggested that calmodulins in tomato vegetative tissues respond to a variety of biotic and abiotic stimuli, such as wounding or systemin treatment [11] and pathogen attack [12]. We have observed that the expression levels of six *SlCaMs* in mature green tomato fruit are stimulated by mechanical injury and *B. cinerea* infection. Among the six genes, *SlCaM2* was the most responsive to both pathogen infection and wounding. Further, expressions of all *SlCaMs* were upregulated by SA and MeJA which occurs 24 and 8 h after treatment, respectively. However, *SlCaM2* also had a detectable peak as early as one hour after treatment by SA and MeJA. In general, *SlCaMs* in tomato fruit are early wound-responsive genes, but late responsive genes to pathogen attack, JA and SA. *SlCaMs* may regulate the wound response and disease resistance via SA- and/or JA-dependent signaling in fruit. Based on its gene expression level, *SlCaM2* is the major *CaM* gene in tomato fruit that responds to mechanical injury and pathogen attack. Transient expression of *SlCaM2* into mature green tomato fruit significantly increases resistance to *B. cinerea.* Conversely, reducing its expression facilitates pathogen growth in the host. Since *SlCaM2* is an SA-responsive gene, it will be interesting to test whether increasing its expression also confers resistance to biotrophic pathogens (*i.e.*, *Cladosporium fulvum*). Moreover, it will be of great utility to identify target(s) of *SlCaM2* to improve overall fruit quality by strengthening tolerance to wounding and concomitantly increasing disease resistance to fungal postharvest plant pathogens.

Acknowledgments

We thank Ernest Paroczay and Ziying Zou for their excellent technical support, Verneta Gaskins for assisting with the pathogen infection assays and Liqun Du for providing pDL28F plasmid. Use of a company or product name by the U.S. Department of Agriculture does not imply approval or recommendation of the product to the exclusion of others that may also be suitable. This research was funded by USDA-ARS NP306 project no. 1245-43000-012-00D.

Author Contributions

H. Peng performed gene expression experiments and statistical analysis of the expression data, and helped to evaluate the data. W. Jurick designed and performed pathogen inoculation and decay experiments. T. Yang established all the experiments, analyzed data and wrote the manuscript. H. Peng and W. Jurick edited the manuscript.

Conflicts of Interest

The authors declare no conflict of interest.

References

1. Dodd, A.N.; Kudla, J.; Sanders, D. The language of calcium signaling. *Annu. Rev. Plant Biol.* **2010**, *61*, 593–620.
2. Batistic, O.; Kudla, J. Analysis of calcium signaling pathways in plants. *BBA Gen Subj.* **2012**, *1820*, 1283–1293.
3. DeFalco, T.A.; Bender, K.W.; Snedden, W.A. Breaking the code: Ca^{2+} sensors in plant signalling. *Biochem. J.* **2010**, *425*, 27–40.
4. Reddy, A.S.N.; Ali, G.S.; Celesnik, H.; Day, I.S. Coping with stresses: Roles of calcium- and calcium/calmodulin-regulated gene expression. *Plant Cell* **2011**, *23*, 2010–2032.
5. Yang, T.; Poovaiah, B.W. Calcium/calmodulin-mediated signal network in plants. *Trends Plant Sci.* **2003**, *8*, 505–512.
6. Bouche, N.; Yellin, A.; Snedden, W.A.; Fromm, H. Plant-specific calmodulin-binding proteins. *Annu. Rev. Plant Biol.* **2005**, *56*, 435–466.
7. Landoni, M.; de Francesco, A.; Galbiati, M.; Tonelli, C. A loss-of-function mutation in calmodulin2 gene affects pollen germination in *Arabidopsis thaliana*. *Plant Mol. Biol.* **2010**, *74*, 235–247.
8. Zhang, W.; Zhou, R.G.; Gao, Y.J.; Zheng, S.Z.; Xu, P.; Zhang, S.Q.; Sun, D.Y. Molecular and genetic evidence for the key role of AtCaM3 in heat-shock signal transduction in *Arabidopsis*. *Plant Physiol.* **2009**, *149*, 1773–1784.
9. Heo, W.D.; Lee, S.H.; Kim, M.C.; Kim, J.C.; Chung, W.S.; Chun, H.J.; Lee, K.J.; Park, C.Y.; Park, H.C.; Choi, J.Y.; *et al.* Involvement of specific calmodulin isoforms in salicylic acid-independent activation of plant disease resistance responses. *Proc. Natl. Acad. Sci. USA* **1999**, *96*, 766–771.
10. Hodges, R.J.; Buzby, J.C.; Bennett, B. Postharvest losses and waste in developed and less developed countries: Opportunities to improve resource use. *J. Agric. Sci.* **2011**, *149*, 37–45.

11. Bergey, D.R.; Ryan, C.A. Wound- and systemin-inducible calmodulin gene expression in tomato leaves. *Plant Mol. Biol.* **1999**, *40*, 815–823.
12. Zhao, Y.; Liu, W.; Xu, Y.P.; Cao, J.Y.; Braam, J.; Cai, X.Z. Genome-wide identification and functional analyses of calmodulin genes in *Solanaceous* species. *BMC Plant Biol.* **2013**, *13*, doi:10.1186/1471-2229-13-70.
13. Robert-Seilaniantz, A.; Grant, M.; Jones, J.D.G. Hormone crosstalk in plant disease and defense: More than just jasmonate-salicylate antagonism. *Annu. Rev. Phytopathol.* **2011**, *49*, 317–343.
14. Chao, W.S.; Gu, Y.Q.; Pautot, V.; Bray, E.A.; Walling, L.L. Leucine aminopeptidase RNAs, proteins, and activities increase in response to water deficit, salinity, and the wound signals systemin, methyl jasmonate, and abscisic acid. *Plant Physiol.* **1999**, *120*, 979–992.
15. Vlot, A.C.; Dempsey, D.A.; Klessig, D.F. Salicylic acid, a multifaceted hormone to combat disease. *Annu. Rev. Phytopathol.* **2009**, *47*, 177–206.
16. Asghari, M.; Aghdam, M.S. Impact of salicylic acid on post-harvest physiology of horticultural crops. *Trends Food Sci. Technol.* **2010**, *21*, 502–509.
17. Tornero, P.; Gadea, J.; Conejero, V.; Vera, P. Two PR-1 genes from tomato are differentially regulated and reveal a novel mode of expression for a pathogenesis-related gene during the hypersensitive response and development. *Mol. Plant Microbe Interact.* **1997**, *10*, 624–634.
18. Uquillas, C.; Letelier, I.; Blanco, F.; Jordana, X.; Holuigue, L. NPR1-independent activation of immediate early salicylic acid-responsive genes in *Arabidopsis*. *Mol. Plant Microbe Interact.* **2004**, *17*, 34–42.
19. Van Loon, L.C.; Rep, M.; Pieterse, C.M.J. Significance of inducible defense-related proteins in infected plants. *Annu. Rev. Phytopathol.* **2006**, *44*, 135–162.
20. Rohwer, C.L.; Erwin, J.E. Jasmonate-induced changes in polyphenol oxidase, peroxidase, and proteinase inhibitors in horticultural species. *Hortscience* **2006**, *41*, 1072–1072.
21. Yang, T.B.; Peng, H.; Whitaker, B.D.; Jurick, W.M. Differential expression of calcium/calmodulin-regulated SlSRs in response to abiotic and biotic stresses in tomato fruit. *Physiol. Plantarum.* **2013**, *148*, 445–455.
22. Kazan, K.; Manners, J.M. Jasmonate signaling: Toward an integrated view. *Plant Physiol.* **2008**, *146*, 1459–1468.
23. Saftner, R.A.; Bai, J.H.; Abbott, J.A.; Lee, Y.S. Sanitary dips with calcium propionate, calcium chloride, or a calcium amino acid chelate maintain quality and shelf stability of fresh-cut honeydew chunks. *Postharvest. Biol. Technol.* **2003**, *29*, 257–269.
24. Abbott, J.A.; Conway, W.S.; Sams, C.E. Postharvest calcium-chloride infiltration affects textural attributes of apples. *J. Am. Soc. Hortic. Sci.* **1989**, *114*, 932–936.
25. Park, S.; Cheng, N.H.; Pittman, J.K.; Yoo, K.S.; Park, J.; Smith, R.H.; Hirschi, K.D. Increased calcium levels and prolonged shelf life in tomatoes expressing *Arabidopsis* H^+/Ca^{2+} transporters. *Plant Physiol.* **2005**, *139*, 1194–1206.
26. Ritenour, M.A.; Stoffella, P.J.; He, Z.L.; Narciso, J.A.; Salvatore, J.J. Postharvest calcium chloride dips of whole tomato fruit reduce postharvest decay under commercial conditions. *Hortscience* **2006**, *41*, 1016–1017.

27. Saftner, R.A.; Conway, W.S.; Sams, C.E. Postharvest calcium infiltration alone and combined with surface coating treatments influence volatile levels, respiration, ethylene production, and internal atmospheres of "Golden Delicious" apples. *J. Am. Soc. Hortic. Sci.* **1999**, *124*, 553–558.
28. Poovaiah, B.W.; Shekhar, V.C. Effects of calcium infiltration of golden delicious apples on fruit firmness and senescence. *Hortscience* **1978**, *13*, 357–357.
29. Sams, C.E.; Conway, W.S.; Abbott, J.A.; Lewis, R.J.; Benshalom, N. Firmness and decay of apples following postharvest pressure infiltration of calcium and heat-treatment. *J. Am. Soc. Hortic. Sci.* **1993**, *118*, 623–627.
30. Conway, W.S.; Sams, C.E. The effects of postharvest infiltration of calcium, magnesium, or strontium on decay, firmness, respiration, and ethylene production in apples. *J. Am. Soc. Hortic. Sci.* **1987**, *112*, 300–303.
31. Lecourieux, D.; Raneva, R.; Pugin, A. Calcium in plant defence-signalling pathways. *New Phytol.* **2006**, *171*, 249–269.
32. Nimchuk, Z.; Eulgem, T.; Holt, B.F., 3rd; Dangl, J.L. Recognition and response in the plant immune system. *Annu. Rev. Genet.* **2003**, *37*, 579–609.
33. Fu, Z.Q.; Dong, X. Systemic acquired resistance: Turning local infection into global defense. *Annu. Rev. Plant Biol.* **2013**, *64*, 839–863.
34. Mur, L.A.J.; Kenton, P.; Atzorn, R.; Miersch, O.; Wasternack, C. The outcomes of concentration-specific interactions between salicylate and jasmonate signaling include synergy, antagonism, and oxidative stress leading to cell death. *Plant Physiol.* **2006**, *140*, 249–262.
35. Koornneef, A.; Leon-Reyes, A.; Ritsema, T.; Verhage, A.; den Otter, F.C.; van Loon, L.C.; Pieterse, C.M.J. Kinetics of salicylate-mediated suppression of jasmonate signaling reveal a role for redox modulation. *Plant Physiol.* **2008**, *147*, 1358–1368.
36. Abd el Rahman, T.; el Oirdi, M.; Gonzalez-Lamothe, R.; Bouarab, K. Necrotrophic pathogens use the salicylic acid signaling pathway to promote disease development in tomato. *Mol. Plant Microbe Interact.* **2012**, *25*, 1584–1593.
37. El Oirdi, M.; Abd el Rahman, T.; Rigano, L.; el Hadrami, A.; Rodriguez, M.C.; Daayf, F.; Vojnov, A.; Bouarab, K. *Botrytis cinerea* manipulates the antagonistic effects between immune pathways to promote disease development in tomato. *Plant Cell* **2011**, *23*, 2405–2421.
38. Wang, L.; Tsuda, K.; Truman, W.; Sato, M.; Nguyen, L.V.; Katagiri, F.; Glazebrook, J. CBP60g and SARD1 play partially redundant critical roles in salicylic acid signaling. *Plant J.* **2011**, *67*, 1029–1041.
39. Zhang, Y.; Xu, S.; Ding, P.; Wang, D.; Cheng, Y.T.; He, J.; Gao, M.; Xu, F.; Li, Y.; Zhu, Z.; *et al.* Control of salicylic acid synthesis and systemic acquired resistance by two members of a plant-specific family of transcription factors. *Proc. Natl. Acad. Sci. USA* **2010**, *107*, 18220–18225.
40. Wang, L.; Tsuda, K.; Sato, M.; Cohen, J.D.; Katagiri, F.; Glazebrook, J. *Arabidopsis* CaM binding protein CBP60g contributes to MAMP-induced SA accumulation and is involved in disease resistance against *Pseudomonas syringae*. *PLoS Pathogens* **2009**, *5*, e1000301.
41. Du, L.Q.; Ali, G.S.; Simons, K.A.; Hou, J.G.; Yang, T.B.; Reddy, A.S.N.; Poovaiah, B.W. Ca^{2+}/calmodulin regulates salicylic-acid-mediated plant immunity. *Nature* **2009**, *457*, 1154–1158.
42. Poovaiah, B.W.; Du, L.Q.; Wang, H.Z.; Yang, T.B. Recent advances in calcium/calmodulin-mediated signaling with an emphasis on plant-microbe interactions. *Plant Physiol.* **2013**, *163*, 531–542.

43. Leon, J.; Rojo, E.; Titarenko, E.; Sanchez-Serrano, J.J. Jasmonic acid-dependent and -independent wound signal transduction pathways are differentially regulated by Ca^{2+}/calmodulin in *Arabidopsis thaliana*. *Mol. Genet. Gen.* **1998**, *258*, 412–419.
44. Dombrowski, J.E.; Bergey, D.R. Calcium ions enhance systemin activity and play an integral role in the wound response. *Plant Sci.* **2007**, *172*, 335–344.
45. Qiu, Y.J.; Xi, J.; Du, L.Q.; Suttle, J.C.; Poovaiah, B.W. Coupling calcium/calmodulin-mediated signaling and herbivore-induced plant response through calmodulin-binding transcription factor AtSR1/CAMTA3. *Plant Mol. Biol.* **2012**, *79*, 89–99.
46. Laluk, K.; Prasad, K.V.S.K.; Savchenko, T.; Celesnik, H.; Dehesh, K.; Levy, M.; Mitchell-Olds, T.; Reddy, A.S.N. The calmodulin-binding transcription factor signal responsive1 is a novel regulator of glucosinolate metabolism and herbivory tolerance in *Arabidopsis*. *Plant Cell Physiol.* **2012**, *53*, 2008–2015.
47. Galon, Y.; Nave, R.; Boyce, J.M.; Nachmias, D.; Knight, M.R.; Fromm, H. Calmodulin-binding transcription activator (CAMTA) 3 mediates biotic defense responses in *Arabidopsis*. *FEBS Lett.* **2008**, *582*, 943–948.
48. Sargent, S.A.; Moretti, C.L. Tomato. In *Agricultural Handbook Number 66 (HB66)*, Draft-Updated August 2014; USDA-ARS, USA. Available online: http://www.ba.ars.usda.gov/hb66/tomato.pdf (accessed on 11 August 2014).
49. Yan, H.J.; Jurick, W.M., II; Lou, Y.G.; Gaskins, V.L.; Kim, Y.-K. First report of pyrimethanil resistance in *Botrytis cinerea* from stored apples in pennsylvania. *Plant Dis.* **2014**, *98*, 999.
50. Untergrasser, A.; Cutcutache, I.; Koressaar, T.; Ye, J.; Faircloth, B.C.; Remm, M.; Rozen, S.G. Primer 3—New capabilities and interfaces. *Nucleic Acids Res.* **2012,** *40*, e115.
51. Orzaez, D.; Mirabel, S.; Wieland, W.H.; Granell, A. Agroinjection of tomato fruits. A tool for rapid functional analysis of transgenes directly in fruit. *Plant Physiol.* **2006**, *140*, 3–11.

9

Cell Wall Metabolism in Response to Abiotic Stress

Hyacinthe Le Gall, Florian Philippe, Jean-Marc Domon, Françoise Gillet, Jérôme Pelloux and Catherine Rayon *

EA3900-BIOPI, Biologie des Plantes et Innovation, Université de Picardie Jules Verne, 80039 Amiens, France; E-Mails: hyacinthe.le.gall@u-picardie.fr (H.L.G.); florian.philippe@u-picardie.fr (F.P.); jean-marc.domon@u-picardie.fr (J.-M.D.); francoise.gillet@u-picardie.fr (F.G.); jerome.pelloux@u-picardie.fr (J.P.)

* Author to whom correspondence should be addressed; E-Mail: catherine.rayon@u-picardie.fr

Academic Editor: Philip Harris

Abstract: This review focuses on the responses of the plant cell wall to several abiotic stresses including drought, flooding, heat, cold, salt, heavy metals, light, and air pollutants. The effects of stress on cell wall metabolism are discussed at the physiological (morphogenic), transcriptomic, proteomic and biochemical levels. The analysis of a large set of data shows that the plant response is highly complex. The overall effects of most abiotic stress are often dependent on the plant species, the genotype, the age of the plant, the timing of the stress application, and the intensity of this stress. This shows the difficulty of identifying a common pattern of stress response in cell wall architecture that could enable adaptation and/or resistance to abiotic stress. However, in most cases, two main mechanisms can be highlighted: (i) an increased level in xyloglucan endotransglucosylase/hydrolase (XTH) and expansin proteins, associated with an increase in the degree of rhamnogalacturonan I branching that maintains cell wall plasticity and (ii) an increased cell wall thickening by reinforcement of the secondary wall with hemicellulose and lignin deposition. Taken together, these results show the need to undertake large-scale analyses, using multidisciplinary approaches, to unravel the consequences of stress on the cell wall. This will help identify the key components that could be targeted to improve biomass production under stress conditions.

Keywords: plant cell wall; abiotic stress; water deficit; temperature; salt; flooding; heavy metals; light; air pollutants

1. Introduction

During their entire life cycle, plants are exposed to various environmental stresses. Two major categories can be distinguished: abiotic stress, which encompasses a variety of unfavorable environmental conditions, such as drought, submergence, salinity, heavy metal contamination and nutrient deficiency, and biotic stress caused by infectious living organisms, such as bacteria, viruses, fungi, or nematodes. Both types of stress negatively affect the productivity and survival of plants [1]. Abiotic stress reduces agricultural yield, so that novel crop genotypes adapted to environmental stress need to be developed [2–4]. Restriction of plant growth by retarding cell extension is often the earliest visible effect of stress. Plants have developed a range of morphological, physiological and biochemical mechanisms that enable them to avoid and/or tolerate stress factors and survive [5]. Reactive oxygen species (ROS) homeostasis is changed in response to stress; they are regarded as molecules causing damage to cells at high concentration as well as ubiquitous signaling molecules at low concentration, thus participating in recognizing and responding to stress factors [6]. Transduction of the signal modifies gene and protein expression levels leading to physiological responses. ROS are produced in most plant subcellular locations including the plasma membrane, mitochondria, nucleus, chloroplast and cell wall [7].

The plant cell wall determines cell size and shape through the mechanical control of cell expansion, which governs tissue and organ morphology. It can be considered a protective barrier and is a complex structure composed of cellulose microfibrils and non-cellulosic neutral polysaccharides embedded in a physiologically active pectin matrix, cross-linked with structural proteins and, depending on the tissue/organ, with lignin [8,9]. Plant cell walls are divided into a primary wall, which is initiated during cell division and deposited during cell elongation, and a secondary wall, which is deposited within the primary wall of specific cell types with specialized functions. In parallel, depending on the presence/absence of some particular polysaccharides, type I and type II cell walls can be distinguished. Type I primary walls are a framework of cellulose microfibrils, mostly cross-linked with xyloglucans (XG) and embedded in a complex matrix of pectic polysaccharides [8,9]. Unlike type I, type II primary cell walls contain small amounts of pectic polysaccharides and xyloglucans, whereas glucuronoarabinoxylans (GAX) and mixed-linkage (1→3),(1→4)-β-D-glucan (β-glucan) are the main matrix polymers that interlace cellulose microfibrils [10]. Furthermore, ferulic and *p*-coumaric acid arabinosyl esters can cross-link GAX in type II primary walls, as well as in the secondary wall, to stiffen the matrix [11,12]. Cell wall-related proteins are believed to play a central role in modulating cell wall extensibility, which mediates cell enlargement and expansion. These proteins include xyloglucan endo-β-transglucosylases/hydrolases (XET/XTHs; GH16), endo-1,4-β-D-glucanase (EGase, GH9), and expansins (EXPA, EXPB, EXPLA, EXPLB; GH45) [13,14]. Other cell wall-modifying enzymes play a major role in controlling cell wall plasticity/rheology. These include pectin-modifying enzymes such as pectin methylesterase (PME; CE8), polygalacturonase (PG; GH28), pectin/pectate lyase-like (PLL; PL1) and pectin acetylesterase (PAE; CE13) [15].

The massive deposition of cellulose and hemicelluloses inside primary walls gives secondary walls their characteristic thickness [16]. During secondary cell wall formation, monolignols, precursors of lignin, are secreted into the cell wall space and randomly cross-linked through oxidative polymerization [17]. The cross-linking is dependent on the availability of ROS generated by laccases and cell wall peroxidases (PRX). This process reinforces the strength and rigidity of the cell walls and can be a key component of the plant response to environmental factors [17–19]. Cell wall architecture is important in plant resistance to abiotic stress and essential in stress sensing and signal transduction [20]. A summary of state of the art of recent reviews is presented in Table 1.

Table 1. List of references of recent reviews on the effects of abiotic stress on cell wall metabolism. References to the papers are given in brackets.

Reference	Stress	Cell Wall	Data Studied
[21]	C, Fl, L, Wd	Primary cell wall	Transcriptomic and proteomic
[22]	Wd	Primary and secondary cell wall	Physiological, proteomic and cell wall composition
[23]	Ap, HM, L, Wd	Secondary cell wall	Transcriptomic, proteomic and metabolomic
[24]	Wd, S, C, Ap, HM, L	Secondary cell wall	Transcriptomic, proteomic and cell wall composition
[25]	S	Primary cell wall	Transcriptomic and proteomic
[26]	S	Primary and secondary cell wall	Proteomic
[27]	HM	Primary cell wall	Transcriptomic and proteomic
[28]	HM	Primary and secondary cell wall	Transcriptomic and proteomic

Ap: Air pollutants; C: Cold; Fl: Flooding; HM: Heavy Metal; L: Light; S: Salt stress; Wd: Water deficit.

The present review summarizes the most recent insights into the effects of eight abiotic stress factors (drought, flooding, heat, cold, salinity, metals, light irradiance, and air pollutants) on primary and secondary cell wall metabolism. Although callose synthesis and deposition have been observed in response to described abiotic stress factors, this polymer is usually not present in the walls of most cell types and is thus beyond the scope of the present review (For reviews: [29,30]). Although the cuticle composed of cutin and waxes, interacts with cell wall, this review will not extend on the effect of abiotic stress on plant cuticle, and will only focus on cell wall components involved in cell wall architecture [31].

2. Water Deficit

Plant growth requires concerted water uptake and irreversible cell wall expansion to enlarge cells [32]. Drought or dehydration results in water deficit stress. Water deficit is defined as an imbalance between soil water availability and evaporative demand [33]. It is one of the major environmental stresses that reduces plant growth and productivity [34]. It is generally agreed that drought stress generates physiological changes in higher plants, including loss of turgor, osmotic adjustment, and reduced leaf water potential (for reviews: [35,36]). Turgor pressure is a critical factor in regulating cell growth. It is the physical force needed to drive cell enlargement, which notably depends upon the extensibility of the cell wall [37,38]. A low turgor pressure caused by water stress leads to a reduction or cessation of growth by decreasing cell extensibility and cell expansion [39]. Plants exposed to water deficit display morphological changes that are the result of plant cell wall modifications. In Black spruce (*Picea mariana*) saplings, drought stress induced a decrease in cell wall thickness, reflecting a lower carbon allocation to cell wall formation and preventing the adaptation of the hydraulic system to drought [40]. Cell expansion can only occur when turgor pressure is greater than the cell wall yield threshold [33,35]. In winter triticale, water deficit induced leaf rolling, which was correlated with a higher level of cell wall-bound phenolics [41]. This adaptive mechanism can limit water loss by restricting the leaf transpiration surface [42,43].

In response to stress, an increase in cell wall elasticity (CWE) can contribute to the maintenance of cell turgor or symplast volume [44–46]. CWE can be correlated with plant drought tolerance [46,47], as shown by a study performed on six cultivars of common bean, which highlighted that the cultivars most resistant to drought showed a strong decrease in their elasticity modulus (ε) in association with a higher CWE; this ultimately enabled an osmotic adjustment [47]. In contrast, the highly drought-sensitive cultivars did not show any significant change in their ε and CWE. These variations in ε and CWE in response to water stress presumably reflect differences in wall structure. However, these changes could depend on the species considered, as in *Ziziphus jujuba* and smooth cordgrass (*Spartina alterniflora*) a large increase in ε was observed, thus indicating that rigid cell walls in leaves may be necessary to maintain cell/tissue integrity during rehydration following a period of stress [37,48,49].

A number of studies report that cellulose biosynthesis can be altered in response to water deficit, as shown by the decreased level of cellulose content in several different species including Arabidopsis, tobacco suspension cells, grape leaves and wheat roots [50–53]. However, in other studies, an increased level of UDP-Glc in the expression of *SuSy* (sucrose synthase) and UDP-glucose pyrophosphorylase (*UGPase*) encoding genes was observed in cotton under drought stress, suggesting a potentially higher cellulose biosynthesis [54]. Increased cellulose synthesis could be a means by which cell wall integrity and cell turgor pressure are maintained, thus allowing continuous cell growth under low water potential [55]. Xyloglycan biosynthesis seems to follow a similar trend in the elongation zone of rice roots under water deficit [56]. In that study, *XTH* and xylose isomerase genes encoding two xyloglucan-modifying enzymes were up-regulated in the early stages following stress application, indicating a role of xyloglucan in the maintenance of root growth [56]. Similarly, the overexpression of an *XTH* gene from pepper in transgenic plants confirmed the role of XTH in enabling better drought tolerance [57,58]. This appears to be a common feature of distinct species, as transgenic Arabidopsis plants overexpressing a hot pepper (*Capsicum annuum*, cv. Pukang) *XTH* (*CaXTH3*) exhibited abnormal leaf morphology resulting in a severely wrinkled leaf shape, an increase in the number of small-sized cells in the leaf mesophyll cells and, ultimately, an increased tolerance to severe water deficit. This suggests that CaXTH3 may be involved in cell wall remodeling to strengthen the wall layers, and hence it could participate in the protection of mesophyll cells against water deficit [57]. In another study performed on transgenic tomato plants overexpressing *CaXTH3*, similar results were shown, highlighting the key role of XTH in water-deficit tolerance [58]. This can be further illustrated by a proteomic analysis in maize, which showed that XTH proteins were differentially regulated upon drought stress exposure [59]. In tomato, a drought responsive transcription factor ASR1 (Abscisic Stress Ripening) was identified as interacting with *Solyc08g082650.2.1*, which was annotated as a cellulose synthase-like (*CSL-G*) involved in hemicellulose biosynthesis [55]. Expansin, another cell wall protein involved in cell wall extension, appears to be strongly regulated by water stress deficiency [60]. In maize roots, expansin activity increased by 4-fold (apical region) and by 2-fold (basal region) under low water potential, which could be related to the higher expression of expansin encoding genes in the root apical zone, corresponding to maintaining elongation [61,62]. However, surprisingly, in the basal region of maize roots, the cell wall underwent a stiffening reaction rather than extension in response to expansin, thus indicating that loss of susceptibility to expansin was associated with growth arrest in this area [61,63]. In soybean, the expression of *GmEXPB2*, a gene that encodes a β-expansin (*EXPB*), was induced by mild water deficiency [64]. This gene, which is intrinsically involved in the response of the root system architecture to some abiotic stresses, including water deficiency, could be an interesting target for improving crop production under these conditions. In a similar way, the overexpression of the wheat β-expansin *EXPB23* under the control of a stress inducible promoter (RD29) in transgenic tobacco plant conferred a higher tolerance to drought stress [65]. The potential key role of expansin in drought tolerance can be further illustrated by studies showing that, in rose (*Rosa hybrida*, cv. Samantha), the virus-induced gene silencing of *RhNAC2*, a transcription factor, or *RhEXPA4*, decreased dehydration tolerance during cell expansion of rose petals [66]. The authors suggested that *RhEXPA4* expression might be regulated by *RhNAC2*. Overexpression of *RhEXPA4* in Arabidopsis transgenic plants conferred a strong phenotype with shorter stems, curly leaves, compact inflorescences and strong drought tolerance, indicating the key role of EXP in response to water deficit in the plant [67]. In wheat coleoptiles, PEG treatment, which can be used to modify osmotic potential and thus induce plant water deficit, limited the increase in the amounts of middle- and low-molecular-weight polysaccharides. In contrast, no effect was shown on high-molecular-weight polysaccharides, such as (1→3),(1→4)-β-D-glucan and arabinoxylan, despite growth arrest [68]. These results demonstrated that the cell wall maintained its ability to extensibility.

In parallel with observations on cellulose and hemicellulose, pectins have been shown to play a key role in modulating cell wall structure in response to drought stress. This fits well with the recognized importance of pectin in controlling plant growth and development (for reviews: [38,69]). In particular, the large amount of homogalacturonan (HG), the most abundant pectic polymer in type I cell walls, combined with a precise regulation of its degree of methylesterification, is likely to be a key element in the control of the stiffness and hydration status of the pectic matrix during drought stress. In addition, it was shown that the amount of side chains of RGI and/or RGII determined the hydration status of the cell wall matrix [70]. The comparison of two wheat cultivars, one tolerant and one sensitive to water stress deficiency, shed new light on the role of this polymer. In the tolerant cultivar, the amount of side chains was indeed increased during water stress, with consequent effects on the viscosity status of the cell wall [52,70]. This latter parameter could be an indicator of drought tolerance [52,70]. In some plants, known as resurrection plants, which can recover completely from a fully-dehydrated state, side chains of pectin are highly enriched in arabinose-rich polymers, including pectin-arabinan, arabinogalactan proteins and arabinoxylans [22,71,72]. The presence of arabinan-rich pectin would prevent water loss during desiccation. In parallel, pectin-degrading enzymes, including PG, can be down-regulated by water stress with consequences on cell wall integrity and cell expansion [53,73]. Alternatively, PG could be involved in controlling cellular water relationships [53,74]. In another study on date palm fruit, the level of methylesterification and *O*-acetylation of xyloglucan was reduced up to 75% and 38% respectively 232 days after pollination, indicating cell wall remodeling of fruits in a water deficit environment [71]. The overexpression of *OsBURP16*, which encodes a putative precursor of PG1β, one of the subunits that regulates PG activity in rice transgenic plants, led to an increase in PG activity and a decrease in pectin content [75,76]. Thus, an increase in the sensitivity of these transgenic rice plants to water stress deficit was shown. In these plants, the intercellular spaces of leaves were larger, suggesting that cell adhesion could be reduced. HG modifications are likely to be an important determinant of stress tolerance as transgenic Arabidopsis, overexpressing a pepper PME inhibitor protein (CaPMEI1), exhibited enhanced tolerance to drought stress during post-germination growth [77].

The secondary cell wall can be strengthened by the incorporation of lignin. Lignification is a complex process involving several different phenolic substrates and enzymes (for a review: [78]). It can occur prematurely to avoid cell wall damage when plants are exposed to a long water stress deficit [23]. Two recent reviews highlighted changes in lignin content and structure following biotic and abiotic stress ([23,24]). Phenylalanine ammonia-lyase (PAL) is a key intermediate at the crossroads of phenolics and lignin synthetic pathways. PAL, which catalyzes the deamination of phenylalanine to trans-cinnamate, was shown to be regulated upon stress exposure. In rice, a proteomic approach showed an increased level of PAL following drought stress [73]. In Grey-maranta (*Ctenanthe setosa*) exposed to long-term drought stress, PAL activity was increased up to 16-fold, with consequent effects on lignin content, ionically wall-bound peroxidase and polyphenol oxidase activities and leaf rolling [79]. The caffeoyl-CoA 3-*O*-methyl-transferase (*CCoAOMT*) encoding gene, which converts caffeoyl-CoA into feruloyl-CoA and thus plays a crucial role in the synthesis of monolignols, was up-regulated in roots of wild water melon subjected to drought [80]. The increase in cinnamoyl-CoA reductase (*CCR*) transcripts in maize roots was associated with a progressive inhibition of wall extensibility and root growth, and may facilitate root acclimation to drying environments [81]. In Eucalyptus, drought stress caused disruption of lignin deposition in leaves and increased the S/G unit ratio [82]. In roots of rice seedlings, the increase in peroxidase activity was correlated with cell wall stiffening, which could be involved in the regulation of root growth reduction caused by water deficit [83]. In white clover leaf, a low water potential enhanced the activation of guaiacol peroxidase and coniferyl alcohol peroxidase, which was closely correlated with an increase in lignin content [84]. Similarly, in the leaf elongation zone of darnel ryegrass (*Lolium temulentum* L.), a grass plant, an increase in cell wall peroxidase activity was associated with a decrease in cell expansion during drought stress, due to phenolic cross-linkage formation between cell wall components [85]. However, the association of free phenolic compounds with the cell wall can be an indicator of plant resistance to drought stress [41,81,86]. The introduction of phenolic compounds linked to the cell wall matrix could generate a stiffening of the cell wall, which might alter cell wall elongation after rehydration. The wall is thus more compact, tight, and less permeable to water. This could allow the plant to maintain leaf or root turgor under a low water potential due to the hydrophobic stabilizing properties of these molecules, thus preventing water loss to the apoplast [41,81,86]. In contrast to free phenolics, cross-linked phenolic compounds and lignin could trigger wall rigidification and growth arrest in the later stage of drought stress, which could lead to a loss of productivity, especially for crops [56]. A Diagram summarizing the plant cell wall response to water deficit is presented in Figure 1.

Although much progress has been made over recent years in understanding the consequences of water deficit on cell wall structure and dynamics, much remains to be investigated before targeting key cell wall polymers that could be useful in engineering drought-resistant crops.

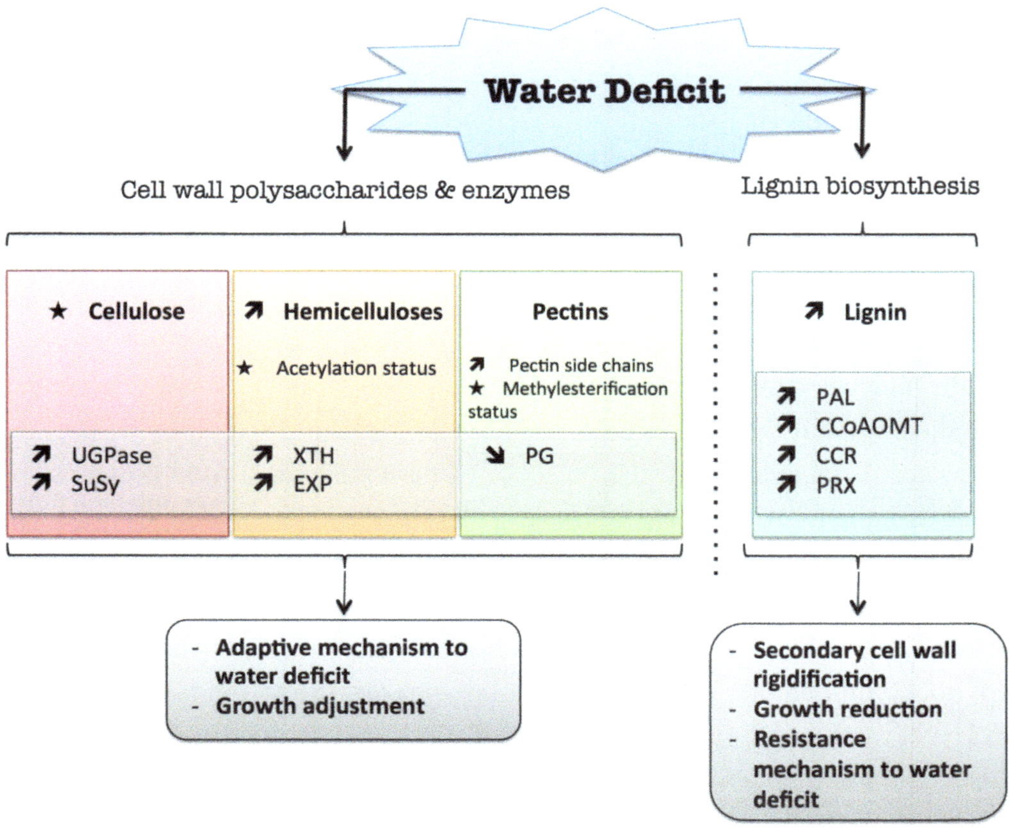

Figure 1. Diagram summarizing the plant cell wall response to water deficit. The schematic presentation is deduced from the results of different studies reported in the review. Arrow (↗) means increased abundance and arrow (↘) means decreased abundance of the molecules. Star (★) means contrasting data on the gene/protein or the molecule studied according to the literature cited in the review. Cell wall (CW); xyloglucan endo-β-transglucosylases/hydrolases (XET/XTH); expansin (EXP); sucrose synthase (SuSy); UDP-glucose pyrophosphorylase (UGPase); polygalacturonase (PG); pectin methylesterase (PME); phenylalanine ammonia-lyase (PAL); caffeoyl-CoA 3-*O*-methyl-transferase (CCoAOMT); cinnamoyl-CoA reductase (CCR); cell wall peroxidases (PRX).

3. Flooding

The occurrence of floods, which is often associated with climate change, has increased over recent decades. Flooding stress in plants results in a lower oxygen level in the root zone due to this molecule's low diffusion rate in water [87]. In soybean seedlings and wheat roots exposed to flooding stress, various deleterious effects have been observed, such as suppression or reduction of hypocotyl and root elongation, and suppression of lateral root development. These effects are dependent upon the flooding intensity (water volume and duration) and the extent of recovery periods [88–90]. In Arabidopsis root, flooding affected the rapid hyponastic movement of the leaf, which was related to differential growth between the abaxial and adaxial leaf parts [91]. The hyponastic movement could be a locally restricted expansion of the cells in the proximal, abaxial region while adaxial cells remained largely unaffected. *EXP* and *XTH*, whose expression is modulated by SHYG (a NAC transcription factor that regulates 1-aminocyclopropane-1-carboxylic acid oxidase, a key enzyme in ethylene biosynthesis under waterlogging conditions) could contribute to cell wall remodeling, enabling differential growth [91]. Flooding alters cell wall polysaccharides. FTIR analysis performed on maize seedlings exposed to flooding showed a degradation of cell wall polysaccharides and a decrease in pectin content [92]. A similar trend was observed in azuki bean seedling (*Vigna angularis*), where a decrease in cell wall sugar content was observed, correlated with cell wall thinning [93]. Furthermore, the xyloglucan content had a lower molecular mass and the apoplastic pH increased, which might inhibit the activity of expansin [93]. Cell wall polysaccharides play an essential role in maintaining the seedling establishment of an Amazonian plant, *Himatanthus sucuuba*, in floodplains. In the flood zone, the seed endosperm of this tree is richer in mannan while species grown in non-flooded areas contain galactomannan and arabinogalactan, indicating that the presence of galactose on mannan in addition of pectin makes the seed tissues more water-permeable. Thus, the presence of pure mannan slowed down water uptake in seeds adapted to environments with plenty of water [94]. Proteomic analysis performed on soybean seedlings revealed that proteins involved in cell wall metabolism were highly regulated in response to flooding stress [90]. For instance, polygalacturonase inhibitor 2-like protein, expansin-like B1-like and rhamnogalacturonate lyase B-like proteins were flooding-responsive proteins common to all the organs studied (root tip, root without tip and hypocotyl). Polygalacturonase inhibitor 2-like protein and EXPLB1 protein levels were greatly increased, while the abundance of other cell wall-related proteins, including cellulose synthase-interactive protein 1 and CAD, decreased, which was dependent on the severity of the flooding. In wheat roots under flooding stress, similar effects to those observed with soybean have been reported. Most down-regulated proteins, such as β-glucanase, β-glucosidase, and β-galactosidase, are related to cell wall metabolism and structural modification [88]. It has been suggested that these down-regulated proteins might be correlated with the inhibition of cell wall elongation under flooding stress, which significantly suppressed the growth of wheat plants.

These alterations in protein abundance are regulated by transcriptional changes since RT-qPCR analysis showed that flooding treatment resulted in the down-regulation of genes encoding cellulose synthase in soybean roots; this could have consequences on cellulose synthesis [95]. Similar trends were observed in woody plants such as gray poplar where the expression of genes encoding cellulose synthases, required for secondary cell wall formation, was strongly down-regulated (up to 44-fold) in roots, while those involved in primary cell wall biosynthesis were less drastically affected (up 4.4-fold). These effects were dependent upon the period of flooding exposure. The primary wall would therefore be a key target for mediating cellular integrity in response to flooding [96]. In addition, other genes encoding proteins involved in cell wall degradation, including cellulase, β-(1,4) glucanase, pectate lyase and polygalacturonase, were down-regulated suggesting the reduced enzyme activity of these proteins could alter the cell wall structure with consequent effects on root growth, leading to growth arrest and suppression of lateral root formation [95]. Another transcriptome analysis performed on soybean seedlings, showed differential mRNA abundance of PGIP and expansin encoding genes. The discrepancy between the inhibition of cell elongation of soybean seedling under flooding and the relative increase in the abundance of EXPLB1 could indicate that flooding recruits proteins with distinct functions in the cell wall. To a certain extent, PGIP, expansin and rhamnoglacturonate lyase B could be flooding stress indicator proteins, which could trigger defense systems [90].

Flooding alters lignification as shown by the decrease in lignin deposition, assessed by phloroglucinol staining, being suppressed in roots and hypocotyls of soybean seedlings under flooding stress [97].

In parallel, proteomic analysis showed a down-regulation of an extracellular superoxide dismutase (Cu-Zn), which might play a role in the regulation of lignification under flooding. Germin-like protein with oxalate oxidase or superoxide dismutase activity produces H_2O_2, which could be used in cross-linking reactions within the cell wall. Four germin-like protein precursors were down-regulated during flooding, thus indicating a reduction in the linking of polysaccharides and proteins within the cell wall, as well as in lignification [97]. In addition, in poplar, the expression of genes encoding cell wall degradation enzymes, such as genes encoding lignin biosynthesis enzymes, including PAL, trans-cinnamate 4-hydroxylase (C4H), 4 coumarate CoA-ligase (4CL), ferulate 5-hydroxylase (F5H), COMT, and CCoAOMT, were repressed indicating the alteration of lignification biosynthesis under flooding.

These data are mostly derived from genomic or proteomic approaches and provide valuable knowledge about the role of the plant cell wall in response to oxygen deprivation (Figure 2). However, very few data are currently available at the biochemical level on cell wall polymers, cell wall enzyme activities and signaling processes that regulate this response.

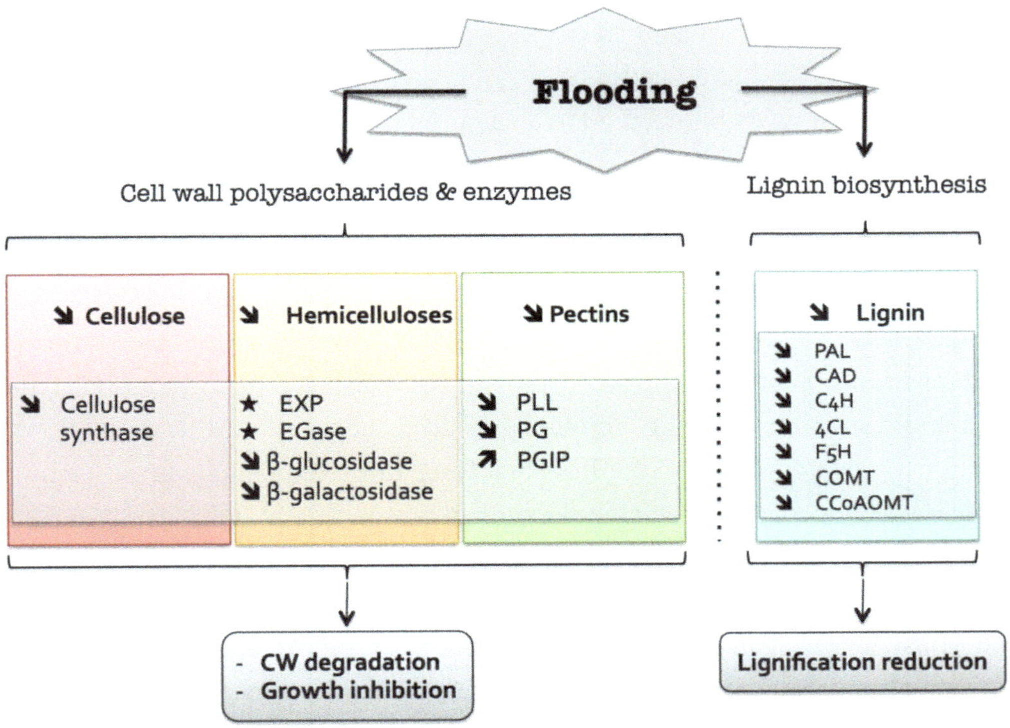

Figure 2. Overview of the plant cell wall response to flooding. The schematic presentation is deduced from the results of different studies reported in the review. Arrow (➚) means increased abundance and arrow (➘) means decreased abundance of the molecules. Star (★) means contrasting data on the gene/protein or the molecule studied according to the literature cited in the review. Cell wall (CW); expansin (EXP); endoglucanase (EGase); polygalacturonase (PG); polygalacturonase inhibitor protein (PGIP); pectin/pectate lyase-like (PLL); phenylalanine ammonia-lyase (PAL); hydroxycinnamyl alcohol dehydrogenase (CAD); trans-cinnamate 4-hydroxylase (C4H); 4 coumarate CoA-ligase (4CL); ferulate 5-hydroxylase (F5H); caffeate *O*-methyltransferase (COMT); caffeoyl-CoA 3-*O*-methyl-transferase (CCoAOMT).

4. Heat

In the context of global warming, an increase in temperature is likely to be a key abiotic stress to which plants will have to adjust to survive [98]. Heat stress is often considered an increase in temperature beyond a threshold level for a period of time that induces irreversible damage to plant growth (*i.e.*, 10–15 °C above the basal temperature). The upper threshold temperature depends on the plant species and genotypes within the species (for a review: [98]). Heat stress is an important agricultural problem in many areas, which can lead to a drastic reduction in economic yield, especially for cereals [99,100]. The application of transitory or constantly high temperatures causes a range of morpho-anatomical, physiological and biochemical changes in plants, particularly observed in Arabidopsis, maize and wheat [101–104]. These affect plant growth and development but in a different way, depending on the developmental stage at which plants are exposed [99]. Various studies have demonstrated that roots are more sensitive to heat stress, suggesting that high soil temperature is more detrimental than high air temperature for whole-plant growth [105]. To maintain growth and productivity, specific tolerance mechanisms must be implemented by plants [106]. Although the cell wall is not the structure primarily altered under heat shock, some studies have revealed that it could be affected at various levels. Cell wall composition and sugar metabolism were compared between maize genotypes during corn expansion, as a target for genetic improvement during heat stress [107]. In sweet corn plants, sugar concentrations in both hemicellulose and cellulose fractions of cobs decreased under stress treatment. The hemicellulose fraction in the shank from the first ear also decreased. In contrast, in dent corn plants, there was no reduction in sugar content, except in the cellulose fraction. Overall, this suggested that synthesis of cell wall components was affected by heat stress and differed between genotypes. High temperature treatment enhanced growth of vegetative plant parts but reduced husk expansion, by suppressing cob extensibility through impairing hemicellulose and cellulose synthesis following a reduction in photosynthesis supply. Therefore, the reduction in grain yield after high temperature treatment was due to effects on sink rather than on source activities [107]. In coffee leaves, heat treatment (37 °C) generated changes in cell wall polymers, as illustrated by a ~40% increased level in leaf hemicelluloses and a 50% decreased level in pectin, with contrasting effects depending on the sugars considered [108]. For instance, in the pectic fraction, arabinose and galactose contents were increased by 40% and 44% respectively, whereas uronic acid and rhamnose decreased by 60% and 40% respectively. An increase in arabinose and galactose contents could be related to a larger amount of type II arabinogalactan, which could contribute to the rigidification of the cell wall by either oxidative cross-linking or water retention. In contrast, the decrease in rhamnose and uronic contents probably reflected changes in the structural organization of cell wall pectin. In the non-cellulosic fraction, the increased xylose content (~30%) was associated with a decrease in arabinose (~33%), thus indicating a lower substitution of arabinose in the xylose backbone of xylan under heat stress. Furthermore, the analysis of leaf apoplastic proteomes of *Coffea arabica* plants reflected these changes in cell wall metabolism in a changing environment related to the greenhouse effect [109]. In contrast, in wheat grain, which displays a primary type II cell wall, arabinoxylan content increased under heat stress [110]. Overall, this shows that the plant response might be dependent upon its cell wall type. Cell-wall related gene expression can be altered in response to heat stress as well. A transcriptomic study performed on Chinese cabbage (*Brassica rapa* L.) showed that some cell wall-related genes could play a role in the acquisition of thermotolerance [111]. Indeed, several genes, encoding proteins from the XTH family, were up-regulated following heat treatment. Other genes, encoding cell wall enzymes or proteins as diverse as arabinogalactan protein, β-glucosidase, cellulose synthase, expansin, extensin, glycosyl transferase, pectin esterase and xylosidase, were up-regulated up to 2–3 fold following their first exposure to permissive high temperature conditions (37 °C). This corresponds to acclimation to a non-lethal temperature to acquire thermotolerance, *i.e.*, the ability for the plant to survive an otherwise lethal high temperature. These results support the idea that thermotolerance acquisition partially requires cell wall-remodeling enzymes, which could be involved in the biosynthesis or *in muro* modification of major cell wall polymers [111]. The authors suggested that the functional significance of XTH during this period could correspond to establishing secondary cell wall deposition, after cell elongation has ceased. Following high temperature exposure, these proteins may have a role in wall strengthening, in addition to that classically observed for cell expansion, which might help plants to adapt to high temperature. XTH could be involved in the reinforcement of connections between the primary and secondary cell walls, thus increasing XG polymerization. In another study, performed on grapevine during fruit ripening, an increased level of *XET* transcripts was measured following heat stress [112]. This could be related to the adaptation of berry volume to temperature and the need for more flexible cell walls. As observed for water stress, the fine-tuning of the cellulose-hemicellulose network appears to be a key component of heat tolerance. In *Agrostis* grass species, expansin is one of the major genes that is up-regulated in response to heat stress at 40 °C [113]. Cell walls with increased levels of expansin may therefore be loosened, relaxed, and become more elastic, which could help maintain cellular functions during heat stress. The overexpression of a grass expansin gene, *PpEXP1* from *Poa pratensis*, in tobacco plants led to a better heat tolerance (35 °C), with less structural damage to cells. This confirms that expansin plays an important role in plant adaptation to heat stress [114]. However, it appears specific to species and/or expansin isoforms, as in *Brassica napus* seedlings subjected to heat stress, where the expression of *EXPA5* was down-regulated up to 10-fold compared to the control [115]. Similar results were observed in poplar (*Populus simonii)*, where thirteen genes encoding *EXP* were down-regulated [116]. Other cell wall-remodeling enzymes are regulated under heat stress at gene expression level, such as a PME gene, *PME35*, which was reduced nearly 10-fold in *Brassica napus* [115].

In addition to its effects on the primary cell wall, heat stress can mediate changes in secondary cell wall metabolism. Synthesis of lignin can be altered during the application of elevated temperature. For instance, in *Coffea* leaves, heat stress increased G and S monolignol contents while a decrease in H monolignol was observed, indicating a structural change in lignin [108]. A proteomic analysis on rice seedlings, growing under various high temperatures, highlighted a differential expression of proteins involved in lignin biosynthesis, including PRX and PAL, with consequent effects on the inhibition of lignification [117]. In contrast to these results, a gene encoding a laccase was up-regulated in Chinese cabbage under heat treatment [113]. This enzyme is suspected of being involved in lignin biosynthesis, based on its ability to oxidize lignin precursors. This result is in accordance with observations in strawberry where elevated temperature possibly regulated the lignin biosynthetic pathway through increased PRX activity [118].

Heat causes changes in cell wall metabolism and this is an important physiological mechanism of heat tolerance (Figure 3). However, much more work is needed to gain a clearer understanding of the cell wall at the physiological, genetic, and biochemical levels of plant heat tolerance.

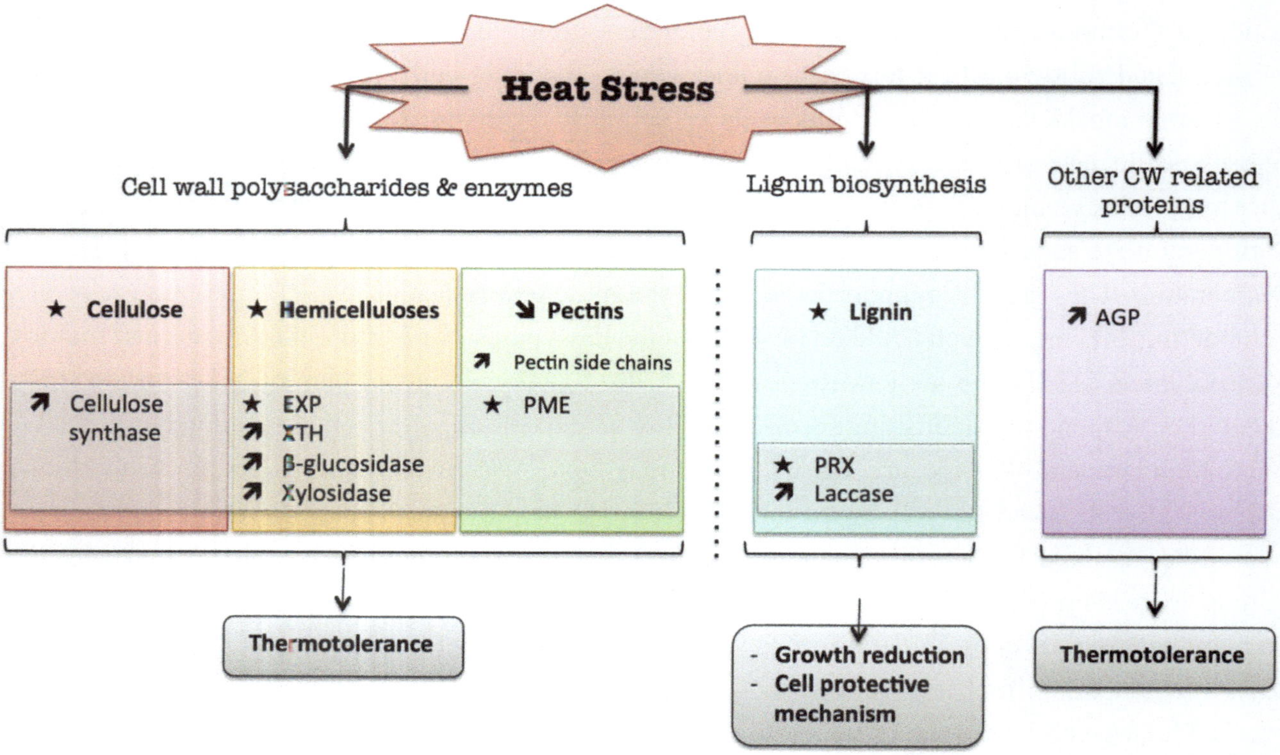

Figure 3. Diagram summarizing the plant cell wall response to heat. The schematic presentation is deduced from the results of different studies reported in the review. Arrow (↗) means increased abundance and arrow (↘) means decreased abundance of the molecules. Star (★) means contrasting data on the gene/protein or the molecule studied according to the literature cited in the review. Cell wall (CW); xyloglucan endo-β-transglucosylases/hydrolases (XET/XTH); expansin (EXP); pectin methylesterase (PME); cell wall peroxidases (PRX); arabinogalactan protein (AGP).

5. Cold

Cold stress, which includes both chilling and freezing injuries, modifies gene expression and plant metabolism with consequent effects on many biological functions [119–122]. The aerial parts of the plant are affected, including leaf expansion, wilting, chlorosis, and necrosis [123,124]. Root growth can also be altered under cold stress [125]. During drastic cold exposure, freezing can induce extracellular ice formation, which leads to cell dehydration and may result in cell collapse [126]. This could be related to the mechanical constraint generated by the enlargement of ice crystals on the cell wall and the plasma membrane, ultimately leading to cell rupture [123,127]. In relation to global warming, the seasons are often altered with changes in temperature and occurrence of freezing stress that can appear without any previous chilling period [128]. A sudden shift to extreme temperature can induce severe physiological damage, which is more important for native plants of a warm habitat cultivated in a temperate climate, most of which are of agronomic interest (maize, soybean, tomato, *etc.*). Such stress can cause a marked decrease in production and significant economic loss [124]. The challenge is thus to improve crops to winter hardiness [129]. In the past twenty years, major efforts have been directed towards understanding the plant response to chilling (0–15 °C) and frost (<0 °C) temperatures [121,127,130–133].

Temperate plants are often chilling-tolerant. They have also developed mechanisms to increase their ability to withstand freezing temperatures following a period of low but non-freezing temperatures [134–137]. This process, called cold acclimation, is a multigenic and quantitative trait that is associated with complex physiological and biochemical changes [119,137] (for a review: [138]). Aside from other factors such as membrane integrity and protoplast properties, plant resistance to cold stress often depends on tissue morphology and cell wall mechanical properties [139–142]. It has been proposed that cell wall rigidity may be an important factor in cell resistance to freeze-induced dehydration [143].

Cold acclimation induces changes in cell wall polysaccharide composition and in the activities of cell wall-modifying enzymes. In a suspension culture of grape cells or in apple, cold acclimation increased cell wall strength and decreased the pore-size of the cell wall [143]. Pectins appear to be a key element of the plant response to cold stress, as shown by a number of studies on various species. In oilseed rape leaves, the cold response was associated with an increase in pectin content [142,144]. This was also observed in pea epicotyls during cold acclimation (2 °C), with a higher level of arabinosyl residues, and in bromeliad (*Nidularium minutum*), a tropical plant, where pectin content increased in plants grown at 10 °C and 15 °C compared to those grown at 25 °C [145,146]. In a frost-tolerant pea genotype, cold acclimation (10 °C day/2 °C night) was accompanied by an increase in homogalacturonan (HG), xylogalacturonan and highly branched rhamnogalacturonan I with branched and unbranched (1→5)-α-arabinans and (1→4)-β-galactans. HG was more esterified in the frost-tolerant pea genotype. The increased cold tolerance in the frost genotype might be related to an increased synthesis of arabinan and galactan side chains and HG methylesterification status, which may act as a gelling component, or regulators of pore size as well as a cold protectant following frost exposure [147]. Changes in the activities of pectin-modifying enzymes have been shown for several species in variable growth conditions in response to cold. For instance, increased PME activity was observed in cold-acclimated oilseed rape plants, which correlated with an increased rigidity of the cell wall in the leaf. This rigidity could be related to a demethylesterification of pectin, which generates free carboxyl groups, and Ca^{2+} ion-mediated cross-linking of these carboxyl groups to create a stiff pectate gel [142]. In contrast, in two chicory root varieties subjected to low temperature in field experiment, PME activity was differentially affected [148]. PME activity was higher during summer and then decreased consistently in both varieties (Nausica and Arancha). However, the decrease in PME activity was stronger in the Nausica variety, which is more cold-sensitive, compared to the Arancha variety where PME activity almost recovered by the end of the cold season [148]. Unlike in oilseed rape, PME activity was not increased in a frost-tolerant pea genotype [147]. During pollen germination, the expression levels of many genes responsible for pollen germination and tube growth were not influenced by cold stress, in particular PME, which were strongly expressed under normal conditions and had steady-state levels of expression in cold-treated samples [149]. This discrepancy could indicate distinct genotype-, species- or tissue-dependent mechanisms of temperature control of PME activity. PG is also regulated in response to cold stress. In young rice leaves of *Ubi::OsBURP16* transgenic plants, which overexpressed a beta-subunit of PG, reduced cell adhesion caused by pectin degradation, which affected cell wall integrity, was associated with a decreased tolerance to cold stress [76]. Cold treatment does not only affect pectin-remodeling enzymes but also pectin-associated receptors or WAKs (wall-associated kinases). WAKs are required for cell elongation and development and were shown to play a role in the plant/pathogen interaction [38,150–153]. An up-regulation of *WAK* genes was observed in rice seedling and a cold-tolerant maize genotype under cold treatment [154]. WAKs could bind either oligogalacturonides or pectin in the cell wall in response to cold stress, indicating a potential role of the cell wall as a cold stress sensor [154].

This modification in pectin structure is often accompanied by changes in neutral sugar composition in vegetative tissues [126,145]. An increase in hemicellulose content could enhance cell wall stiffening and prevent cell collapse caused by dehydration [144,155]. A reduction in xylose and glucose contents in non-cellulosic components was observed in cold-acclimated winter oilseed rape leaves [144]. In contrast, in the winter/dormant stems of poplar, xylose and glucose contents increased and the cellulose content increased by almost 3-fold [156]. Unlike a frost-tolerant pea genotype, a frost-sensitive pea cultivar accumulated substantial amounts of (1→4)-β-xylans and glucuronoxylans, but not pectins, indicating a premature end to cell growth and an initiation of vascularization [147]. Some alterations in the type II cell wall have been shown upon cold exposure. In Miscanthus plants grown at the juvenile stage, an increase in (1→3),(1→4)-β-D-glucan content was observed during cold acclimation (12 °C day/12 °C night) while in wheat roots cultivated over a low temperature acclimation (2 °C) period, hemicellulose synthesis was modified [155,157]. Hemicellulose synthesis increased during the first 24 h following application of stress. In contrast, in the first hour of low temperature exposure, a decrease in the proportion of β-glucans and branched glucuronoarabinoxylans occurred while an increase in the activities of β-fucosidase and β-glucosidase was observed. The activities of glycosidases could account for the transient accumulation of GXAG (oligosaccharin) in winter wheat tissues during the first hours of cold acclimation at 2 °C [158]. Such an oligosaccharin could be produced as a result of the increased hemicellulose turnover and could be physiologically active oligosaccharides that could stimulate the acquisition of freezing tolerance in a winter variety of *Triticum aestivum* L. [155]. Moreover, the authors hypothesized that GXAG could increase cell receptivity to ABA signaling [155,158].

The observed changes in cell wall composition correlate with a strong up-regulation of genes encoding proteins involved in cell wall modification [155]. During cold acclimation, *EXP* and *XTH* gene expression was down-regulated in sweet potato while *EXP* genes were up-regulated in Arabidopsis and rice, showing a contrasting regulation of these genes at low temperatures [125,159,160]. In a distinct experiment, extensin and *XTH* encoding genes were down-regulated in Arabidopsis under cold response [161]. The functional role of XTH was shown in Arabidopsis transgenic plants, overexpressing *AtXTH21*, which had an improved freezing tolerance [162]. Similarly, in rice, a xyloglucan endotransglucosylase (*OsXET9*) was highly induced under cold stress, especially in seedlings [163]. Transcriptomic data in cotton showed an up-regulation of genes encoding a cellulose synthase, *PME3*, and *MUR4*, a UDP-D-xylose 4-epimerase, an enzyme that catalyzes UDP-xylose-4-epimerase to UDP-L-arabinose, in response to cold [164]. In potato plants, the expression of xylose isomerase, which catalyzes the conversion of D-xylulose into D-xylose, a constituent of non-cellulosic polysaccharides, was increased [165]. Overall, cold stress had a drastic effect on the regulation of genes encoding cellulose synthase, as illustrated by the results obtained on cotton, in which cellulose synthases were down-regulated by 2-fold upon cold treatment [164]. In poplar during the winter dormancy period, cellulose synthesis was increased in stems compared to summer stems [156,166], but most cellulose synthase genes were surprisingly down-regulated, which might suggest an unknown control mechanism of cellulose deposition in the winter cell wall of stems [156]. In contrast, in a rice cold-tolerant genotype, cellulose synthase genes were up-regulated by 5 to 8-fold after 48h of cold exposure [154]. The effects of cold stress can also target other cell wall compounds such as arabinogalactan proteins (AGPs), which are located in plant cell walls and belong to a super family of highly glycosylated hydroxyproline-rich glycoproteins. In cotton, the expression of a non-classic AGP which contains a predicted signal peptide, a short AGP domain of seven amino acids, a His-stretch, a Pro-rich domain and a PAC (PRP-AGP containing Cys) domain, was found to be up-regulated in response to cold stress [167]. Transgenic Arabidopsis plants overexpressing *GhAGP31* confirmed that GhAGP31 improved the freezing tolerance of Arabidopsis seedlings [167].

Lignification and changes in lignin content can occur during growth at low temperature as well [23]. Lignin synthesis can be enhanced during cold acclimation and participate in cell wall modification by strengthening it, thus preventing freezing damage and cell collapse. A large increase in CAD (hydroxycinnamyl alcohol dehydrogenase) activity under cold stress was observed in frost-sensitive or cold-tolerant Miscanthus genotypes, but it was largest in the frost-tolerant clone *M. sinensis* August Feder [157]. Furthermore, PAL enzyme activity was higher in the frost-tolerant genotype during cold acclimation [157]. In *Brassica napus*, PAL activity, as well as the rate of accumulation of different phenolic compounds in leaves, depended on the range of low temperature applied to the plant [168–170]. The accumulation of ferulic acid observed in these leaves could increase cell wall rigidity and cell resistance to mechanical stress, thus explaining the growth delay by limited cell expansion [23,170,171]. These data suggest that the phenolic pathway activated by cold stress participates in protecting against freezing damage [23,157]. Lignin biosynthesis genes are affected in the cold response. It was shown that *CAD* gene expression was induced in sweet potato or Arabidopsis in response to cold [161,172]. Other genes involved in lignin biosynthesis, including *PAL*, *CCR*, *COMT* (caffeate *O*-methyltransferase), and *CCoAOMT*, were shown to be up-regulated in pea and *Phaseolus vulgaris* in response to cold [161,164,173,174]. This is concomitant with the increase in cell wall strengthening in response to low temperature stress.

All these data show that the cell wall structure changes in response to cold (Figure 4). Although it is likely that the cell wall could modulate different aspects of the cold response, an investigation of the cell wall as a primary effector is still required.

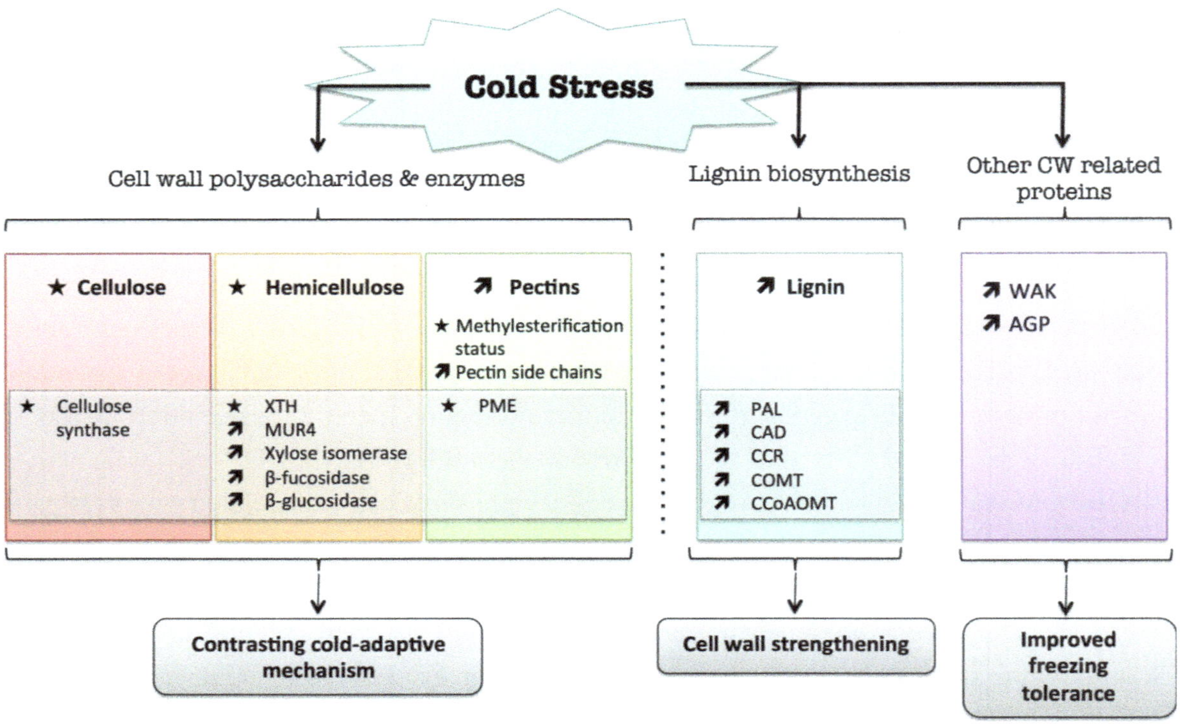

Figure 4. Overview of the plant cell wall response to cold. The schematic presentation is deduced from the results of different studies reported in the review. Arrow (➚) means increased abundance and arrow (➘) means decreased abundance of the molecules. Star (★) means contrasting data on the gene/protein or the molecule studied according to the literature cited in the review. Cell wall (CW); xyloglucan endo-β-transglucosylases/hydrolases (XET/XTH); pectin methylesterase (PME); UDP-D-xylose 4-epimerase (MUR4); phenylalanine ammonia-lyase (PAL); hydroxycinnamyl alcohol dehydrogenase (CAD); caffeate *O*-methyltransferase (COMT); caffeoyl-CoA 3-*O*-methyl-transferase (CCoAOMT); cinnamoyl-CoA reductase (CCR); cell wall peroxidases (PRX); arabinogalactan protein (AGP); wall-associated kinase (WAK).

6. Salt

Another major abiotic stress is due to soil salinity, which can affect plant productivity. More than 6% of the world's total land and approximately 20% of irrigated land is affected by this abiotic stress [175]. Increasing salinization of arable land is expected to have drastic effects, with up to 30% land loss within the next 15 years [106]. Salt stress has two origins: the deposition of oceanic salts by wind and rain and the erosion of rocks that release soluble salts [175,176]. The major effect of salt stress is a decrease in available soil water due to the reduction in osmotic potential of the soil solution leading to water deficit-related plant growth impairment [177–179]. Plants display shoot growth inhibition during the first phase of salt stress, partly due to a loss of cell wall extensibility. In maize, reduced cell wall extensibility was related to the formation and accumulation of diferulic acid cross-links in the cell [180]. Changes in cell wall structure appear to be a common feature of the response to salt stress. For instance, in coffee plants grown with 150 mM NaCl during 25 days, the palisade and spongy parenchyma exhibited thinner cells, which might be related to decreasing water content in leaf cells [181]. Such physiological response was reported in salt-sensitive plants, called glycophytes. This group includes a number of cultivated plants such as rice, soybean and maize [180,182,183]. The resistance of glycophytes to lethal salinity can be improved by short exposure to low levels of salt stress, a process known as salt acclimation [177,184]. This acclimation mechanism has been described and shown to involve changes in cell wall-related proteins [185,186]. Candidates that could participate in salt acclimation are receptor-like protein kinases (RLKs), primary cell wall "sensors", such as WAKs [187]. Although in abiotic stress the function of this protein is not so well defined, induction of *WAKL4* (WAK-like kinase 4) gene expression was observed in Arabidopsis roots under high salt concentration (750 mM NaCl) [188]. These data suggest that the main salt stress perception-to-signaling event could occur at the cell wall-plasma membrane interface. Following salt stress perception, the plant response was shown to involve the regulation of a number of primary cell wall protein-encoding genes, through the recruitment of specific transcription factors. For example, *AtMYB41* or the rice R2R3-type MYB transcription factor *OsMPS* (MULTIPASS) were activated by salt stress (100 mM NaCl) and shown to regulate the biosynthesis of cell wall-related genes [189,190]. *OsMPS* is transiently expressed in vegetative and reproductive tissues and negatively regulates the expression of expansin and endoglucanase genes under salt stress. The hormone-dependent regulation of *OsMPS* expression allowed root growth adaptation to changing environmental conditions while *AtMYB41* was shown to down-regulate the expression of *EXP* or *XTH* encoding genes and up-regulate the expression of *HRGP* when Arabidopsis plants were exposed to 200 mM NaCl [191]. There appear to be distinct responses when species and/or salt-stress conditions are compared. For instance, *RhEXPA4* expression in rose was increased by salt while transgenic Arabidopsis plants overexpressing *RhEXPA4* displayed a compact phenotype (shorter stems, curly leaves, compact inflorescence) and were salt-tolerant [67]. Similar results were observed with the overexpression of wheat expansin, *TaEXPB23*, in transgenic tobacco plants [192]. However, Arabidopsis transgenic plants overexpressing *AtEXP3* or *AtEXP-B1* were salt-sensitive following exposure from 250 mM to 300 mM NaCl treatment [193]. In physic nut roots (*Jatropha curcas*), *EXP* genes were both up-regulated and down-regulated while *XET* genes were up-regulated during different periods of salt exposure at 100 mM NaCl [194]. In parallel, other structural cell wall proteins, including AGP and proline-rich protein (PRP), were affected by salt [25,195]. The expression of rice and poplar (*Populus euphratica*) genes encoding AGP and PRP was up-regulated in response to salt [196–199]. The glycine-rich proteins (GRP), another class of cell wall structural proteins, appear to be involved in salt tolerance. Genes encoding these proteins were induced under salt exposure, with consequent effects on seedling growth and seed germination [200]. MsGRP, a GRP from *Medicago sativa*, probably participates in iron or water transport to maintain cellular homeostasis [200]. *SOS6*, which encodes a cellulose synthase-like protein *AtCSLD5*, plays a role in the regulation by decreasing ROS content under salt stress. The Arabidopsis *sos6* mutant displayed a decreased level of arabinose, rhamnose and galacturonic acid sugar contents under salt stress suggesting that SOS6 is required to maintain cell wall integrity in these conditions [201]. In the primary cell wall of glycophytes, other classes of proteins and polysaccharides are modified upon salt stress treatment. These include RGP (reversibly glycosylated polypeptide), such as RGP1 in pea roots (*Pisum sativum*), or glycosyl hydrolase 1 (GH1 family), GRP in Arabidopsis root, which are up-regulated under salt treatment [200,202–204]. In potato, an increased expression of a UDP-glucose-4-epimerase (UGE), which interconverts UDP-glucose and UDP-galactose, was observed under various abiotic stresses such as high salinity [165]. Moreover, in red-osier dogwood seedlings, a salt-adapted plant, the hemicellulose composition and the lignification status were not altered but an increase in the rhamnose content in the pectic fraction was observed, accompanied by a decrease in plant growth [205]. In addition, cellulose and hemicellulose displayed a steady-state level while a slight decrease in uronic acid was observed in root of a salt-tolerant soybean genotype in response to salt stress. In the root elongation zone of the salt-sensitive soybean genotype, a significant decrease in the cell wall total sugar content was shown, with a more severe decrease (~70%–80%) in the abundance of pectin, compared to that observed in the salt-tolerant one which only displayed a 30% decrease [206]. The higher abundance of pectin in the salt-tolerant cultivar could explain its salt tolerance. A similar study performed on three maize hybrids with contrasting salt tolerance showed a decrease in cellulose concentration in all genotypes and an accumulation of non-methylated uronic acid, which was more pronounced in the salt-sensitive maize genotype [207]. This accumulation was delayed in the salt-tolerant maize genotype, particularly in the youngest shoots, which favored their elongation due to the smaller increase in non-methylated uronic acid [207]. Unlike maize, highly methylated pectin, as shown by immunolabeling with JIM7, increased within the cell wall of Aspen petiole cells [208]. This was associated with an increase in the elasticity modulus, indicating a decrease in cell wall plasticity to maintain the turgor pressure necessary for plant growth. According to these authors, the increase in highly methylated pectin could prevent water molecules from binding to pectin leading to a less hydrated cell wall, and thus favoring water dehydration in the plant during salt stress.

Salt stress has been shown to impact secondary cell wall formation and structure, as revealed by an altered lignin biosynthesis. Salt treatment increased root lignification and the number of lignified vessels [182,209]. Many enzymes involved in the lignin biosynthesis pathway were altered upon salt treatment. For instance, in tomato, *SlPAL5* gene expression was increased after treatment with 200 mM of NaCl while the expression of PAL was decreased in rice [210,211]. Proteomic studies have also revealed that COMT and CCoAOMT are salt-responsive proteins [26]. This was confirmed at the activity level as in rice roots, COMT enzyme activity was decreased in salt stress conditions [211]. However, the CCoAOMT enzyme was salt-induced in the roots of Arabidopsis and rice [212,213]. The action of these enzymes might help reduce the bypass water flow that allows Na^+ ions to enter rice roots *via* an apoplastic pathway [214].

The response of halophyte plants to salt stress appears to differ from that of glycophytes. Indeed, in these plants, cell homeostasis and osmotic adjustment are maintained, with consequences on the synthesis of proteins and cell wall pectins [215]. For instance, in *Sonneratia alba* (a halophyte plant), a higher calcium concentration was measured suggesting that the plant could sequester and efficiently use Ca^{2+} within cell walls to increase their rigidity through pectin egg-box formation [216]. Like in glycophytes, WAK protein could be a salt-stress sensor. In favor of this hypothesis, the results obtained in *Nitraria sphaerocarpa* showed that the expression of *WAK4* was increased after long exposure to salt [217]. Salt also has consequences on proteins that are involved in the biosynthesis or remodeling of cell wall polymers. In roots of *Salicornia europaea*, genes encoding cell wall proteins of the primary cell wall, including UDP-L-rhamnose synthase and cellulose synthases, were down-regulated while other genes, encoding pectin methylesterase inhibitor proteins, were increased under saline conditions [218].

In conclusion, the cell wall plays an important role in salt tolerance, especially in the detection of salt stress. However, the cell wall modifications are different in glycophytes and halophytes. In glycophytes, there is an increase in primary and secondary cell walls whereas in halophytes only the secondary cell wall structure is altered (Figure 5).

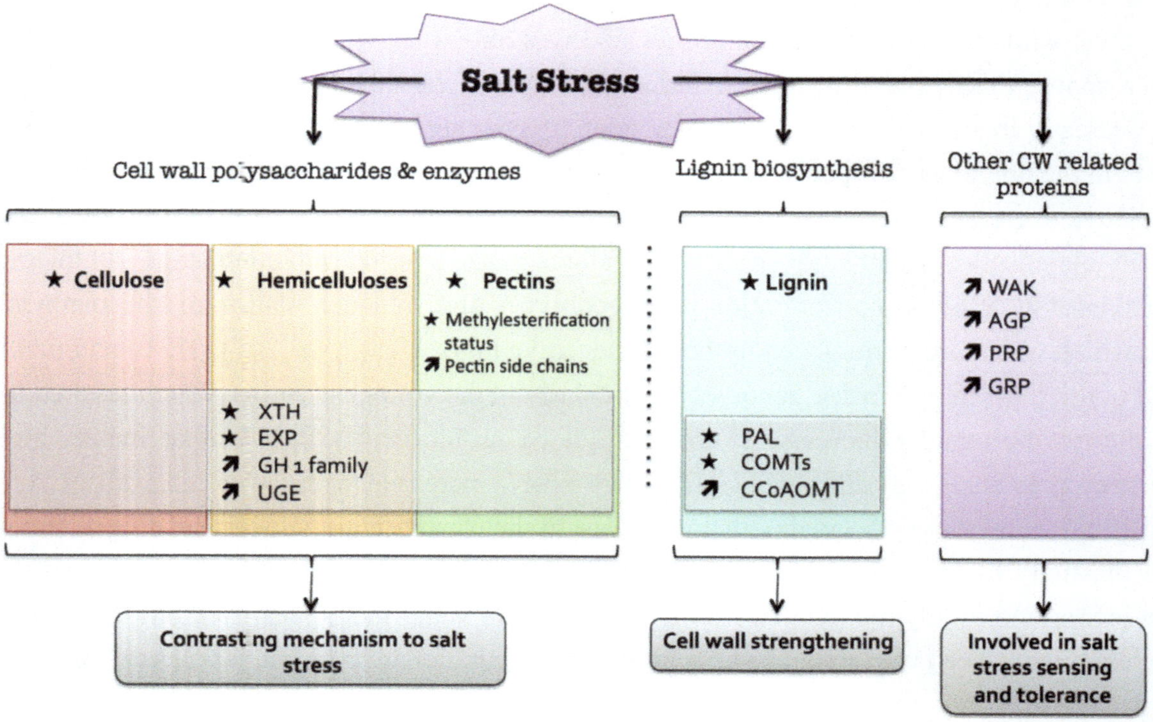

Figure 5. Diagram summarizing the plant cell wall response to salt. The schematic presentation is deduced from the results of different studies reported in the review. Arrow (➚) means increased abundance and arrow (➘) means decreased abundance of the molecules. Star (★) means contrasting data on the gene/protein or the molecule studied according to the literature cited in the review. Cell wall (CW); xyloglucan endo-β-transglucosylases/hydrolases (XET/XTH); glycosyl hydrolase family (GH); reversibly glycosylated polypeptide (RGP); phenylalanine ammonia-lyase (PAL); caffeate *O*-methyltransferase (COMT); caffeoyl-CoA 3-*O*-methyl-transferase (CCoAOMT); arabinogalactan protein (AGP); wall-associated kinase (WAK); proline-rich protein (PRP); glycine-rich protein (GRP).

7. Heavy Metals

Heavy metal (HM) pollution has become an increasingly serious environmental problem over recent decades due to its toxicity, leading to many detrimental effects on living organisms, from plants to animals [219,220]. For instance, heavy metal exposure was shown to reduce root growth and to enhance formation of lateral roots [221–223]. When considering plant mineral nutrition, metals are often categorized into those that are essential for many physiological processes, e.g., copper (Cu), zinc (Zn) and iron (Fe), and non-essential ones, such as lead (Pb), cadmium (Cd), aluminum (Al) and mercury (Hg). An excess, as well as a deficiency, of the former can cause harmful effects in plant cells, while the latter may be toxic even at relatively low concentrations [224]. It is also possible to classify them as redox-active metals (e.g., Cu and Fe), which catalyze the formation of hydroxyl radicals by directly participating in the Haber-Weiss reaction, or metals without a redox capacity (e.g., Cd, Pb, Hg, Zn) which inactivate the cellular antioxidant pool and disrupt the metabolic balance, thus enhancing the load of ROS [225,226]. This affects the distribution of calcium (Ca^{2+}), Ca^{2+}-binding protein activity, signaling processes and MAPK pathways [227,228].

The cell wall represents a physical barrier against the entry of HM into the symplastic compartment. One strategy of plant cells for coping with HM is to remove them from the protoplast by sequestration into extra-cytoplasmic compartments such as the cell wall [27,28]. This protects the most sensitive sites within the protoplast from HM toxicity [27,229,230]. Plant cell wall polysaccharides play a crucial role in HM binding and accumulation in the cell wall, although other compounds, such as proteins, amino acids and phenolics, can also take part in this process [27,231]. One of the advantages of precipitating metals within the cell wall is their strong metabolic inactivation [232].

Pectin, together with hemicellulose, plays a crucial role in binding HM [233,234]. It was shown that the presence of pectin containing boron cross-linked rhamnogalacturonan II in the cell wall promoted the formation of an Al-complex, preventing damage and the irreversible inhibition of growth [235,236]. The presence of exogenous silicon increased cell wall affinity for Zn in maize roots and rice seedlings, thereby reducing its uptake and translocation [237,238]. The degree of methylesterification of pectins appears to contribute to HM tolerance [239]. Some amounts of heavy metal bound to low-methylesterified pectins in the cell wall can enter the protoplast together with this compound by endocytosis, in particular in meristematic dividing cells [229,240]. A striking example of metal accumulation in the cell wall was observed in the common reed (*Phragmites australis)*, which is considered to be a plant with a high detoxification and phytoremediation potential. Zn concentrations followed a gradient with the sequence: intercellular space >cell wall > vacuole > cytoplasm, indicating that most Zn was accumulated within the apoplast [232]. The ability to bind divalent metal cations depends on the number of functional groups, such as –COOH, –OH, and –SH present in the cell wall [241]. The main pectic domain responsible for binding divalent and trivalent metal cations is HG. The HG fraction, particularly those with a low degree of methylesterification, recognized by JIM 5 antibodies containing some amounts of free carboxyl groups, can bind divalent and trivalent metal ions [242]. It is known that low methylesterified HG can interact with calcium and form an egg-box structure in the cell wall. Calcium ions can be replaced by trivalent Al^{3+}, and also by divalent HM cations, including Pb^{2+}, Cu^{2+}, Cd^{2+} and Zn^{2+} [243]. For instance, under short-term Al supply, Al accumulated primarily in the root apoplast where Al^{3+} strongly binds to the negatively-charged binding sites provided by non-methylesterified pectin in the cell wall [239]. In addition, in flax hypocotyls, Cd^{2+} caused reorganization in the distribution of pectin epitopes related to PME activities [244]. Enhanced JIM 5 and 2F4 labeling at the junctions of the inner tissues indicated that the presence of blockwise de-esterified HGA might oppose cell separation under Cd stress [245,246]. This was accompanied by an increase in PME and peroxidase enzyme activities, indicating a reinforcement of cell junctions leading to better tissue cohesion and Cd sequestration [246]. An increase in cell wall polysaccharides confers on the plant the capacity to store metal in the cell wall. In the common bean root apex, the reduction in pectin in the cell wall following PEG treatment is likely to change cell wall structure, thus affecting its porosity and inhibiting Al accumulation [247]. In another study, Al induced pectin biosynthesis in the roots of an Al-tolerant rye cultivar, but reduced it in the sensitive one [248]. This was associated with a decreased level in root PME gene expression in the tolerant rye genotype, although Al treatment increased PME activity in both cultivars.

The degree of methylesterification and acetylation of pectin decreased its affinity for HM. In a copper-tolerant plant, *Silene paradoxa*, a reduction in the root pectin content, together with an increase in its methylesterification status, was observed, in contrast to the copper-sensitive cultivar, indicating a Cu-tolerant/excluder phenotype [249]. In *Medicago truncatula*, the silencing of two Al-induced genes, pectin acetylesterase and annexin, in hairy roots slightly increased the sensitivity of the plant to Al, suggesting that these genes play a role in Al resistance [250].

In addition to pectin-remodeling enzymes, other cell wall enzymes are regulated following HM exposure. For instance, in Arabidopsis, XET enzyme activity was inhibited under exposure to Al, which was mostly bound to hemicellulose [251]. This was associated with root growth inhibition. Interestingly, transgenic poplar overexpressing *XET* had a higher Cd tolerance by reducing Cd uptake and accumulation in roots. This was accompanied by a higher xyloglucan degradation activity, leading to a reduction in XG content in the root cell wall [252].

The structure of the cell wall has an effect on the translocation of HM within xylem. In rice plant, mutation in a cellulose synthase gene (*CESA9*) modified the structure of the secondary cell wall, which further reduced translocation of Cd within xylem and its accumulation inside the plant [253]. In *Elsholtzia splendens*, a hyperaccumulator of Cu, excess Cu altered the composition and distribution of cell wall sugars in roots [254]. Cu enhanced cellulose, hemicellulose and pectin contents with a particular increase in galacturonic acids and KDO levels during HM exposure. In *Sedum alfredii Hance*, an HM-hyperaccumulator, transcriptome analysis showed that cell wall biosynthesis genes encoding *CesA* and *XTH* were up-regulated [255]. In addition to the above-mentioned study, several transcriptome analyses have been performed on plants exposed to HM, revealing that cell wall-related genes are altered. A large-scale rice transcriptome analysis under arsenate (As(V)) stress showed changes in gene expression including 40 cell wall-related genes among 637 transcripts [222]. Of these, 13 were up-regulated and 27 down-regulated. Down-regulated genes were cellulose synthase like A (*CslA*), xyloglucan galactosyltransferases (*XGT*), xyloglucan xylosyltransferases, galactomannan galactosyltransferases (*XXT*), glycosyl transferase (GT8, pectin synthesis), xyloglucan endotransglycosylases/hydrolases (*XTH*), β-galactosidases (βGAL), glycoside hydrolases 9 (GH9) and polygalacturonases (GH28). The down-regulation of these genes could limit cell expansion, as root growth is inhibited under arsenic stress. In *Medicago truncatula*, cell wall-modifying genes, including pectin esterase and pectin esterase precursors, were up-regulated following treatment with Al [250]. Furthermore, microarray analysis performed on rice exposed to metal stress showed that WAK proteins were up-regulated in rice subjected to As(V) stress, indicating that WAK may be involved in heavy metal stress tolerance [222]. The presence of heavy metals in the cell wall does not always result in cell wall stiffening and inhibition of cell elongation. For example, the presence of Cu and Fe can lead to polysaccharide scission, promoting cell elongation. Pectin and xyloglucan can be broken down by hydroxyl radical (•OH) production generated by hydrogen peroxide in the presence of ascorbate and a transient metal ion [256]. Thus, (•OH) free radicals produced in the cell wall under metal stress could act as a site-specific oxidant, targeted to play a role in the cell wall loosening process supporting cell expansion [257].

Lignin biosynthesis is induced under HM exposure. Cu had a positive effect on the biosynthesis of lignin and increased activities of several enzymes, including PRX and laccases, which carry out the polymerization of monolignol precursors of lignin [258]. In addition, in *Panax ginseng* exposed to a higher level of Cu, PAL and CAD enzyme activities were induced. This was correlated with an accumulation of lignin, indicating a protective response to higher levels of Cu [259]. Furthermore, transcriptome analysis of rice seedling revealed that the expression of 6 lignin biosynthesis genes was increased in rice roots under As(V) exposure. A previous study indicated that lignification might be an important step in root growth reduction in plants exposed to heavy metal stress. Therefore, As(V) may have a significant effect on the inhibition of root elongation by regulating cell wall organization. The ROS production induced by As(V) exposure may contribute to cell wall rigidity and inhibition of rice root growth. A similar trend has been observed for *Medicago truncatula* exposed to Al stress, where genes encoding *4CL* and *COMT* were up-regulated in response to Al stress in root tips [250]. It was proposed that the extent of root growth inhibition in response to Al was closely correlated with the extent of lignin deposition. It is conceivable that the Al-induced increase in activity of peroxidases triggered lignin production resulting in cell wall stiffening and reduced root growth in *M. truncatula* [250]. In a Cd-hyperaccumulator plant, *Sedum alfredii Hance*, genes encoding laccase and *C4H* were not induced but expressed at a higher basal level compared with the non-hyperaccumulating genotype [255].

Taken together, these data show a key role for cell wall composition particularly that of pectins, in the tolerance of plants to heavy metals, and could open new perspectives to engineer HM-resistant genotypes (Figure 6).

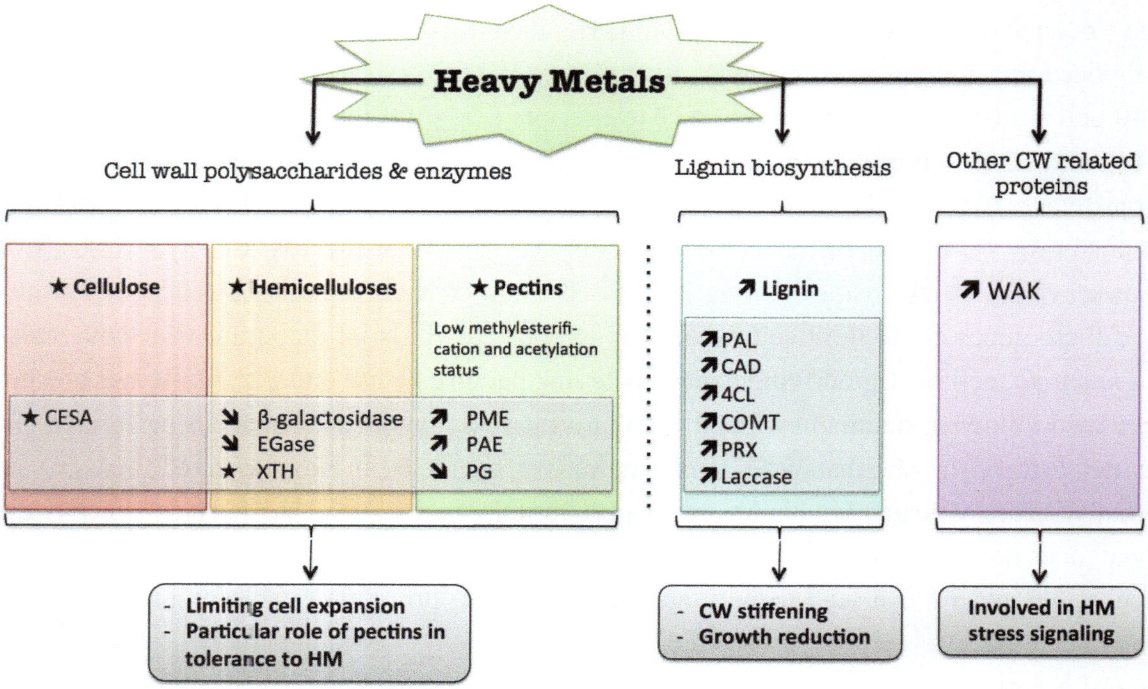

Figure 6. Diagram summarizing the plant cell wall response to heavy metals. The schematic presentation is deduced from the results of different studies reported in the review. Arrow (↗) means increased abundance and arrow (↘) means decreased abundance of the molecules. Star (★) means contrasting data on the gene/protein or the molecule studied according to the literature cited in the review. Cell wall (CW); xyloglucan endo-β-transglucosylases/hydrolases (XET/XTH); endoglucanase (EGase); pectin methylesterase (PME); pectin acetylesterase (PAE); polygalacturonase (PG); phenylalanine ammonia-lyase (PAL); hydroxycinnamyl alcohol dehydrogenase (CAD); 4 coumarate CoA-ligase (4CL); caffeate *O*-methyltransferase (COMT); cell wall peroxidases (PRX).

8. Light

Although light, which contains Ultra Violet (UV), visible and far-red wavelengths, is essential for plants to ensure normal growth, its excess can cause cellular damage. Plants have therefore developed means of protection, notably through changes in the cell wall.

8.1. Shade Avoidance

Plant-to-plant competition is an example of environment sensing and response. A shade avoidance system (SAS) is present in several plant species upon the detection of shade signals from neighboring plants, notably within dense populations [260]. Following sensing through the absorption of light by photoreceptors, it enables morphological responses and differential growth to reach the upper canopy [261]. A reduction in blue and red light (red to far-red ratio) is an important factor that is sensed by the plant photoreceptors that trigger stem and petiole elongation, primarily through cellular expansion [260]. This elongation is a consequence of cell wall loosening mediated by cell wall-modifying enzymes such as expansins and XTHs [262].

A microarray analysis performed on Arabidopsis seedlings identified two *EXP* genes (N37536 and R29778) whose expression was modified by changes in light quality [263]. Similarly, in *Stellaria longipes*, five expansin genes, among the 12 expansin sequences identified from the internodes, were up-regulated relative to the controls in response to changes in light regimes [264]. The key role of the fine-tuning of the cellulose-hemicellulose network in response to light can be further illustrated by the regulation of *XTH*. In Arabidopsis, the expression of five *XTH* genes (*XTH9*, *-15*, *-16*, *-17*, and *-19*) out of 33 was increased by low red to far-red light conditions. In parallel, the expression of *XTH16*, *-17*, *-19* and *-22* was significantly increased in response to green shade which mimics white light conditions in a dense canopy (combined reductions of blue, R/FR, and total light intensity). The functional role of XTH in the shade response was shown by characterizing knockout mutants for two of these *XTH* genes (*XTH15* and *XTH17*), which revealed reduced or absent shade avoidance responses to these light signals. This ultimately demonstrated that cell wall loosening is a key element in the plant response to shade avoidance [265].

8.2. Light Irradiance

As observed in pea and lettuce, light is involved in the regulation of cell wall sugar content in plants, which depends on the tissue or organ considered. [266,267]. In etiolated pea seedling, stem elongation is inhibited by about 75% following both blue and red light exposure as compared to dark controls. The fast inhibitory response was mediated by blue light while the long one was induced by red light. The degree of growth rate inhibition was increased by longer exposure to red light [268]. That inhibition was due to changes of viscoelastic properties of the cell wall.

During Bartlett pear (*Pyrus communis* L.) ripening, cell wall catabolism occurred in both sun- and shade-grown fruits, but pectin solubilization was clearly delayed in sun-exposed fruit. This was associated with a decreased removal of RG I-arabinan side chains rather than reduced depolymerization of the entire polymer. After ripening, sun-exposed pears remained firmer, and there was not a great impact on pectin molecular weight. At harvest, however, a higher proportion of water-solubilized uronic acids and alkali-solubilized neutral sugars and a larger mean molecular size of tightly bound glucans were found in sun-exposed pears [269]. Light radiation altered genes encoding proteins involved in cell wall modifications [155]. The overexpression of a blue light photoreceptor, *BnCRY1*, in *Brassica napus* reduced seedling elongation. In parallel, the expression of cell wall related genes including *XTH* and *PME* showed a 2.5- and 4-fold decreased level respectively in that transgenic line [270].

The lignin biosynthetic pathway is regulated by the photoperiod and light quality [24]. Changes in light conditions have been shown to be associated with an increase in lignin content in a number of different developmental processes and species. *Ebenus cretica* L. seedlings grown in the light displayed 2.5 times more lignin than those grown in the dark [271]. The observed alteration in leaf morphology in response to changing light conditions could thus be the consequence of modifications of the structural components of lignin. Similar results were observed in coffee leaves exposed to sunlight, where the lignin content in leaves has been positively linked to radiation, with an increase in the bulk elasticity modulus that could be explained by the higher content of lignin when compared with shade leaves [24]. Light intensity has also been shown to exert an effect on the activity of enzymes involved in the lignin biosynthetic pathway. Analyses of orchid plants exposed to different light intensities, after transfer from *in vitro* to soil, revealed that PAL and CAD enzyme activities were induced in leaves [272]. In

Arabidopsis, 7000 genes exhibited altered expression when plants were subjected to high light intensity. Among them, 110 showed an increased expression, and several of them, including *4CL*, are involved in the lignin biosynthetic pathway [273].

8.3. UV Light

In recent years, there has been an increasing interest in the biological effects of UV light (230–400 nm) on plants. UV-B radiation (280–315 nm) reduced leaf elongation in Antarctic grass (*Deschampsia antarctica*) and in lettuce, when compared with plants exposed to reduced UV-B. This could be partly explained by the accumulation of insoluble hydroxycinnamic acids (*p*-coumaric, caffeic and ferulic acids) cross-linking the cell wall polysaccharide network and cell wall PRX thus limiting the expansion of epidermal cells and leaf elongation [267,274]. Among the effects of UV-B, cell wall elasticity was reduced in grapevine cells [275]. In cell wall isolates from pea leaves or on citrus fruit pectin, UV-A (315–400 nm) or UV-B produced hydroxyl radical ($^{\bullet}OH$) in presence of hydrogen peroxide. Pectin is able to transform ($^{\bullet}OH$) to superoxide ($O_2^{\bullet -}$), which may have an intracellular impact, since it can pass cellular membrane [276]. ($^{\bullet}OH$) released could contribute to methane production from galacturonic acid methyl ester groups, and also, at a lower level, from acetate groups, [276,277]. This indicates that pectin might be involved in the UV radiation signaling pathway as a source of superoxide [276,278]. In tomato pericarp, UV-C irradiation (230–280 nm) exposure during postharvest retarded cell wall disassembly. This was notably related to the down-regulation of transcriptional expression or through the inhibition of the activities of many cell wall-degrading enzymes, including PME, PG and cellulase, during ripening [279,280].

Changes in cell wall structure and composition under UV light could produce mechanical barriers and protective substances such as secondary compounds including lignin [281]. In cucumber, cotyledons exposed to UV-B radiation lignin accumulated in cell walls of the parenchyma of tracheid elements and in the intercellular space [282,283]. Phenylpropanoids, such as hydroxycinnamic acids and their derivatives (sinapate esters), as well as flavonoids, might play complementary roles in UV-B protection. In addition, in grapevine leaves, the expression of genes of the lignin biosynthetic pathway, such as *PAL*, *C4H*, *4CL* and *CCoAOMT*, were all up-regulated in response to UV-B radiation while genes associated with cell wall loosening were down-regulated [284,285].

These rather scarce data show that light intensity and quality influence cell wall architecture, particularly lignin, which could attenuate radiation and thus protect plant tissues and organs. However, the mechanism of cell wall regulation by light remains to be investigated. An overview of the plant cell wall response to light is presented in Figure 7.

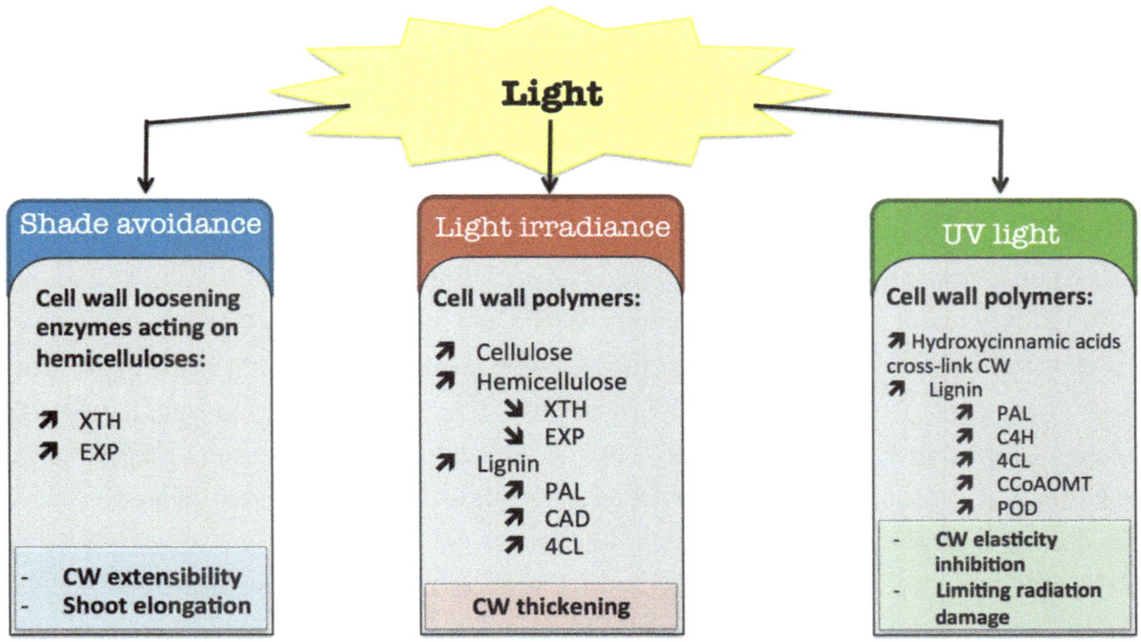

Figure 7. Diagram summarizing the plant cell wall response to light. The schematic presentation is deduced from the results of different studies reported in the review. Arrow (↗) means increased abundance and arrow (↘) means decreased abundance of the molecules. Star (★) means contrasting data on the gene/protein or the molecule studied according to the literature cited in the review. Cell wall (CW); xyloglucan endo-β-transglucosylases/hydrolases (XET/XTH); expansin (EXP); phenylalanine ammonia-lyase (PAL); hydroxycinnamyl alcohol dehydrogenase (CAD); 4 coumarate CoA-ligase (4CL); caffeoyl-CoA 3-*O*-methyl-transferase (CCoAOMT); trans-cinnamate 4-hydroxylase (C4H); cell wall peroxidases (PRX).

9. Air Pollutants

According to the US environment protection Agency (EPA) the main air pollutants are ozone, particle, carbon monoxide, nitrogen oxide, sulfur dioxide and lead. Among these factors the main pollutants affecting ecosystems are carbon dioxide (CO_2) and ozone (O_3) [286]. CO_2 is the primary greenhouse gas emitted through human activities (U.S. environment protection Agency, EPA) and is the substrate for photosynthesis. Its increased concentration in the atmosphere can stimulate plant growth photosynthesis and the accumulation of plant biomass in the short-term. But, elevated atmospheric CO_2 is accompanied by a decrease in oxygen levels which thus can reduce photorespiration, a major source of hydrogen peroxide. This may then reduce abiotic stress response signaling pathway [287]. The tropospheric ozone concentrations (O_3) are more than double compared to the levels measured over a century ago [288]. (O_3) are high enough to cause damage to crops and thus decrease crop yields [289]. Unlike CO_2, O_3 is rapidly converted to ROS. It then interferes with basic physiological functions by accelerating foliar damage, senescence and reducing photosynthesis [290]. This chapter only reports data on the effects of ozone exposure and elevated atmospheric CO_2 on cell wall metabolism since these two main air pollutants for which a number of data is available.

9.1. Ozone (O_3)

Ozone (O_3) is considered the most important consequence of air pollution and is harmful to plants, animals and humans [291]. The cell wall has been shown to be an early target of ozone [292]. The plant response to ozone is thus dependent on the species, the age of the plant, the season, the time of day and, ultimately, ozone concentration [293,294]. For instance, tobacco was more sensitive to O_3 than deciduous trees while juvenile Manna ash (*Fraxinus ornus* L.) trees were more sensitive than mature trees [295]. O_3 induces mainly leaf and root injuries, which can lead to biomass reduction, as observed in tobacco, poplar and birch [291,295,296]. In Manna ash (*Fraxinus ornus* L.), one of the main structural impacts of O_3 was punctures within the cell wall, as observed on leaf mesophyll cells [295]. The development of symptoms was accompanied by the thickening of Manna ash (*Fraxinus ornus* L.) palisade mesophyll cell walls, which appears to be a common feature in several species [295,297]. Cell wall thickening is accompanied by an accumulation of cellulose layers in woody plant species including, *Fagus sylvatica*, *Acer pseudoplatanus*, which reduces the inner space of the cells [297,298]. Cell wall thickening facilitates detoxification processes and confers greater mechanical resistance to cell collapse [297,299]. In Manna ash, pectins and other cell wall polysaccharides are likely to be key components of this cell wall thickening, as shown by the cytochemical cell wall composition of stomata, which displayed more cellulose and lignin in the outer cell wall layer and more pectin in the inner cell wall layer [293,295]. This might contribute to maintaining the mechanical properties of these cells [295]. In ragweed (*Ambrosia artemisiifolia* L.) pollen, FTIR analysis showed a decrease in phenolic compounds and an increase in pectin components (rhamnogalacturonic acid, arabinan) under elevated ozone [300]. Notably, a decrease in acetylester groups, indicating a de-esterification of pectin in response to ozone, was shown. Transcriptome analysis further revealed that the expression of pectin methylesterase and pectate lyase encoding genes was strongly induced in ragweed pollen in response to ozone. Although Amb a1, an allergen found in *Ambrosia artemisiifolia* pollen, which is a pectate lyase, was highly expressed in response to ozone, no significant difference was measured at the protein level, as tested by an ELISA experiment [300] Cell wall sugar content increased under ozone exposure in strawberry plant, which was related to increased pectin content (100% more) after a short treatment (7 days) rather than long exposure (2 months). These changes were primarily due to water-soluble pectin suggesting that either pectin degradation was decreased or *de novo* synthesis increased. In parallel, long-term exposure reduced non-soluble pectin by 60% and 80% in more and less O_3-tolerant strawberry cultivars, respectively. This appeared to be related to changes in the amount of low methoxy pectin content, which increased in the tolerant genotype and decreased in the sensitive one.

O_3 can be used to delay fruit ripening during cold storage, due to its ability to inhibit ethylene oxidizing capacity. In kiwifruit, O_3 treatment at 0 °C reduced cell wall disassembly of the fruits during ripening at room temperature [301]. In that study, the swelling capacity of the cell wall associated with the solubilization of pectin was reduced when fruit was treated with O_3. The treatment induced lower PG and endo-1,4-β-glucanase activities, enzymes involved in pectin and hemicellulose degradation, compared with the control environment. In contrast to the above-mentioned findings, in O_3-sensitive aspen, leaves were thinner compared to the tolerant genotype, concomitant with a reduction in cell walls. The average thinner cell walls in mesophyll cells in the sensitive aspen clone could decrease cell protection from O_3 present in the apoplast [302]. At the transcriptional level, in an O_3-sensitive rice cultivar, genes encoding cell wall degradation enzymes, including cellulase and β-galactosidase, were up-regulated. Other genes encoding cell wall-related proteins, such as cell wall invertase inhibitor or glycine-rich protein, were similarly up-regulated while the expression of genes encoding proteins involved in cell wall biosynthesis, such as *XET* and *COBRA*, was down-regulated [303]. In contrast, in the O_3-tolerant rice genotype, a different transcriptional regulation is likely to occur, with an increased expression of genes encoding some expansin and xylanase inhibitors.

O_3 also has an impact on wood formation. In poplar, stems were more affected than leaves, with a reduction in stem diameter and the anatomy of the tension wood [296]. O_3 reduced cambial growth and xylem differentiation, which was associated with a reduction in cellulose and lignin biosynthesis [296]. This was accompanied by changes in the expression of specific genes, and by a decrease in activity of enzymes involved in cellulose (SuSy, UGPase) and lignin formation (PAL, CAD) [296]. In poplar exposed to ozone, changes in lignin structure occurred, with enrichment in condensed lignin by a higher proportion of H-units near the foliar necrotic area [304]. CAD activity was rapidly and strongly increased (10-fold) whatever the foliar developmental stage, while PAL enzyme activity was increased in old and mid-aged leaves [304]. In non-woody plants, at the transcriptome level, *PAL* mRNA, a key enzyme involved in the phenylpropanoid pathway including lignin biosynthesis, was rapidly and transiently induced reaching a 3-fold higher level compared to control Arabidopsis or parsley plants [305,306]. In contrast, O_3 was also found to modify cell wall components by depolymerization of lignin, which released small phenolic compounds that could be used as elicitor molecules [307].

Cell wall peroxidases might regulate the level of ROS by scavenging H_2O_2 to polymerize lignin and cross-link cell wall proteins and polysaccharides [308]. In sunflower leaves exposed to ozone, cell wall PRX was induced thus leading to cell wall stiffening and contributing to slowing down O_3 penetration into the cell [308,309]. Furthermore, PRX displayed higher activity with syringyl compared with guaiacyl molecules. In addition, PRX activity in strawberry leaves decreased in the less tolerant cultivar while it increased in the ozone-tolerant genotype under short term ozone exposure, indicating a significant contribution of cell wall PRX to the detoxification of ozone in the apoplast [310] These data, which are mostly derived from cytochemical analysis and enzyme activity measurements, implies a contribution of cell wall proteins to apoplastic ozone detoxification. In addition, some genomic approaches provided valuable knowledge about the role of the plant cell wall in response to ozone (Figure 8). However, very few data are currently available at the biochemical level on cell wall polymers, cell wall enzyme activities and signaling processes that regulate this response.

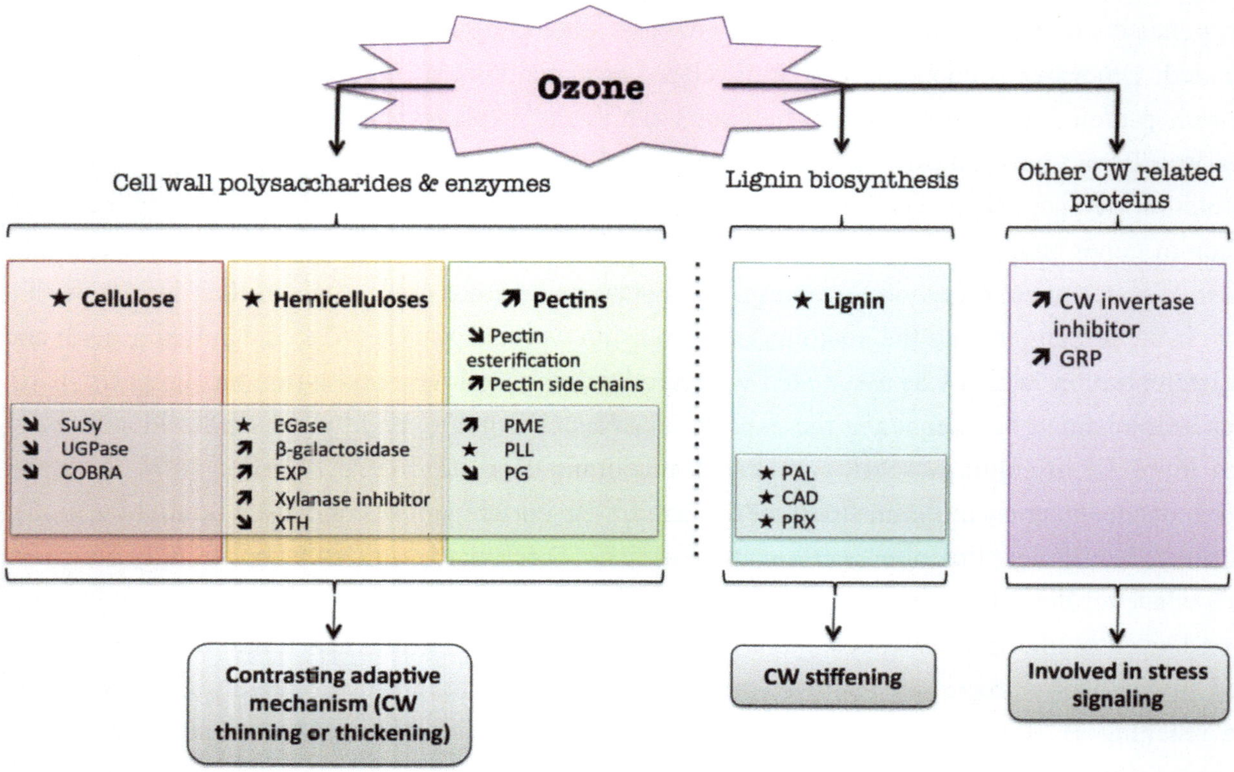

Figure 8. Diagram summarizing the plant cell wall response to ozone. The schematic presentation is deduced from the results of different studies reported in the review. Arrow (➚) means increased abundance and arrow (➘) means decreased abundance of the molecules. Star (★) means contrasting data on the gene/protein or the molecule studied according to the literature cited in the review. Cell wall (CW); xyloglucan endo-β-transglucosylases/hydrolases (XET/XTH); expansin (EXP); sucrose synthase (SuSy); UDP-glucose pyrophosphorylase (UGPase); endoglucanase (EGase); pectin methylesterase (PME); pectin/pectate lyase-like (PLL); polygalacturonase (PG); phenylalanine ammonia-lyase (PAL); hydroxycinnamyl alcohol dehydrogenase (CAD); glycine-rich protein (GRP); cell wall peroxidases (PRX).

9.2. CO_2

The atmospheric CO_2 level has increased by 30% over the last 250 years and is forecast to rise by 1%–2% per year [311]. This has an impact on plant development as C3 crops show a significant increase in photosynthesis and growth in response to rising CO_2 while C4 plants, including maize, Miscanthus, and sugarcane differ in their response [312,313].

In woody plants, including poplar, growing under elevated CO_2, trees have a larger leaf size, increased stem and branch growth and root biomass [314–320]. CO_2 responsiveness was found to be dependent on genotype. The unresponsive aspen clone grew slower than the responsive one, suggesting different strategies used in carbon partitioning [317]. A significant decrease in cell wall thickness was observed in the CO_2-responsive clone [317]. During the late season, the CO_2-unresponsive clone displayed a cell wall thickening in the leaves, which is likely to be related to an increased level of UDP-sugar pyrophosphorylase transcripts. In contrast, a long term exposure (11 years) to an elevated CO_2, decreased cell wall thickness in different aspen genotypes and in sweetgum (*Liquidambar styraciflua* L.), suggesting that a short term exposure may not provide a realistic view of a long term exposure [319,320]. In addition, no significant changes in the cell wall sugar content were observed [320]. In herbaceous plants, including Arabidopsis, grown under elevated CO_2 conditions, a rise in whole plant biomass (+20%) was observed before flowering. UDP-Glc dehydrogenase, a key sugar in the nucleotide interconversion pathway, necessary for cell wall polysaccharide biosynthesis, was increased in these young plants. However, although the CO_2 concentration increased the dry weight of the cell wall, it had no significant effect on cell wall composition as observed in sweetgum [320,321]. In Arabidopsis, an increase in plasmodesmata formation in the midrib of the first leaves exposed to elevated CO_2 was observed. This might be related to a higher sucrose translocation in the sieve elements [322]. In barley, under ambient CO_2, cell wall elasticity was decreased compared to elevated CO_2 levels, which could be considered as a positive mechanism for maintaining cell turgor and preventing cell wall rupture [323]. In addition, high CO_2 was shown to have an effect on post-harvest biochemical and textural properties of white asparagus spears, based on changes in cell wall components [324]. For instance, during post-harvest storage, high CO_2 levels inhibited the excessive accumulation of secondary wall components, including cellulose and lignin, at room temperature (20 °C), thus decreasing the toughening of asparagus spears. Pectin content was less influenced by elevated CO_2 compared to secondary cell wall components. In contrast, at lower temperature (10 °C) and elevated CO_2 atmosphere, no significant changes in cell wall composition and distribution were observed compared with cold applied on its own. Thickening of cell walls and lignification could be delayed by elevated CO_2 at room temperature. This could explain why elevated CO_2 partially inhibited PAL activity, as observed in green asparagus grown in an oxygen modified atmosphere [325]. In arrowhead, shoot elongation was stimulated by elevated CO_2. This growth phenotype was associated with higher transcript levels of genes encoding *EXP* and *XTH* genes, both of which are known to control cell wall loosening during cell growth [326]. MicroRNAs which are important regulators of gene expression are altered under long-term elevated CO_2 in Arabidopsis. For instance, miRPAL2 which regulates, cellulose synthase genes including *AtCesA10*, pectin biosynthesis gene including galacturonosyltransferase-like 9 or glucosyltransferase family 8 (GT8) is up-regulated, indicating an alteration of plant cell wall biosynthesis under elevated CO_2 [327].

Another, transcriptome analysis performed on old aspen trees after 12 years of exposure to elevated CO_2, showed a significantly up-regulatation of genes encoding enzymes involved in cell wall loosening and expansion (*XTH*, *EXP*). This indicates a radial growth expansion in response to elevated CO_2 [318]. In addition, other genes involved in cellulose biosynthesis, non-cellulosic and pectin biosynthesis had increased transcript levels. The overall average expression of these genes was 33% higher in the vascular cambium zone (VCZ), which reflects a more intensive cell wall biosynthesis there. Pectin-remodeling enzymes, including pectate lyase, pectin acetylesterase and pectin methylesterase, were highly up-regulated in response to elevated CO_2, and a *PME* orthologous to *AtPME3* was up-regulated by at least 25% in the VCZ. Similar patterns of expression were shown for a putative pectin acetylesterase gene and a number of NDP-sugar epimerases (UDP-glucuronate 4-epimerase, UDP-glucose 4-epimerase, and UDP-arabinose 4-epimerase (MUR4)). Overall, this indicates a significant increase in pectin and hemicellulose biosynthesis in vascular meristematic cells in response to growth under elevated CO_2. In contrast, genes involved in lignin biosynthesis were mostly down-regulated in the VCZ (24%) and in leaves (11.4%) indicating that down-regulation of lignin biosynthesis is important for maintaining cell growth in this zone [318]. In another atudy, performed on Aspen leaves of both unresponsive and CO_2 responsive genotypes revealed a distinct response between them during both early and late growing seasons. During the early season, the expression of genes encoding cell wall glycosyltransferase (GT43) was increased in both clones in elevated CO_2 conditions. In contrast, some other genes encoding cell wall-remodeling enzymes were differentially regulated. For instance, genes encoding enzymes involved in secondary cell wall metabolism, including cell wall *PRX*, were up-regulated in the CO_2-unresponsive clone. Furthermore, a strong up-regulation of NAC protein (for NAM, no apical meristem), a transcription factor known to regulate secondary cell wall synthesis genes in fibers, was observed in the sensitive clone leaves. This would suggest that NAC could control the carbon deposition in the leaf secondary cell wall thickening of the sensitive clone in response to elevated CO_2 [317]. In contrast, in the genotype that displayed a high responsiveness to elevated CO_2, genes encoding enzymes involved in lignin biosynthesis, including *CCoAOMT* and laccase, were down-regulated in leaves during the early growing season. No significant changes were observed in the late growing season in that genotype. A similar trend was observed for fascilin-like arabinogalactan protein and pectin methylesterase genes. The responsive clone is likely to have a strategy that maintains the ability to grow while the sensitive one devotes more energy to passive defense, *i.e.*, lignin biosynthesis and cell wall thickening.

All these data confirm that atmospheric pollution has an impact on plant cell wall metabolism (Figures 8 and 9). However, these data are scarce and studies need to be developed to supply plants adapted to such environments in the future.

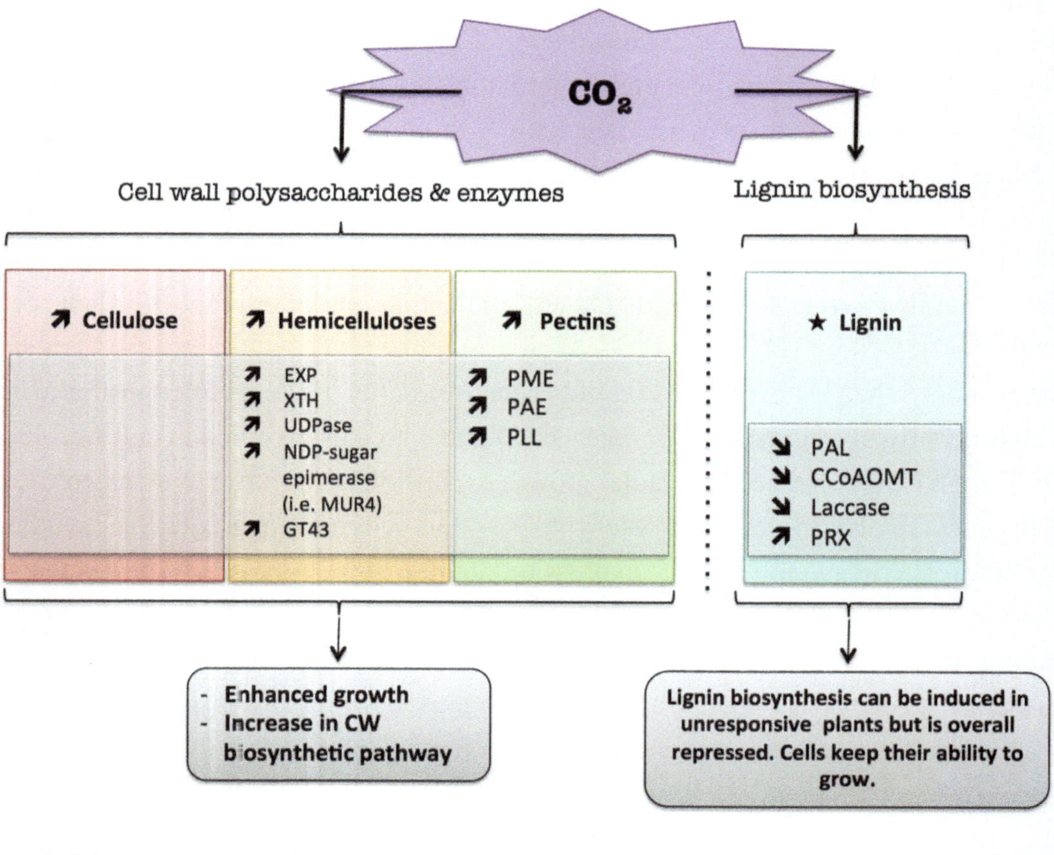

Figure 9. Diagram summarizing the plant cell wall response to elevated CO_2. The schematic presentation is deduced from the results of different studies reported in the review.

Arrow (🡽) means increased abundance and arrow (🡾) means decreased abundance of the molecules. Star (★) means contrasting data on the gene/protein or the molecule studied according to the literature cited in the review. Cell wall (CW); xyloglucan endo-β-transglucosylases/hydrolases (XET/XTH); expansin (EXP); UDP-glucose pyrophosphorylase (UGPase); pectin methylesterase (PME); pectin acetylesterase (PAE); pectin/pectate lyase-like (PLL); glycosyltransferase family 43 (GT43); phenylalanine ammonia-lyase (PAL); caffeoyl-CoA 3-*O*-methyl-transferase (CCoAOMT); cell wall peroxidases (PRX).

10. Conclusions

The present review outlined the impact of various abiotic stresses on the plant cell wall. Overall, cell wall architecture is affected by abiotic stress, but the extent of the available data differs depending on the stress considered. Although changes in the cell wall in response to drought or cold have been quite well documented, data concerning some other stresses, including flooding and air pollutants, are scarcer. Furthermore, the data available in the literature are more related to transcriptome and proteome analysis than to the detailed chemical composition of the cell wall in response to stress. Stress exposure, depending on the species and stress intensity, appears to result in both plant cell wall loosening and tightening. One possible strategy is thus to improve the viscoelastic properties of the primary wall by increasing levels of cell wall remodeling and biosynthesis enzymes, and by modulating other wall loosening agents, including pectin, thus contributing to increasing the hydration status of the plant and maintaining turgor pressure for growth (Figure 10A). This could enable a relative growth status to be maintained. Another option is to modulate the viscoelastic properties by reinforcing the secondary wall with cellulose deposition and non-cellulosic components. Tightening which corresponds to a decrease in cell wall extensibility and/or turgor pressure is often accompanied by lignin biosynthesis in response to stress, except during flooding, when genes and/or enzymes involved in lignin biosynthesis are repressed/inhibited (Figure 10B). However, as the review reports, the stress effects on plant cell wall architecture depend on the plant species, plant genotype, and the age of the plant, as shown under ozone exposure or in response to cold. It also relies on the timing and intensity of the stress, as shown under flooding. Clearly, cell wall adjustment under abiotic stress is an important phenomenon in plant adaptation.

Although much progress has been made over the last decade in understanding the consequences of abiotic stress on cell wall metabolism, a multidisciplinary approach including physiology, biochemistry, proteomics and genetics based on one species for each given stress could provide more consistent results to elucidate the complex mechanism of the cell wall in its response to abiotic stress. Research needs to be pursued to engineer plants with cell walls adapted to global warming in the future.

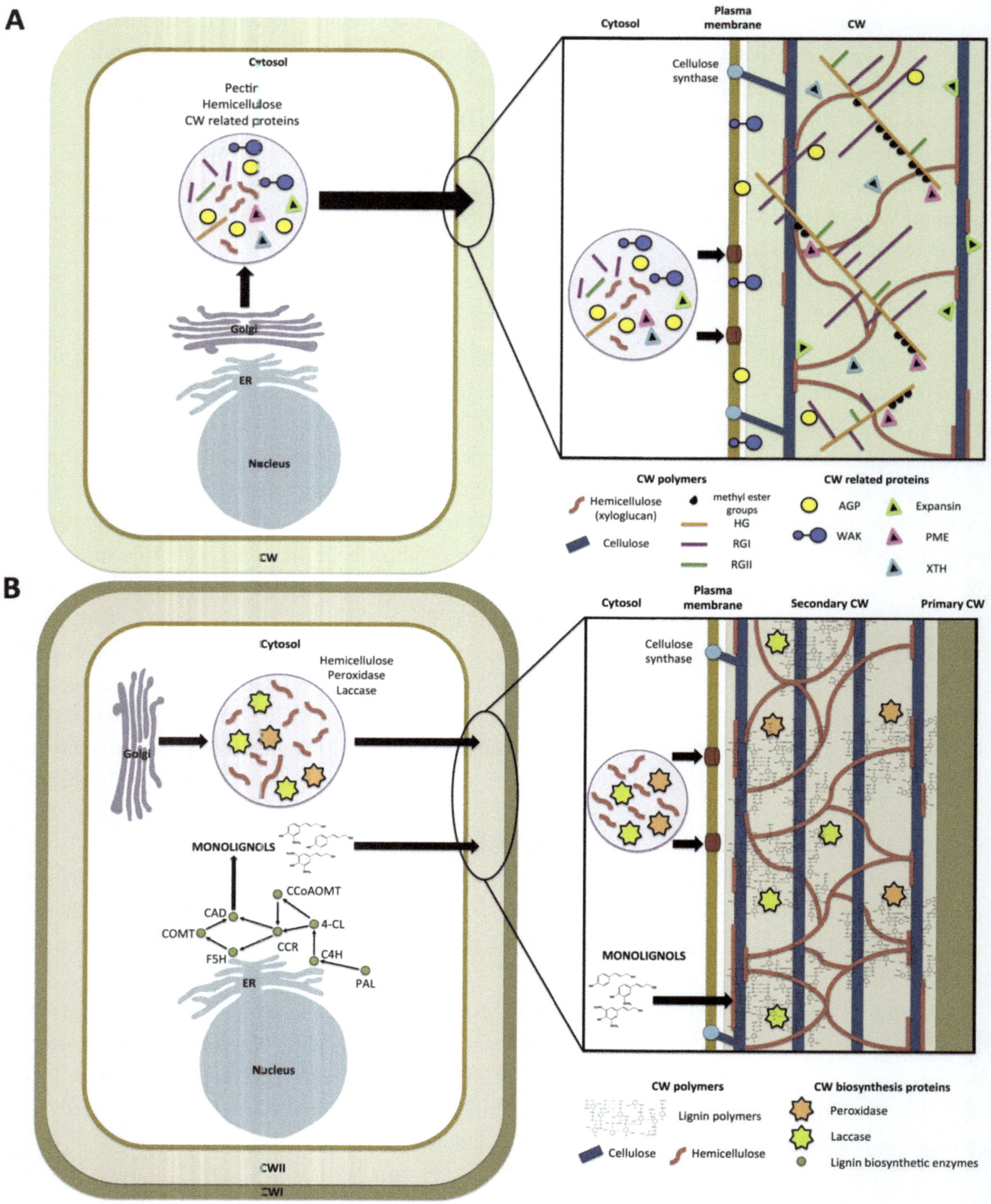

Figure 10. Model depicting changes in the plant cell wall mechanism in response to abiotic stress. (**A**) Key points of the dynamics of type I primary cell wall (CW) during abiotic stress exposure are shown. Cell wall polymer biosynthesis takes place at different locations within the cell. During cell wall formation, hemicelluloses and pectin are synthesized in the Golgi

apparatus of the plant cell and then secreted to the apoplastic space while cellulose synthesis occurs at the plasma membrane. The organization and interactions of wall components is coordinated with dynamic assembly in the apoplast and rearrangement occurs during wall extension [328]. During growth under non-optimal environmental conditions, the primary cell wall composition is altered. In several abiotic stresses, cellulose, hemicellulose and pectin contents are changed. Different adaptive mechanisms can be observed, which depend on the species considered, the organ, the tissue, the age, and the exposure intensity of the stress. In most "tolerant" species, the biosynthesis of cellulose and xyloglucan, which is the most abundant non-cellulosic component of type I primary walls, is induced. This is associated with an up-regulation of the expression of genes encoding *EXP* (expansin) and *XTH* (xyloglucan endo-β-transglucosylases/hydrolases). In parallel, levels of rhamnogalacturonan I (RGI) enriched in arabinan and/or galactan side chains are increased. The level of methylesterification of homogalacturonan (HG), regulated by PME (pectin methylesterase), is reduced. All these modifications of the cell wall architecture lead to a relative maintenance of cell wall extensibility, adapted to coping with a particular abiotic stress. For species that are "sensitive" to a given abiotic stress, cell wall degradation is observed, related to a decrease in cell wall polysaccharide content, an increase in cell wall hydrolases (EGase) and a decrease in cell wall biosynthesis and remodeling enzymes (PME, PG, SuSy). Abiotic stress can also alter wall-associated kinases (WAK), which are required for cell elongation and development. In plants that display tolerance to abiotic stress, the expression of genes encoding *WAK* are up-regulated, which suggests a perception of the stress at the cell wall/plasma membrane interface through the detection of released plant cell wall fragments [187,188,222]. Genes encoding other cell wall proteins, including arabinogalactan protein (AGP), proline-rich protein (PRP) and glycine-rich protein (GRP) are induced in response to abiotic stress, which could covalently link with pectin or hemicellulose thus contributing to the strengthening of the wall [329]. Another possibility is that HRGP is a signal molecule in plant defense [330]; (**B**) Key points of the secondary cell wall (CWII) during abiotic stress exposure are shown. The model is adapted from [78]. To cope with abiotic stress, some plants adopt a strategy that consists of increasing the secondary cell wall thickening by hemicellulose and cellulose synthesis [147,295]. This is often associated with a rigidification of the secondary wall by lignin deposition [23,182,282]. Monolignols, which are the building blocks of lignin, are synthesized from phenylalanine through the general phenylpropanoid and monolignol-specific pathways. This involves cytosolic enzymes including phenylalanine ammonia-lyase (PAL), 4 coumarate CoA-ligase (4CL), cinnamoyl-CoA reductase (CCR), caffeate *O*-methyltransferase (COMT), and caffeoyl-CoA 3-*O*-methyl-transferase (CCoAOMT) as well as ER membrane-anchored proteins including trans-cinnamate 4-hydroxylase (C4H), and ferulate 5-hydroxylase (F5H). The monolignols are then transported to the cell wall where they are polymerized by apoplastic peroxidase (PRX) and laccases into lignin [117,144]. This other adaptive mechanism, based on a decrease in cell wall expansion and cell extensibility, can limit water loss and prevent cell collapse due to dehydration.

Acknowledgments

This work was funded by grants from the Agence Nationale de la Recherche (GALAPAGOS, ANR-12-BSV5-0001 and NOSTRESSWALL, ANR-12-ADAP-0011) and the Conseil Regional de Picardie (PhD studentship awarded to F.P.). The funding of the Institut Universitaire de France (IUF) to J.P. is gratefully acknowledged.

Author Contributions

All authors contributed to this review.

Conflicts of Interest

The authors declare no conflict of interest.

References

1. Shaik, R.; Ramakrishna, W. Machine learning approaches distinguish multiple stress conditions using stress-responsive genes and identify candidate genes for broad resistance in rice. *Plant Physiol.* **2014**, *164*, 481–495.
2. Wu, Q.; Shu, P.-C.; Cai, T.; Fan, X.-W.; Li, Y.-Z. Existence of abiotic stress-responsive genes within the regions of QTLs controlling maize grain yield: One of the root causes for unstability of the QTLs? *S. Afr. J. Bot.* **2014**, *93*, 231–241.
3. Wang, Y.; Frei, M. Stressed food—The impact of abiotic environmental stresses on crop quality. *Agric. Ecosyst. Environ.* **2011**, *141*, 271–286.
4. Dolferus, R. To grow or not to grow: A stressful decision for plants. *Plant Sci.* **2014**, *229*, 247–261.
5. Potters, G.; Jansen, M.A.K.; Guisez, Y.; Pasternak, T. Stress drives plant cells to take the road towards embryogenesis. In *Floriculture, Ornamental and Plant Biotechnology, Advances and Topical Issues*; Teixeira da Silva, J.A., Ed.; Global Science Books Ltd.: London, UK, 2006; Volume 2, pp. 289–294.
6. Wrzaczek, M.; Brosché, M.; Kangasjärvi, J. ROS signaling loops—Production, perception, regulation. *Curr. Opin Plant Biol.* **2013**, *16*, 575–582.
7. Gill, S.S.; Tuteja, N. Reactive oxygen species and antioxidant machinery in abiotic stress tolerance in crop plants. *Plant Physiol. Biochem.* **2010**, *48*, 909–930.
8. McCann, M.; Roberts, K. Architecture of the primary cell wall. In *The Cytoskeletal Basis of Plant Growth and Form*; Llcyd, C.W., Ed.; Academic Press: London, UK, 1991; pp. 109–129.
9. Carpita, N.C.; Gibeaut, D.M. Structural models of primary cell walls in flowering plants: Consistency of molecular structure with the physical properties of the walls during growth. *Plant J.* **1993**, *3*, 1–30.
10. Carpita, N.C.; McCann, M.C. Maize and sorghum: Genetic resources for bioenergy grasses. *Trends Plant Sci.* **2008**, *13*, 415–420.
11. Carpita, N.C. Structure and biogenesis of the cell walls of grasses. *Annu. Rev. Plant Physiol. Plant Mol. Biol.* **1996**, *47*, 445–476.

12. Grabber, J.H.; Ralph, J.; Lapierre, C.; Barrière, Y. Genetic and molecular basis of grass cell-wall degradability. I. Lignin–cell wall matrix interactions. *C. R. Biol.* **2004**, *327*, 455–465.
13. Sampedro, J.; Cosgrove, D.J. The expansin superfamily. *Genome Biol.* **2005**, *6*, 242.
14. Eklöf, J.M.; Brumer, H. The XTH gene family: An update on enzyme structure, function, and phylogeny in xyloglucan remodeling. *Plant Physiol.* **2010**, *153*, 456–466.
15. Sénéchal, F.; Wattier, C.; Rustérucci, C.; Pelloux, J. Homogalacturonan-modifying enzymes: Structure, expression, and roles in plants. *J. Exp. Bot.* **2014**, *65*, 5125–5160.
16. Liepman, A.H.; Wightman, R.; Geshi, N.; Turner, S.R.; Scheller, H.V. Arabidopsis a powerful model system for plant cell wall research. *Plant J.* **2010**, *61*, 1107–1121.
17. Vanholme, R.; Demedts, B.; Morreel, K.; Ralph, J.; Boerjan, W. Lignin Biosynthesis and Structure. *Plant Physiol.* **2010**, *153*, 895–905.
18. Tronchet, M.; Balagué, C.; Kroj, T.; Jouanin, L.; Roby, D. Cinnamyl alcohol dehydrogenases-C and D, key enzymes in lignin biosynthesis, play an essential role in disease resistance in Arabidopsis. *Mol. Plant Pathol.* **2010**, *11*, 83–92.
19. Hamann, T. Plant cell wall integrity maintenance as an essential component of biotic stress response mechanisms. *Front. Plant Sci.* **2012**, *3*, 77.
20. Seifert, G.J.; Blaukopf, C. Irritable walls: The plant extracellular matrix and signaling. *Plant Physiol.* **2010**, *153*, 467–478.
21. Sasidharan, R.; Voesenek, L.A.; Pierik, R. Cell wall modifying proteins mediate plant acclimatization to biotic and abiotic stresses. *Crit. Rev. Plant Sci.* **2011**, *30*, 548–562.
22. Moore, J.P.; Vicré-Gibouin, M.; Farrant, J.M.; Driouich, A. Adaptations of higher plant cell walls to water loss: Drought *vs.* desiccation. *Physiol. Plant* **2008**, *134*, 237–245.
23. Moura, J.C.M.S.; Bonine, C.A.V.; de Oliveira Fernandes Viana, J.; Dornelas, M.C.; Mazzafera, P. Abiotic and biotic stresses and changes in the lignin content and composition in plants. *J. Integr. Plant Biol.* **2010**, *52*, 360–376.
24. Cabane, M.; Afif, D.; Hawkins, S. Lignins and abiotic stresses. *Adv. Bot. Res.* **2013**, *61*, 220.
25. Zagorchev, L.; Kamenova, P.; Odjakova, M. The role of plant cell wall proteins in response to salt stress. *Sci. World J.* **2014**, Article ID 764089.
26. Zhao, Q.; Zhang, H.; Wang, T.; Chen, S.; Dai, S. Proteomics-based investigation of salt-responsive mechanisms in plant roots. *J. Proteomics* **2013**, *82*, 230–253.
27. Krzeslowska, M. The wall cell in plant cell response to trace metals: Polysaccharide remodelling and its role in defense strategy. *Acta Physiol. Plant.* **2011**, *33*, 35–51.
28. Ovečka, M.; Takáč, T. Managing heavy metal toxicity stress in plants: Biological and biotechnological tools. *Biotechnol. Adv.* **2014**, *32*, 73–86.
29. Chen, X.Y.; Kim, J.Y. Callose synthesis in higher plants. *Plant Signal. Behav.* **2009**, *4*, 489–492.
30. Stass, A.; Horst, W.J. Callose in abiotic stress. In *Chemistry, Biochemistry and Biology of (1→3)-β-Glucans and Related Polysaccharides*; Bacic, A., Fincher, G.B., Stone, B.A., Eds.; Academic: New York, NY, USA, 2009; pp. 499–524.
31. Guzmán, P.; Fernández, V.; Graça, J.; Cabral, V.; Kayali, N.; Khayet, M.; Gil, L. Chemical and structural analysis of *Eucalyptus globulus* and *E. camaldulensis* leaf cuticles: A lipidized cell wall region. *Front. Plant Sci.* **2014**, *5*, doi:10.3389/fpls.2014.00481.

32. Cosgrove, D.J.; Li, Z.C. Role of expansin in cell enlargement of oat coleoptiles (analysis of developmental gradients and photocontrol). *Plant Physiol.* **1993**, *103*, 1321–1328.
33. Tardieu, F.O.; Granier, C.; Muller, B. Water deficit and growth. Co-ordinating processes without an orchestrator? *Curr Opin. Plant Biol.* **2011**, *14*, 283–289.
34. Zlatev, Z.; Cebola Lidon, F. An overview on drought induced changes in plant growth, water relations and photosynthesis. *Emir. J. Food Agric.* **2012**, *24*, 57–72.
35. Shao, H.-B.; Chu, L.-Y.; Jaleel, C.A.; Zhao, C.-X. Water-deficit stress-induced anatomical changes in higher plants. *C. R. Biol.* **2008**, *331*, 215–225.
36. Lisar, S.Y.S.; Motafakkerazad, R.; Hossain, M.M.; Rahman, I.M.M. Water stress in plants: Causes, effects and responses. In *Water Stress*; Rahman, M., Hasegawa, H., Eds.; InTech: Rijeka, Croatia, 2012; pp. 1–14.
37. Tardieu, F. Any trait or trait-related allele can confer drought tolerance: Just design the right drought scenario. *J. Exp. Bot.* **2011**, *63*, 25–31.
38. Wolf, S.; Greiner, S. Growth control by cell wall pectins. *Protoplasma* **2012**, *249*, 169–175.
39. Tardieu, F.; Parent, B.; Caldeira, C.F.; Welcker, C. Genetic and physiological controls of growth under water deficit. *Plant Physiol.* **2014**, *164*, 1628–1635.
40. Balducci, L.; Deslauriers, A.; Giovannelli, A.; Beaulieu, M.; Delzon, S.; Rossi, S.; Rathgeber, C.B.K. How do drought and warming influence survival and wood traits of *Picea mariana* saplings? *J. Exp. Bot.* **2014**, *4*, 1–13.
41. Hura, T.; Hura, K.; Dziurka, K.; Ostrowska, A.; Bączek-Kwinta, R.; Grzesiak, M. An increase in the content of cell wall-bound phenolics correlates with the productivity of triticale under soil drought. *J. Plant Physiol.* **2012**, *169*, 1728–1736.
42. Izanloo, A.; Condon, A.G.; Langridge, P.; Tester, M.; Schnurbusch, T. Different mechanisms of adaptation to cyclic water stress in two South Australian bread wheat cultivars. *J. Exp. Bot.* **2008**, *59*, 3327–3346.
43. Beikircher, B.; de Cesare, C.; Mayr1, S. Hydraulics of high-yield orchard trees: A case study of three *Malus domestica* cultivars. *Tree Physiol.* **2013**, 33, 1296–1307.
44. Chartzoulakis, K.; Patakas, A.; Kofidis, G.; Bosabalidis, A.; Nastou, A. Water stress affects leaf anatomy, gas exchange, water relations and growth of two avocado cultivars. *Sci. Hortic.* **2002**, *95*, 39–50.
45. Saito, T.; Terashima, I. Reversible decreases in the bulk elastic modulus of mature leaves of deciduous *Quercus* species subjected to two drought treatments. *Plant Cell Environ.* **2004**, *27*, 863–875.
46. De Diego, N.; Sampedro, M.C.; Barrio, R.J.; Saiz-Fernandez, I.; Moncalean, P.; Lacuesta, M. Solute accumulation and elastic modulus changes in six radiata pine breeds exposed to drought. *Tree Physiol.* **2013**, *33*, 69–80.
47. Martínez, J.P.; Silva, H.; Ledent, J.F.; Pinto, M. Effect of drought stress on the osmotic adjustment, cell wall elasticity and cell volume of six cultivars of common beans (*Phaseolus vulgaris* L.). *Eur. J. Agron.* **2007**, *26*, 30–38.
48. Clifford, S.C.; Arndt, S.K.; Corlett, J.E.; Joshi, S.; Sankhla, N.; Popp, M.; Jones, H.G. The role of solute accumulation, osmotic adjustment and changes in cell wall elasticity in drought tolerance in *Ziziphus mauritiana* (Lamk.). *J. Exp. Bot.* **1998**, *49*, 967–977.

49. Hessini, K.; Martínez, J.P.; Gandour, M.; Albouchi, A.; Soltani, A.; Abdelly, C. Effect of water stress on growth, osmotic adjustment, cell wall elasticity and water-use efficiency in *Spartina alterniflora*. *Environ. Exp. Bot.* **2009**, *67*, 312–319.
50. Iraki, N.M.; Bressan, R.A.; Hasegawa, P.M.; Carpita, N.C. Alteration of the physical and chemical structure of the primary cell wall of growth-limited plant cells adapted to osmotic stress. *Plant Physiol.* **1989**, *91*, 39–47.
51. Sweet, W.J.; Morrison, J.C.; Labavitch, J.M.; Matthews, M.A. Altered synthesis and composition of cell wall of grape (*Vitis vinifera* L.) leaves during expansion and growth inhibiting water deficit. *Plant Cell Physiol.* **1990**, *31*, 407–414.
52. Piro, G.; Leucci, M.R.; Waldron, K.; Dalessandro, G. Exposure to water stress causes changes in the biosynthesis of cell wall polysaccharides in roots of wheat cultivars varying in drought tolerance. *Plant Sci.* **2003**, *165*, 559–569.
53. Bray, E.A. Genes commonly regulated by water-deficit stress in *Arabidopsis thaliana*. *J. Exp. Bot.* **2004**, *55*, 2331–2341.
54. Zheng, M.; Meng, Y.; Yang, C.; Zhou, Z.; Wang, Y.; Chen, B. Protein expression changes during cotton fiber elongation in response to drought stress and recovery. *Proteomics* **2014**, *14*, 1776–1795.
55. Ricardi, M.M.; Gonzalez, R.M.; Zhong, S.; Dominguez, P.G.; Duffy, T.; Turjanski, P.G.; Salter, J.D.S.; Alleva, K.; Carrari, F.; Giovannoni, J.J. Genome-wide data (ChIP-seq) enabled identification of cell wall-related and aquaporin genes as targets of tomato ASR1, a drought stress-responsive transcription factor. *BMC Plant Biol.* **2014**, *14*, 29.
56. Yang, L.; Wang, C.C.; Guo, W.D.; Li, X.B.; Lu, M.; Yu, C.L. Differential expression of cell wall related genes in the elongation zone of rice roots under water deficit. *Russ. J. Plant Physiol.* **2006**, *53*, 390–395.
57. Cho, S.K.; Kim, J.E.; Park, J.A.; Eom, T.J.; Kim, W.T. Constitutive expression of abiotic stress-inducible hot pepper *CaXTH3*, which encodes a xyloglucan endotransglucosylase/hydrolase homolog, improves drought and salt tolerance in transgenic *Arabidopsis* plants. *FEBS Lett.* **2006**, *580*, 3136–3144.
58. Choi, J.Y.; Seo, Y.S.; Kim, S.J.; Kim, W.T.; Shin, J.S. Constitutive expression of *CaXTH3*, a hot pepper xyloglucan endotransglucosylase/hydrolase, enhanced tolerance to salt and drought stresses without phenotypic defects in tomato plants (*Solanum lycopersicum* cv. Dotaerang). *Plant Cell Rep.* **2011**, *30*, 867–877.
59. Zhu, J.; Alvarez, S.; Marsh, E.L.; LeNoble, M.E.; Cho, I.J.; Sivaguru, M.; Chen, S.; Nguyen, H.T.; Wu, Y.; Schachtman, D.P.; *et al*. Cell wall proteome in the maize primary root elongation zone. II. Region-specific changes in water soluble and lightly ionically bound proteins under water deficit. *Plant Physiol.* **2007**, *145*, 1533–1548.
60. Jones, L.; McQueen-Mason, S. A role for expansins in dehydration and rehydration of the resurrection plant *Craterostigma plantagineum*. *FEBS Lett.* **2004**, *559*, 61–65.
61. Wu, Y.; Sharp, R.E.; Durachko, D.M.; Cosgrove, D.J. Growth maintenance of the maize primary root at low water potentials involves increases in cell-wall extension properties, expansin activity, and wall susceptibility to expansins. *Plant Physiol.* **1996**, *111*, 765–772.

62. Wu, Y.; Thorne, E.T ; Sharp, R.E.; Cosgrove, D.J. Modification of expansin transcript levels in the maize primary root at low water potentials. *Plant Physiol.* **2001**, *126*, 1471–1479.
63. Wu, Y.; Cosgrove, D.J. Adaptation of roots to low water potentials by changes in cell wall extensibility and cell wall proteins. *J. Exp. Bot.* **2000**, *51*, 1543–1553.
64. Guo, W.; Zhao, J.; Li, X.; Qin, L.; Yan, X.; Liao, H. A soybean β-expansin gene *GmEXPB2* intrinsically involved in root system architecture responses to abiotic stresses. *Plant J.* **2011**, *66*, 541–552.
65. Li, F.; Han, Y.; Feng, Y.; Xing, S.; Zhao, M.; Chen, Y.; Wang, W. Expression of wheat expansin driven by the *RD29* promoter in tobacco confers water-stress tolerance without impacting growth and development. *J. Biotechnol.* **2013**, *163*, 281–291.
66. Dai, F.; Zhang, C.; Jiang, X.; Kang, M.; Yin, X.; Lu, P.; Zhang, X.; Zheng, Y.; Gao, J. *RhNAC2* and *RhEXPA4* are involved in the regulation of dehydration tolerance during the expansion of rose petals. *Plant Physiol* **2012**, *160*, 2064–2082.
67. Lü, P.; Kang, M.; Jiang, X.; Dai, F.; Gao, J.; Zhang, C. *RhEXPA4*, a rose expansin gene, modulates leaf growth and confers drought and salt tolerance to Arabidopsis. *Planta* **2013**, *237*, 1547–1559.
68. Wakabayashi, K.; Hoson, T.; Kamisaka, S. Changes in amounts and molecular mass distribution of cell-wall polysaccharides of wheat (*Triticum aestivum* L.) coleoptiles under water. *J. Plant Physiol.* **1997**, *151*, 33–40.
69. Peaucelle, A.; Braybrook, S.; Höfte, H. Cell wall mechanics and growth control in plants: The role of pectins revisited. *Front Plant Sci.* **2012**, *3*, 1–6.
70. Leucci, M.R.; Lenucci, M.S.; Piro, G.; Dalessandro, G. Water stress and cell wall polysaccharides in the apical root zone of wheat cultivars varying in drought tolerance. *J. Plant Physiol.* **2008**, *165*, 1168–1180.
71. Gribaa, A.; Dardelle, F.; Lehner, A.; Rihouey, C.; Burel, C.; Ferchichi, A.; Driouich, A.; Mollet, J.C. Effect of water deficit on the cell wall of the date palm (*Phoenix dactylifera* "Deglet nour", Arecales) fruit during development. *Plant Cell Environ.* **2013**, *36*, 1056–1070.
72. Moore, J.P.; Nguema-Ona, E.; Fagerström, A.; Fangel, J.U.; Willats, W.G.T.; Hugo, A.; Vivier, M.A. Profiling the main cell wall polysaccharides of grapevine leaves using high throughput methods. *Carbohydr. Polym.* **2014**, *99*, 190–198.
73. Pandey, A.; Rajamani, U.; Verma, J.; Subba, P.; Chakraborty, N.; Datta, A.; Chakraborty, S.; Chakraborty, N. Identification of extracellular matrix proteins of rice (*Oryza sativa* L.) involved in dehydration-responsive network: A proteomic approach. *J. Proteome Res.* **2010**, *9*, 3443–3464.
74. Atkinson, R.G.; Schröder, R.; Hallett, I.C.; Cohen, D.; MacRae, E. Overexpression of polygalacturonase in transgenic apple trees leads to a range of novel phenotypes involving changes in cell adhesion. *Plant Physiol.* **2002**, *129*, 122–133.
75. Tucker, G.A.; Robertson, N.G.; Grierson, D. The conversion of tomato fruit polygalacturonase isoenzyme 2 into isoenzyme 1 *in vitro*. *Eur. J. Biochem.* **1981**, *115*, 87–90.
76. Liu, H.; Ma, Y.; Chen, N.A.; Guo, S.; Liu, H.; Guo, X.; Chong, K.; Xu, Y. Overexpression of stress-inducible *OsBURP16*, the β subunit of polygalacturonase 1, decreases pectin content

and cell adhesion and increases abiotic stress sensitivity in rice. *Plant Cell Environ.* **2013**, *37*, 1144–1158.

77. An, S.H.; Sohn, K.H.; Choi, H.W.; Hwang, I.S.; Lee, S.C.; Hwang, B.K. Pepper pectin methylesterase inhibitor protein CaPMEI1 is required for antifungal activity, basal disease resistance and abiotic stress tolerance. *Planta* **2008**, *228*, 61–78.
78. Wang, Y.; Chantreau, M.; Sibout, R.; Hawkins, S. Plant cell wall lignification and monolignol metabolism. *Front. Plant Sci.* **2013**, *4*, 220.
79. Terzi, R.; Güler, N.S.; Çaliskan, N.; Kadioglu, A. Lignification response for rolled leaves of *Ctenanthe setosa* under long-term drought stress. *Turk. J. Biol.* **2013**, *37*, 614–619.
80. Yoshimura, K.; Masuda, A.; Kuwano, M.; Yokota, A.; Akashi, K. Programmed proteome response for drought avoidance/tolerance in the root of a C3 xerophyte (wild watermelon) under water deficits. *Plant Cell Physiol.* **2007**, *49*, 226–241.
81. Fan, L. Progressive inhibition by water deficit of cell wall extensibility and growth along the elongation zone of maize roots is related to increased lignin metabolism and progressive stelar accumulation of wall phenolics. *Plant Physiol.* **2006**, *140*, 603–612.
82. Moura-Sobczak, J.; Souza, U.; Mazzafera, P. Drought stress and changes in the lignin content and composition in Eucalyptus. *BMC Proc.* **2011**, *5*, 103.
83. Lin, C.C.; Kao, C.H. Osmotic stress-induced changes in cell wall peroxidase activity and hydrogen peroxide level in roots of rice seedlings. *Plant Growth Regul.* **2002**, *37*, 177–184.
84. Lee, B.R.; Kim, K.Y.; Jung, W.J.; Avice, J.C.; Ourry, A.; Kim, T.H. Peroxidases and lignification in relation to the intensity of water-deficit stress in white clover (*Trifolium repens* L.). *J. Exp. Bot.* **2007**, *58*, 1271–1279.
85. Bacon, M.A.; Thompson, D.S.; Davies, W.J. Can cell wall peroxidase activity explain the leaf growth response of *Lolium temulentum* L. during drought? *J. Exp. Bot.* **1997**, *48*, 2075–2085.
86. Hura, T.; Hura, K.; Ostrowska, A.; Grzesiak, M.; Dziurka, K. The cell wall-bound phenolics as a biochemical indicator of soil drought resistance in winter triticale. *Plant Soil Environ.* **2013**, *59*, 189–195.
87. Christianson, J.A.; Llewellyn, D.J.; Dennis. E.S.; Wilson, I.W. Globalgene expression responses to waterlogging in roots and leaves of cotton (*Gossypium hirsutum* L.). *Plant Cell Physiol.* **2010**, *51*, 21–37.
88. Kong, F.J.; Oyanagi, A.; Komatsu, S. Cell wall proteome of wheat roots under flooding stress using gel-based and LC MS/MS-based proteomics approaches. *Biochim. Biophys.* **2010**, *1804*, 124–136.
89. Salavati, A.; Khatoon, A.; Nanjo, Y.; Komatsu, S. Analysis of proteomic changes in roots of soybean seedlings during recovery after flooding. *J. Proteomics* **2012**, *75*, 878–893.
90. Nanjo, Y.; Nakamura, T.; Komatsu, S. Identification of indicator proteins associated with flooding injury in soybean seedlings using label-free quantitative proteomics. *J. Proteome Res.* **2013**, *12*, 4758–4798.
91. Rauf, M.; Arif, M.; Fisahn, J.; Xue, G.P.; Balazadeh, S.; Mueller-Roeber, B. NAC transcription factor SPEEDY HYPONASTIC GROWTH regulates flooding-induced leaf movement in *Arabidopsis*. *Plant Cell.* **2013**, *25*, 4941–4955.

92. Vitorino, P.G.; Alves, J.D.; Magalhães, P.C.; Magalhães, M.M.; Lima, L.C.O.; Oliveira, L.E.M. Flooding tolerance and cell wall alterations in maize mesocotyl during hypoxia. *Pesq. Agropec. Bras.* **2001**, *36*, 1027–1035.
93. Ooume, K.; Inoue, Y.; Soga, K.; Wakabayashi, K.; Fujii, S.; Yamamoto, R.; Hoson, T. Cellular basis of growth suppression by submergence in azuki bean epicotyls. *Ann. Bot.* **2009**, *103*, 325–332.
94. Da Silva Ferreira, C.S.; Piedade, M.T.F.; Tine, M.A.S.; Rossatto, D.R.; Parolin, P.; Buckeridge, M.S. The role of carbohydrates in seed germination and seedling establishment of *Himatanthus sucuuba*, an Amazonian tree with populations adapted to flooded and non-flooded conditions. *Ann. Bot.* **2009**, *104*, 1111–1119.
95. Nanjo, Y.; Maruyama, K.; Yasue, H.; Yamaguchi-Shinozaki, K.; Shinozaki, K.; Komatsu, S. Transcriptional responses to flooding stress in roots including hypocotyl of soybean seedlings. *Plant Mol. Biol.* **2011**, *77*, 129–144.
96. Kreuzwieser, J.; Hauberg, J.; Howell, K.A.; Carroll, A.; Rennenberg, H.; Millar, A.H.; Whelan, J. Differential response of gray poplar leaves and roots underpins stress adaptation during hypoxia. *Plant Physiol.* **2009**, *149*, 461–473.
97. Komatsu, S.; Kobayashi, Y.; Nishizawa, K.; Nanjo, Y.; Furukawa, K. Comparative proteomics analysis of differentially expressed proteins in soybean cell wall during flooding stress. *Amino Acids* **2010**, *39*, 1435–1449.
98. Hall, A.E. Breeding for heat tolerance. *Plant Breed. Rev.* **1992**, *10*, 129–168.
99. Wahid, A.; Gelani, S.; Ashraf, M.; Foolad, M.R. Heat tolerance in plants: An overview. *Environ. Exp. Bot.* **2007**, *61*, 199–223.
100. Dolferus, R.; Ji, X.; Richards, R.A. Abiotic stress and control of grain number in cereals. *Plant Sci.* **2011**, *181*, 331–341.
101. Crafts-Brandner, S.J. Sensitivity of Photosynthesis in a C4 Plant, Maize, to Heat Stress. *Plant Physiol.* **2002**, *129*, 1773–1780.
102. Larkindale, J.; Knight, M.R. Protection against Heat Stress-Induced Oxidative Damage in Arabidopsis Involves Calcium, Abscisic Acid, Ethylene, and Salicylic Acid. *Plant Physiol.* **2002**, *128*, 682–695.
103. Kotak, S.; Larkindale, J.; Lee, U.; von Koskull-Döring, P.; Vierling, E.; Scharf, K.-D. Complexity of the heat stress response in plants. *Curr. Opin. Plant Biol.* **2007**, *10*, 310–316.
104. Hossain, A.; Sarker, M.; Saifuzzaman, M.; Teixeira da Silva, J.A.; Lozovskaya, M.V.; Akhter, M.M. Evaluation of growth, yield, relative performance and heat susceptibility of eight wheat (*Triticum aestivum* L.) genotypes grown under heat stress. *Int. J. Plant Prod.* **2013**, *7*, 615–636.
105. Liu, X.; Huang, B. Root physiological factors involved in creeping bentgrass response to high soil temperatures. *Environ. Exp. Bot.* **2005**, *53*, 233–245.
106. Wang, W.; Vinocur, B.; Altman, A. Plant responses to drought, salinity and extreme temperatures: Towards genetic engineering for stress tolerance. *Planta* **2003**, *218*, 1–14.
107. Suwa, R.; Hakata, H.; Hara, H.; El-Shemy, H.A.; Adu-Gyamfi, J.J.; Nguyen, N.T.; Kanai, S.; Lightfoot, D.A.; Mohapatra, P.K.; Fujita, K. High temperature effects on photosynthate

partitioning and sugar metabolism during ear expansion in maize (*Zea mays* L.) genotypes. *Plant Physiol. Biochem.* **2010**, *48*, 124–130.

108. Lima, R.B.; Santos, dos, T.B.; Vieira, L.G.E.; de Lourdes Lúcio Ferrarese, M.; Ferrarese-Filho, O.; Donatti, L.; Boeger, M.R.T.; de Oliveira Petkowicz, C.L. Heat stress causes alterations in the cell-wall polymers and anatomy of coffee leaves (*Coffea arabica* L.). *Carbohydr. Polym.* **2013**, *93*, 135–143.
109. Guerra-Guimarães, L.; Vieira, A.; Chaves, I.; Pinheiro, C.; Queiroz, V.; Renaut, J.; Ricardo, C.P. Effect of greenhouse conditions on the leaf apoplastic proteome of *Coffea arabica* plants. *J. Proteomics* **2014**, *104*, 128–139.
110. Rakszegi, M.; Lovegrove, A.; Balla, K.; Láng, L.; Bedő, Z.; Veisz, O.; Shewry, P.R. Effect of heat and drought stress on the structure and composition of arabinoxylan and β-glucan in wheat grain. *Carbohydr. Polym.* **2014**, *102*, 557–565.
111. Yang, K.A.; Lim, C.J.; Hong, J.K.; Park, C.Y.; Cheong, Y.H.; Chung, W.S.; Lee, K.O.; Lee, S.Y.; Cho, M.J.; Lim, C.O. Identification of cell wall genes modified by a permissive high temperature in Chinese cabbage. *Plant Sci.* **2006**, *171*, 175–182.
112. Rienth, M.; Torregrosa, L.; Luchaire, N.; Chatbanyong, R.; Lecourieux, D.; Kelly, M.T.; Romieu, C. Day and night heat stress trigger different transcriptomic responses in green and ripening grapevine (*Vitis vinifera*) fruit. *BMC Plant Biol.* **2013**, *14*, 108–108.
113. Xu, J.; Belanger, F.; Huang, B. Differential gene expression in shoots and roots under heat stress for a geothermal and non-thermal *Agrostis* grass species contrasting in heat tolerance. *Environ. Exp. Bot.* **2008**, *63*, 240–247.
114. Xu, Q.; Xu, X.; Shi, Y.; Xu, J.; Huang, B. Transgenic Tobacco Plants Overexpressing a Grass *PpEXP1* Gene Exhibit Enhanced Tolerance to Heat Stress. *PLoS One* **2014**, *9*, e100792.
115. Yu, E.; Fan, C.; Yang, Q.; Li, X.; Wan, B.; Dong, Y.; Wang, X.; Zhou, Y. Identification of Heat Responsive Genes in *Brassica napus* Siliques at the Seed-Filling Stage through Transcriptional Profiling. *PLoS One* **2014**, *9*, e101914.
116. Song, Y.; Ci, D.; Tian, M.; Zhang, D. Comparison of the physiological effects and transcriptome responses of *Populus simonii* under different abiotic stresses. *Plant Mol. Biol.* **2014**, *86*, 139–156.
117. Han, F.; Chen, H.; Li, X.J.; Yang, M.F.; Liu, G.S.; Shen, S.H. A comparative proteomic analysis of rice seedlings under various high-temperature stresses. *Biochim. Biophys. Acta* **2009**, *1794*, 1625–1634.
118. Gulen, H.; Eris, A. Effect of heat stress on peroxidase activity and total protein content in strawberry plants. *Plant Sci.* **2004**, *166*, 739–744.
119. Stitt, M.; Hurry, V. A plant for all seasons: Alterations in photosynthetic carbon metabolism during cold acclimation in *Arabidopsis*. *Curr. Opin. Plant Biol.* **2002**, *5*, 199–206.
120. Goulas, E.; Schubert, M.; Kieselbach, T.; Kleczkowski, L.A.; Gardeström, P.; Schröder, W.; Hurry, V. The chloroplast lumen and stromal proteomes of *Arabidopsis thaliana* show differential sensitivity to short- and long-term exposure to low temperature. *Plant J.* **2006**, *47*, 720–734.
121. Chinnusamy, V.; Zhu, J.; Zhu, J.K. Cold stress regulation of gene expression in plants. *Trends Plant Sci.* **2007**, *12*, 444–451.

122. Chang, M.-C.; Chien, W.-F.; Chao, C.-H.; Lu, M.-K. Effects of cold stress on alterations of physiochemical and chemical properties of rice polysaccharides. *Carbohydr. Polym.* **2010**, *80*, 373–376.
123. Mahajan, S.; Tuteja, N. Cold, salinity and drought stresses: An overview. *Arch. Biochem. Biophys.* **2005**, *444*, 139–158.
124. Sanghera, G.S.; Wani, S.H.; Hussain, W.; Singh, N.B. Engineering cold stress tolerance in crop plants. *Curr. Genomics* **2011**, *12*, 30–43.
125. Noh, S.A.; Park, S.H.; Huh, G.H.; Paek, K.-H.; Shin, J.S.; Bae, J.M. Growth retardation and differential regulation of expansin genes in chilling-stressed sweet potato. *Plant Biotechnol. Rep.* **2009**, *3*, 75–85.
126. Stefanowska, M.; Kuras, M.; Kubacka-Zębalska, M.; Kacperska, A. Low Temperature Affects Pattern of Leaf Growth and Structure of Cell Walls in Winter Oilseed Rape (*Brassica napus* L., var. *oleifera* L.). *Ann. Bot.* **1999**, *84*, 313–319.
127. Pearce, R.S. Molecular analysis of acclimation to cold. *Plant Growth Regul.* **1999**, *29*, 47–76.
128. Warren, G.J. Cold stress: Manipulating freezing tolerance in plants. *Curr. Biol.* **1998**, *8*, R514–R516.
129. Janská, A.; Maršík, P.; Zelenková, S.; Ovesná, J. Cold stress and acclimation - what is important for metabolic adjustment? *Plant Biol.* **2010**, *12*, 395–405.
130. Hajela, R.K.; Horvath, D.P.; Gilmour, S.J.; Thomashow, M.F. Molecular cloning and expression of cor (cold regulated) genes in *Arabidopsis thaliana*. *Plant Physiol.* **1990**, *93*, 1246–1252.
131. Kreps, J.A.; Wu, Y.; Chang, H.S.; Zhu, T.; Wang, X.; Harper, J.F. Transcriptome changes for Arabidopsis in response to salt, osmotic, and cold stress. *Plant Physiol.* **2002**, *130*, 2129–2141.
132. Zhu, J.; Dong, C.H.; Zhu, J.K. Interplay between cold-responsive gene regulation, metabolism and RNA processing during plant cold acclimation. *Curr. Opin. Plant Biol.* **2007**, *10*, 290–295.
133. Chen, L.; Zhong, H.; Ren, F.; Guo, Q.-Q.; Hu, X.-P.; Li, X.-B. A novel cold-regulated gene, *COR25*, of *Brassica napus* is involved in plant response and tolerance to cold stress. *Plant Cell. Rep.* **2011**, *30*, 463–471.
134. Janska, A.; Marsik, P.; Zelenkova, S.; Ovesna, J. Cold stress and acclimation-what is important for metabolic adjustment? *Plant Biol.* **2010**, *12*, 395–405.
135. Levitt, J. Chilling, freezing, and high temperature stresses. In *Responses of Plants to Environmental Stresses*, 2nd ed.; Academic Press: Orlando, FL, USA, 1980.
136. Kume, S.; Kobayashi, F.; Ishibashi, M.; Ohno, R.; Nakamura, C.; Takumi, S. Differential and coordinated expression of Cbf and Cor/Lea genes during long-term cold acclimation in two wheat cultivars showing distinct levels of freezing tolerance. *Genes Genet. Syst.* **2005**, *80*, 185–197.
137. Monroy, A.F.; Dryanova, A.; Malette, B.; Oren, D.H.; Farajalla, M.R.; Liu, W.; Danyluk, J.; Ubayasena, L.W.C.; Kane, K.; Scoles, G.J.; *et al.* Regulatory gene candidates and gene expression analysis of cold acclimation in winter and spring wheat. *Plant Mol. Biol.* **2007**, *64*, 409–423.
138. Winfield, M.O.; Lu, C.; Wilson, I.D.; Coghill, J.A.; Edwards, K.J. Plant responses to cold: Transcriptome analysis of wheat. *Plant Biotechnol. J.* **2010**, *8*, 749–771.
139. Fujikawa, S.; Kuroda, K. Cryo-scanning electron microscopic study on freezing behavior of xylem ray parenchyma cells in hardwood species. *Micron* **2000**, *31*, 669–686.
140. Pearce, R.S.; Fuller, M.P. Freezing of barley studied by infrared video thermography. *Plant Physiol.* **2001**, *125*, 227–240.

141. Yamada, T.; Kuroda, K.; Jitsuyama, Y.; Takezawa, D.; Arakawa, K.; Fujikawa, S. Roles of the plasma membrane and the cell wall in the responses of plant cells to freezing. *Planta* **2002**, *215*, 770–778.
142. Solecka, D.; Zebrowski, J.; Kacperska, A. Are involved in cold acclimation and de-acclimation of winter oil-seed rape plants? *Ann. Bot*. **2008**, *101*,521–530.
143. Rajashekar, C.B.; Lafta, A. Cell-wall changes and cell tension in response to cold acclimation and exogenous abscisic acid in leaves and cell cultures. *Plant Physiol.* **1996**, *111*, 605–612.
144. Kubacka-Zebalska, M.; Kacperska, A. Low temperature-induced modifications of cell wall content and polysaccharide composition in leaves of winter oilseed rape (*Brassica napus* L. var. *oleifera* L.). *Plant Sci.* **1999**, *148*, 59–67.
145. Weiser, R.L.; Wallner, S.J.; Waddell, J.W. Cell wall and extensin mRNA changes during cold acclimation of pea seedlings. *Plant Physiol.* **1990**, *93*, 1021–1026.
146. Carvalho, C.P.; Hayashi, A.H.; Braga, M.R.; Nievola, C.C. Biochemical and anatomical responses related to the *in vitro* survival of the tropical bromeliad *Nidularium minutum* to low temperatures. *Plant Physiol. Biochem.* **2013**, *71*, 144–154.
147. Baldwin, L.; Domon, J.M.; Klimek, J.F.; Fournet, F.; Sellier, H.; Gillet, F.; Pelloux, J.; Lejeune-Hénaut, I.; Carpita, N.C.; Rayon, C. Structural alteration of cell wall pectins accompanies pea development in response to cold. *Phytochemistry* **2014**, *104*, 37–47.
148. Thonar, C.; Liners, F.; van Cutsem, P. Polymorphism and modulation of cell wall esterase enzyme activities in the chicory root during the growing season. *J. Exp. Bot.* **2006**, *57*, 81–89.
149. Lee, J.Y. Use of serial analysis of gene expression technology to reveal changes in gene expression in arabidopsis pollen undergoing cold stress. *Plant Physiol.* **2003**, *132*, 517–529.
150. He, Z.H.; Fujiki, M.; Kohorn, B.D. A cell wall-associated, receptor-like protein kinase. *J. Biol. Chem.* **1996**, *271*, 19789–19793.
151. He, Z.H.; Cheeseman, I.; He, D.; Kohorn, B.D. A cluster of five cell wall-associated receptor kinase genes, Wak1–5, are expressed in specific organs of *Arabidopsis*. *Plant Mol. Biol.* **1999**, *39*, 1189–1196.
152. Meier, S.; Ruzvidzo, O.; Morse, M.; Donaldson, L.; Kwezi, L.; Gehring, C. The *Arabidopsis* wall associated kinase-like 10 gene encodes a functional guanylyl cyclase and is co-expressed with pathogen defense related genes. *PLoS One* **2010**, *5*, e8904.
153. Kohorn, B.D.; Kohorn, S.L. The cell wall-associated kinases, WAKs, as pectin receptors. *Front. Plant Sci.* **2012**, *3*, 88–93.
154. Zhang, F.; Huang, L.; Wang, W.; Zhao, X.; Zhu, L.; Fu, B.; Li, Z. Genome-wide gene expression profiling of introgressed indica rice alleles associated with seedling cold tolerance improvement in a japonica rice background. *BMC Genomics* **2012**, *13*, 461.
155. Zabotin, A.I.; Barisheva, T.S.; Zabotina, O.A.; Larskaya, I.A.; Lozovaya, V.V.; Beldman, G.; Voragen, A.G.J. Alterations in cell walls of winter wheat roots during low temperature acclimation. *J. Plant Physiol*. **1998**, *152*, 473–479.
156. Ko, J.H.; Prassinos, C.; Keathley, D.; Han, K.H. Novel aspects of transcriptional regulation in the winter survival and maintenance mechanism of poplar. *Tree Physiol.* **2011**, *31*, 208–225.

157. Domon, J.-M.; Baldwin, L.; Acket, S.; Caudeville, E.; Arnoult, S.; Zub, H.; Gillet, F.; Lejeune-Hénaut, I.; Brancourt-Hulmel, M.; Pelloux, J.; *et al*. Cell wall compositional modifications of *Miscanthus* ecotypes in response to cold acclimation. *Phytochemistry* **2013**, *85*, 51–61.
158. Zabotin, A.I.; Barisheva, T.S.; Trofimova, O.I.; Toroschina, T.E.; Larskaya, I.A.; Zabotina, O.A. Oligosaccharin and ABA synergistically affect the acquisition of freezing tolerance in winter wheat. *Plant Physiol. Biochem.* **2009**, *47*, 854–858.
159. Imin, N.; Kerim, T.; Rolfe, B.G.; Weinman, J.J. Effect of early cold stress on the maturation of rice anthers. *Proteomics* **2004**, *4*, 1873–1882.
160. Yamauchi, Y.; Ogawa, M.; Kuwahara, A.; Hanada, A.; Kamiya, Y.; Yamaguchi, S. Activation of gibberellin biosynthesis and response pathways by low temperature during imbibition of *Arabidopsis thaliana* seeds. *Plant Cell* **2004**, *16*, 367–378.
161. Seki, M.; Narusaka, M.; Ishida, J.; Nanjo, T.; Fujita, M.; Oono, Y.; Kamiya, A.; Nakajima, M.; Enju, A.; Sakurai, T. Monitoring the expression profiles of 7000 *Arabidopsis* genes under drought, cold and high-salinity stresses using a full-length cDNA microarray. *Plant J.* **2002**, *31*, 279–292.
162. Shi, H.; Ye, T.; Zhong, B.; Liu, X.; Jin, R.; Chan, Z. AtHAP5A modulates freezing stress resistance in Arabidopsis through binding to CCAAT motif of AtXTH21. *New Phytol.* **2014**, *203*, 554–567.
163. Dong, J.; Jiang, Y.; Chen, R.; Xu, Z.; Gao, X. Isolation of a novel xyloglucan endotransglucosylase (OsXET9) gene from rice and analysis of the response of this gene to abiotic stresses. *Afr. J. Biotechnol.* **2011**, *10*, 1–11.
164. Zhu, Y.-N.; Shi, D.-Q.; Ruan, M.-B.; Zhang, L.-L.; Meng, Z.-H.; Liu, J.; Yang, W.-C. Transcriptome analysis reveals crosstalk of responsive genes to multiple abiotic stresses in cotton (*Gossypium hirsutum* L.). *PLoS One* **2013**, *8*, e80218.
165. Evers, D.; Legay, S.; Lamoureux, D.; Hausman, J.F.; Hoffmann, L.; Renaut, J. Towards a synthetic view of potato cold and salt stress response by transcriptomic and proteomic analyses. *Plant Mol. Biol.* **2012**, *78*, 503–514.
166. Meng, M.; Geisler, M.; Johansson, H.; Mellerowicz, E.J.; Karpinski, S.; Kleczkowski, L.A. Differential tissue/organ-dependent expression of two sucrose- and cold-responsive genes for UDP-glucose pyrophosphorylase in *Populus*. *Gene* **2007**, *389*, 186–195.
167. Gong, S.Y.; Huang, G.Q.; Sun, X.; Li, P.; Zhao, L.L.; Zhang, D.J.; Li, X.B. GhAGP31, a cotton non-classical arabinogalactan protein, is involved in response to cold stress during early seedling development. *Plant Biol.* **2011**, *14*, 447–457.
168. Solecka, D.; Kacperska, A. Activity of L-phenylalanine ammonia-lyase in winter rape leaves treated with cold. *Acta Biochim. Pol.* **1993**, *40*, 113–118.
169. Solecka, D.; Boudet, A.M.; Kacperska, A. Phenylpropanoid and anthocyanin changes in low-temperature treated winter oilseed rape leaves. *Plant Physiol. Biochem.* **1999**, *37*, 491–496.
170. Solecka, D.; Kacperska, A. Phenylpropanoid deficiency affects the course of plant acclimation to cold. *Physiol. Plant* **2003**, *119*, 253–262.

171. Stefanowska, M.; Kuras, M.; Kacperska, A. Low temperature-induced modifications in cell ultrastructure and localization of phenolics in winter oilseed rape (*Brassica napus* L. var. *oleifera* L.) leaves. *Ann. Bot.* **2002**, *90*, 637–645.
172. Kim, Y.-H.; Jeong, J.C.; Lee, H.-S.; Kwak, S.-S. Comparative characterization of sweet potato antioxidant genes from expressed sequence tags of dehydration-treated fibrous roots under different abiotic stress conditions. *Mol. Biol. Rep.* **2012**, *40*, 2887–2896.
173. Badowiec, A.; Swigonska, S.; Weidner, S. Changes in the protein patterns in pea (*Pisum sativum* L.) roots under the influence of long- and short-term chilling stress and post-stress recovery. *Plant Physiol. Biochem.* **2013**, *71*, 315–324.
174. Badowiec, A.; Weidner, S. Proteomic changes in the roots of germinating *Phaseolus vulgaris* seeds in response to chilling stress and post-stress recovery. *J. Plant Physiol.* **2014**, *171*, 389–398.
175. Munns, R.; Tester, M. Mechanisms of salinity tolerance. *Annu. Rev. Plant Biol.* **2008**, *59*, 651–681.
176. Szabolcs, I. *Salt-Affected Soils*; CRC Press: Boca Raton, FL, USA, 1998; p. 274.
177. Munns, R. Genes and salt tolerance: Bringing them together. *New Phytol.* **2005**, *167*, 645–663.
178. Verslues, P.E.; Agarwal, M.; Katiyar-Agarwal, S.; Zhu, J.; Zhu, J.-K. Methods and concepts in quantifying resistance to drought, salt and freezing, abiotic stresses that affect plant water status. *Plant J.* **2006**, *45*, 523–539.
179. Chinnusamy, V.; Zhu, J.; Zhu, J.-K. Salt stress signaling and mechanisms of plant salt tolerance. *Genet. Eng. (NY)* **2006**, *27*, 141–177.
180. Uddin, M.N.; Hanstein, S.; Faust, F.; Eitenmüller, P.T.; Pitann, B.; Schubert, S. Diferulic acids in the cell wall may contribute to the suppression of shoot growth in the first phase of salt stress in maize. *Phytochemistry* **2014**, *102*, 126–136.
181. De Lima, R.B.; dos Santos, T.B.; Vieira, L.G.E.; de Lourdes Lúcio Ferrarese, M.; Ferrarese-Filho, O.; Donatti, L.; Boeger, M.R.T.; de Oliveira Petkowicz, C.L. Salt stress alters the cell wall polysaccharides and anatomy of coffee (*Coffea arabica* L.) leaf cells. *Carbohydr. Polym.* **2014**, *112*, 686–694.
182. Neves, G.Y.S.; Marchiosi, R.; Ferrarese, M.L.L.; Siqueira-Soares, R.C.; Ferrarese-Filho, O. Root growth inhibition and lignification induced by salt stress in soybean. *J. Agron. Crop. Sci.* **2010**, *196*, 467–473.
183. Horie, T.; Karahara, I.; Katsuhara, M. Salinity tolerance mechanisms in glycophytes: An overview with the central focus on rice plants. *Rice* **2012**, *5*, 1–18.
184. Shabala, S.; Munns, R. Salinity Stress: Physiological constraints and adaptive mechanisms. In *Plant Stress Physiology*; Shabala, S., Ed.; CAB International: Oxford, UK; Oxford, MS, USA, 2012; pp. 59–93.
185. Deinlein, U.; Stephan, A.B.; Horie, T.; Luo, W.; Xu, G.; Schroeder, J.I. Plant salt-tolerance mechanisms. *Trends Plant Sci.* **2014**, *19*, 371–379.
186. Golldack, D.; Li, C.; Mohan, H.; Probst, N. Tolerance to drought and salt stress in plants: Unraveling the signaling networks. *Front. Plant Sci.* **2014**, *5*, 151.
187. Kacperska, A. Sensor types in signal transduction pathways in plant cells responding to abiotic stressors: Do they depend on stress intensity? *Physiol. Plant.* **2004**, *122*, 159–168.

188. Hou, X.; Tong, H.; Selby, J.; Dewitt, J.; Peng, X.; He, Z.-H. Involvement of a cell wall-associated kinase, WAKL4, in *Arabidopsis* mineral responses. *Plant Physiol.* **2005**, *139*, 1704–1716.
189. Lippold, F.; Sanchez, D.H.; Musialak, M.; Schlereth, A.; Scheible, W.-R.; Hincha, D.K.; Udvardi, M.K. *AtMyb41* regulates transcriptional and metabolic responses to osmotic stress in *Arabidopsis*. *Plant Physiol.* **2009**, *149*, 1761–1772.
190. Schmidt, R.; Schippers, J.H.M.; Mieulet, D.; Obata, T.; Fernie, A.R.; Guiderdoni, E.; Mueller-Roeber, B. MULTIPASS, a rice R2R3-type MYB transcription factor, regulates adaptive growth by integrating multiple hormonal pathways. *Plant J.* **2013**, *76*, 258–273.
191. Cominelli, E.; Sala, T.; Calvi, D.; Gusmaroli, G.; Tonelli, C. Over-expression of the *Arabidopsis AtMYB41* gene alters cell expansion and leaf surface permeability. *Plant J.* **2008**, *53*, 53–64.
192. Han, Y.; Li, A.; Li, F.; Zhao, M.; Wang, W. Characterization of a wheat (*Triticum aestivum L.*) expansin gene, *TaEXPB23*, involved in the abiotic stress response and phytohormone regulation. *Plant Physiol. Biochem.* **2012**, *54*, 49–58.
193. Kwon, Y.R.; Lee, H.J.; Kim, K.H.; Hong, S.-W.; Lee, S.J.; Lee, H. Ectopic expression of Expansin3 or Expansinbeta1 causes enhanced hormone and salt stress sensitivity in Arabidopsis. *Biotechnol. Lett.* **2008**, *30*, 1281–1288.
194. Zhang, L.; Zhang, C.; Wu, P.; Chen, Y.; Li, M.; Jiang, H.; Wu, G. Global analysis of gene expression profiles in physic nut (*Jatropha curcas L.*) seedlings exposed to salt stress. *PLoS One* **2014**, *9*, e97878.
195. Kieliszewski, M.J.; Lamport, D.T.A.; Tan, L.; Cannon, M.C. Hydroxyproline-rich glycoproteins: Form and function. In *Annual Plant Reviews*; Wiley-Blackwell: Oxford, UK, 2010; pp. 321–342.
196. Lamport, D.T.A.; Kieliszewski, M.J.; Showalter, A.M. Salt stress upregulates periplasmic arabinogalactan proteins: Using salt stress to analyse AGP function. *New Phytol.* **2006**, *169*, 479–492.
197. Brinker, M.; Brosche, M.; Vinocur, B.; Abo-Ogiala, A.; Fayyaz, P.; Janz, D.; Ottow, E.A.; Cullmann, A.D.; Saborowski, J.; Kangasjarvi, J.; *et al.* Linking the salt transcriptome with physiological responses of a salt-resistant *Populus* species as a strategy to identify genes important for stress acclimation. *Plant Physiol.* **2010**, *154*, 1697–1709.
198. Ma, H.; Zhao, J. Genome-wide identification, classification, and expression analysis of the arabinogalactan protein gene family in rice (*Oryza sativa L.*). *J. Exp. Bot.* **2010**, *61*, 2647–2668.
199. Tseng, I.-C.; Hong, C.-Y.; Yu, S.-M.; Ho, T.-H.D. Abscisic acid- and stress-induced highly proline-rich glycoproteins regulate root growth in rice. *Plant Physiol.* **2013**, *163*, 118–134.
200. Long, R.; Yang, Q.; Kang, J.; Zhang, T.; Wang, H.; Li, M.; Zhang, Z. Overexpression of a novel salt stress-induced glycine-rich protein gene from alfalfa causes salt and ABA sensitivity in *Arabidopsis*. *Plant Cell Rep.* **2013**, *32*, 1289–1298.
201. Zhu, J.; Lee, B.-H.; Dellinger, M.; Cui, X.; Zhang, C.; Wu, S.; Nothnagel, E.A.; Zhu, J.-K. A cellulose synthase-like protein is required for osmotic stress tolerance in Arabidopsis. *Plant J.* **2010**, *63*, 128–140.
202. Kav, N.N.V.; Srivastava, S.; Goonewardene, L.; Blade, S.F. Proteome-level changes in the roots of *Pisum sativum* in response to salinity. *Ann. Appl. Biol.* **2004**, *145*, 217–230.

203. Xu, C.; Sibicky, T.; Huang, B. Protein profile analysis of salt-responsive proteins in leaves and roots in two cultivars of creeping bentgrass differing in salinity tolerance. *Plant Cell Rep.* **2010**, *29*, 595–615.
204. Jiang, Y.; Yang, B.; Harris, N.S.; Deyholos, M.K. Comparative proteomic analysis of NaCl stress-responsive proteins in Arabidopsis roots. *J. Exp. Bot.* **2007**, *58*, 3591–3607.
205. Mustard, J.; Renault, S. Effects of NaCl on water relations and cell wall elasticity and composition of red-osier dogwood (*Cornus stolonifera*) seedlings. *Physiol. Plant.* **2004**, *121*, 265–271.
206. An, P.; Li, X.; Zheng, Y.; Matsuura, A.; Abe, J.; Eneji, A.E.; Tanimoto, E.; Inanaga, S. Effects of NaCl on root growth and cell wall composition of two soya bean cultivars with contrasting salt tolerance. *J. Agro. Crop Sci.* **2014**, *200*, 212–218.
207. Muszyńska, A.; Jarocka, K.; Kurczynska, E.U. Plasma membrane and cell wall properties of an aspen hybrid (*Populus tremula* × *tremuloides*) parenchyma cells under the influence of salt stress. *Acta Physiol. Plant* **2014**, *36*, 1155–1165.
208. Uddin, M.N.; Hanstein, S.; Leubner, R.; Schubert, S. Leaf cell-wall components as influenced in the first phase of salt stress in three maize (*Zea mays* L.) hybrids differing in salt resistance. *J. Agro Crop Sci.* **2013**, *199*, 405–415.
209. Sánchez-Aguayo, I.; Rodríguez-Galán, J.M.; García, R.; Torreblanca, J.; Pardo, J.M. Salt stress enhances xylem development and expression of S-adenosyl-L-methionine synthase in lignifying tissues of tomato plants. *Planta* **2004**, *220*, 278–285.
210. Guo, J.; Wang, M.-H. Characterization of the phenylalanine ammonia-lyase gene (*SlPAL5*) from tomato (*Solanum lycopersicum* L.). *Mol. Biol. Rep.* **2009**, *36*, 1579–1585.
211. Li, X.-J.; Yang, M.-F.; Chen, H.; Qu, L.-Q.; Chen, F.; Shen, S.-H. Abscisic acid pretreatment enhances salt tolerance of rice seedlings: Proteomic evidence. *Biochim. Biophys. Acta* **2010**, *1804*, 929–940.
212. Salekdeh, G.H.; Siopongco, J.; Wade, L.J.; Ghareyazie, B.; Bennett, J. Proteomic analysis of rice leaves during drought stress and recovery. *Proteomics* **2002**, *2*, 1131–1145.
213. Lee, S.; Lee, E.J.; Yang, E.J.; Lee, J.E.; Park, A.R.; Song, W.H.; Park, O.K. Proteomic identification of annexins, calcium-dependent membrane binding proteins that mediate osmotic stress and abscisic acid signal transduction in Arabidopsis. *Plant Cell* **2004**, *16*, 1378–1391.
214. Yeo, A.R.; Flowers, S.A.; Rao, G.; Welfare, K.; Senanayake, N.; Flowers, T.J. Silicon reduces sodium uptake in rice (*Oryza sativa L.*) in saline conditions and this is accounted for by a reduction in the transpirational bypass flow. *Plant Cell Environ.* **1999**, *22*, 559–565.
215. Yang, J.; Yen, H.E. Early salt stress effects on the changes in chemical composition in leaves of ice plant and Arabidopsis. A Fourier Transform Infrared Spectroscopy Study. *Plant Physiol.* **2002**, *130*, 1032–1042.
216. Hayatsu, M.; Suzuki, S.; Hasegawa, A.; Tsuchiya, S.; Sasamoto, H. Effect of NaCl on ionic content and distribution in suspension-cultured cells of the halophyte *Sonneratia alba vs.* the glycophyte *Oryza sativa*. *J. Plant Physiol.* **2014**, *171*, 1385–1391.
217. Chen, J.; Cheng, T.; Wang, P.; Liu, W.; Xiao, J.; Yang, Y.; Hu, X.; Jiang, Z.; Zhang, S.; Shi, J. Salinity-induced changes in protein expression in the halophytic plant *Nitraria sphaerocarpa*. *J. Proteomics* **2012**, *75*, 5226–5243.

218. Fan, P.; Nie, L.; Jiang, P.; Feng, J.; Lv, S.; Chen, X.; Bao, H.; Guo, J.; Tai, F.; Wang, J.; *et al.* Transcriptome Analysis of *Salicornia europaea* under saline conditions revealed the adaptive primary metabolic pathways as early events to facilitate salt adaptation. *PLoS One* **2013**, *8*, e80595.
219. D'Emilio, M.; Caggiano, R.; Macchiato, M.; Ragosta, M.; Sabia, S. Soil heavy metal contamination in an industrial area: Analysis of the data collected during a decade. *Environ. Monit. Asses.* **2012**, *185*, 5951–5964.
220. Kelepertzis, E. Accumulation of heavy metals in agricultural soils of Mediterranean: Insights from Argolida basin, Peloponnese, Greece. *Geoderma* **2014**, *221*, 82–90.
221. Doncheva, S.; Georgieva, K.; Vassileva, V.; Stoyanova, Z.; Popov, N.; Ignatov, G. Effects of succinate on manganese toxicity in pea plants. *J. Plant Nutr.* **2005**, *28*, 47–62.
222. Huang, T.L.; Nguyen, Q.T.T.; Fu, S.F.; Lin, C.Y.; Chen, Y.C.; Huang, H.J. Transcriptomic changes and signalling pathway induced by arsenic stress in rice roots. *Plant Mol. Biol.* **2012**, *80*, 587–608.
223. Feigl, G.; Kumar, D.; Lehotai, N.; Tugyi, N.; Molnor, A.; Ordog, A.; Szepesi, A.; Gemes, K.; Laskay, G.; Erdei, L.; *et al.* Physiological and morphological responses of the root system of Indian mustard (*Brassica jumcea* L. Czern.) and rapeseed (*Brassica napus* L.) to copper stress. *Ecotox. Environ. Safe* **2013**, *91*, 179–189.
224. Appenroth, K.J. What are "heavy metals" in plant sciences? *Acta Physiol. Plant* **2010**, *32*, 615–619.
225. Stohs, S.J.; Bagchi, D. Oxidative mechanisms in the toxicity of metal ions. *Free Radic. Biol. Med.* **1995**, *18*, 321–336.
226. Briat, J.F. Metal-ion-mediated oxidative stress and its control. In *Oxidative Stress in Plants*; Montagu, M., Inzé, D., Eds.; Taylor and Francis: London, UK, 2002; pp. 171–189.
227. Lander, H.M. An essential role for free radicals and derived species in signal transduction. *FASEB J.* **1997**, *11*, 118–124.
228. Opdenakker, K.; Remans, T.; Keunen, E.; Vangronsveld, J.; Cuypers, A. Exposure of *Arabidopsis thaliana* to Cd or Cu excess leads to oxidative stress mediated alterations in MAPKinase transcript levels. *Environ. Exp. Bot.* **2012**, *83*, 53–61.
229. .Krzeslowska, M.; Lenartowska, M.; Mellerowicz, E.J.; Samardakiewicz, S.; Wosny, A. Pectinous cell wall thickenings formation-a response of moss protonema cells to Pb. *Environ. Exp. Bot.* **2009**, *65*, 119–131.
230. Krzeslowska, M.; Lenartowska, M.; Samardakiewicz, S.; Bilski, H.; Wosny, A. Lead deposited in the cell wall of *Funaria hygrometrica* protonema is not stable-a remobilization can occur. *Environ. Pollut.* **2010**, *158*, 325–338.
231. Vuletic, M.; Hadsi-Taskovic Sukalovic, V.; Markovic, K.; Kravic, N.; Vucinic, Z.; Maksimovic, V. Differential response of antioxidative systems of maize roots cell walls to osmotic and heavy metal stress. *Plant Biol.* **2014**, *16*, 88–96.
232. Jiang, X.; Wang, C. Zinc distribution and zinc-binding forms in *Phragmites australis* under zinc pollution. *J. Plant Physiol.* **2008**, *165*, 697–704.
233. Hossain, A.K.M.Z.; Koyama, H.; Hara, T. Growth and cell wall properties of two wheat cultivars differing in their sensitivity to aluminum stress. *J. Plant Physiol.* **2006**, *163*, 39–47.

234. Polec-Powlak, K.; Ruzik, R.; Lipiec, E.; Ciurzynska, M.; Gawronska, H. Investigation of Pb(II) binding to pectin in *Arabidopsis thaliana*. *J. Anal. Atom. Spectrom.* **2007**, *22*, 968–972.
235. Amenos, M.; Corrales, I.; Poschenrieder, C.; Illes, P.; Baluska, F.; Barcelo, J. Different effects of aluminum on the actin cytoskeleton and Brefeldin A-sensitive vesicle recycling in root apex cells of two maize varieties differing in root elongation rate and aluminum tolerance. *Plant Cell Physiol.* **2009**, *50*, 528–540.
236. Horst, W.J.; Wang, Y.; Eticha, D. The role of the root apoplast in Aluminum-induced inhibition of root elongation and in Aluminum resistance of plant. *Rev. Ann. Bot.* **2010**, *106*, 185–197.
237. Vaculík, M.; Lux, A.; Luxová, M.; Tanimoto, E.; Lichtscheidl, I. Silicon mitigates cadmium inhibitory effects in young maize plants. *Environ. Exp. Bot.* **2009**, *67*, 52–58.
238. Gu, J.; Yin, X.; Stomph, T.J.; Wang, H.; Struik, P.C. Physiological basis of genetic variation in leaf photosynthesis among rice (*Oryza sativa* L.) introgression lines under drought and well-watered conditions. *J. Exp. Bot.* **2012**, *63*, 5137–5153.
239. Eticha, D.; Stass, A.; Horst, W.J. Cell-wall pectin and its degree of methylation in the maize root apex: Significance for genotypic differences in aluminium resistance. *Plant Cell Env.* **2005**, *28*, 1410–1420.
240. Baluska, F.; Liners, F.; Hlavacka, A.; Schlicht, M.; van Cutsem, P.; McCurdy, D.W.; Menzel, D. Cell wall pectins and xyloglu-cans are internalised into dividing root cells and accumulate within cell plates during cytokinesis. *Protoplasma* **2005**, *225*, 141–155.
241. Pelloux, J.; Rusterucci, C.; Mellerowicz, E.J. New insight into pectin methylesterase structure and function. *Trends Plant Sci.* **2007**, *12*, 1363–1372.
242. Willats, W.G.T.; Knox, J.P.; Mikkelsen, J.D. Pectin: New insights into an old polymer are starting to gel. *Trends Food Sci. Technol.* **2006**, *17*, 97–104.
243. Dronnet, V.M.; Renard, C.M.G.C.; Axelos, M.A.V.; Thibault, J.F. Heavy metal binding by pectins: Selectivity, quantification and characterization. *Carbohydr. Polym.* **1996**, *30*, 253–263.
244. Paynel, F.; Schaumann, A.; Arkoun, M.; Douchiche, O.; Morvan, C. Temporal regulation of cell wall pectin methylesterase and peroxidase isoforms in cadmium-treated flax hypocotyls. *Ann. Bot.* **2009**, *104*, 1363–1372.
245. Douchiche, O.; Rihouey, C.; Schaumann, A.; Driouich, A.; Morvan, C. Cadmium-induced alterations of the structural features of pectins in flax hypocotyls. *Planta* **2007**, *225*, 1301–1312.
246. Douchiche, O.; Driouich, A.; Morvan, C. Spatial regulation of cell wall structure in response to heavy metal stress: Cadmium-induced alteration of the methyl-esterification pattern of homogalacturonans. *Ann. Bot.* **2010**, *105*, 481–491.
247. Yang, Z-B.; Eticha, D.; Rao, I.M.; Horst, W.J. Alteration of Cell-wall Porosity is Involved in Osmotic Stress-induced Enhancement of Aluminium Resistance in Common Bean (*Phaseolus vulgaris* L.). *J. Exp. Bot.* **2010**, *61*, 3245–3258.
248. El-Moneim, D.A.; Contreras, R.; Silva-Navas, J.S.; Gallego, F.J.; Figueiras, A.M.; Benito, C.; Pectin methylesterase gene and aluminum tolerance in *Secale cereale*. *Environ. Exp. Bot.* **2014**, *107*, 125–133.
249. Colzi, I.; Doumett, S.; Del Bubba, M.; Fornaini, J.; Arnetoli, A.; Gabbrielli, R.; Gonnelli, C. On the role of the cell wall in the phenomenon of copper tolerance in *Silene paradoxa* L. *Environ. Exp. Bot.* **2011**, *72*, 77–83.

250. Chandran, D.; Sharopova, N.; Ivashuta, S.; Gantt, J.S.; VandenBosch, K.A.; Samac, D.A. Transcriptome profiling identified novel genes associated with aluminium toxicity, resistance and tolerance in *Medicago truncatula*. *Planta* **2008**, *228*, 151–166.
251. Yang, Z.B.; Eticha, D.; Rotter, B.; Rao, I.M.; Horst, W.J. Physiological and molecular analysis of polyethylene glycol-induced reduction of aluminium accumulation in the root tips of common bean (*Phaseolus vulgaris*). *New Phytol.* **2011**, *192*, 99–113.
252. Han, Y.; Sa, G.; Sun, J.; Shen, Z.; Zhao, R.; Ding, M.; Deng, S.; Lu, Y.; Zhang, Y.; Shen, X.; Chen, S. Overexpression of *Populus euphratica* xyloglucan endotransglucosylase/hydrolase gene confers enhanced cadmium tolerance by the restriction of root cadmium uptake in transgenic tobacco. *Environ. Exp. Bot.* **2014**, *100*, 74–83.
253. Song, X.Q.; Liu, L.F.; Jiang, Y.J.; Zhang, B.C.; Gao, Y.P.; Liu, X.L.; Lin, Q.S.; Ling, H.Q.; Zhou, Y.H. Disruption of secondary wall cellulose biosynthesis alters cadmium translocation and tolerance in rice plants. *Mol. Plant.* **2013**, *6*, 768–780.
254. Liu, T.; Shen, C.; Wang, Y.; Huang, C.; Shi, J. New insights into regulation of proteome and polysaccharide in cell wall of *Elsholtzia splendens* in response to copper stress. *PLoS One* **2014**, *9*, e109573.
255. Gao, J.; Sun, L.; Yang, X.; Liu, J.X. Transcriptomic analysis of cadmium stress response in the heavy metal hyperaccumulator *Sedum alfredii* hance. *PLoS One* **2013**, *8*, e64643.
256. Tabbì, G.; Fry, S.C.; Bonomo, R.P. ESR study of the non-enzymic scission of xyloglucan by an ascorbate-H2O2-copper system: The involvement of the hydroxyl radical and the degradation of ascorbate. *J. Inorg. Biochem.* **2001**, *84*, 179–187.
257. Kärkönen, A.; Fry, S.C. Effect of ascorbate and its oxidation products on H_2O_2 production in cell-suspension cultures of *Picea abies* and in the absence of cells. *J. Exp. Bot.* **2006**, *57*, 1633–1644.
258. Claus, H. Laccases: Structure, reaction, distribution. *Micron* **2004**, *35*, 93–96.
259. Ali, M.B.; Singh, N.; Shohael, A.M.; Hahn, E.J.; Paek, K.Y. Phenolics metabolism and lignin synthesis in root suspension cultures of *Panax ginseng* in response to copper stress. *Plant Sci.* **2006**, *171*, 147–154.
260. Keuskamp, D.H.; Keller, M.M.; Ballaré, C.L.; Pierik, R. Blue light regulated shade avoidance. *Plant Signal. Behav.* **2012**, 7, 514–517.
261. Franklin, K.A. Shade avoidance. *New Phytol.* **2008**, *179*, 930–944.
262. Sasidharan, R.; Keuskamp, D.H.; Kooke, R.; Voesenek, L.A.; Pierik, R. Interactions between auxin, microtubules and XTHs mediate green shade-induced petiole elongation in *Arabidopsis*. *PLoS One* **2014**, *9*, e90587.
263. Ma, L.; Li, J.; Qu, L.; Hager, J.; Chen, Z.; Zhao, H.; Deng, X.W. Light control of *Arabidopsis* development entails coordinated regulation of genome expression and cellular pathways. *Plant Cell.* **2001**, *13*, 2589–2607.
264. Sasidharan, R.; Chinnappa, C.C.; Voesenek, L.A.C.J.; Pierik, R. The regulation of cell wall extensibility during shade avoidance: A study using two contrasting ecotypes of *Stellaria longipes*. *Plant Physiol.* **2008**, *148*, 1557–1569.
265. Sasidharan, R.; Chinnappa, C.C.; Staal, M.; Elzenga, J.T.; Yokoyama, R.; Nishitani, K.; Voesenek, L.A.; Pierik. R. Light quality-mediated petiole elongation in *Arabidopsis* during

shade avoidance involves cell wall modification by xyloglucan endotransglucosylase/hydrolases. *Plant Physiol.* **2010**, *154*, 978–990.

266. Masuda, Y.; Kamisaka, S.; Yanagisawa, H.; Suzuki, H. Effect of light on growth and metabolic activities in pea seedlings I. changes in cell wall polysaccharides during growth in the dark and in the light. *Biochem. Physiol. Pflanzen.* **1981**, *176*, 23–34.
267. Wargent, J.J.; Gegas, V.C.; Jenkins, G.I.; Doonan, J.H.; Paul, N.D. UVR8 in *Arabidopsis thaliana* regulates multiple aspects of cellular differentiation during leaf development in response to ultraviolet B radiation. *New Phytol.* **2009**, *183*, 315–326.
268. Kigel, J.; Cosgrove, D.J. Photoinhibition of stem elongation by blue and red light. *Plant Physiol.* **1991**, *95*, 1049–1056.
269. Raffo, M.D.; Ponce, N.M.; Sozzi, G.O.; Vicente, A.R.; Stortz, C.A. Compositional changes in "Bartlett" pear (*Pyrus communis* L.) cell wall polysaccharides as affected by sunlight conditions. *J. Agric. Food Chem.* **2011**, *59*, 12155–12162.
270. Sharma, P.; Chatterjee, M.; Burman, N.; Khurana, J.P. Cryptochrome 1 regulates growth and development in *Brassica* through alteration in the expression of genes involved in light, phytohormone and stress signalling. *Plant Cell Environ.* **2014**, *37*, 961–977.
271. Syros, T.D.; Yupsanis, T.A.; Economou, A.S. Expression of peroxidases during seedling growth in *Ebenus cretica* L. as affected by light and temperature treatments. *Plant Growth Regul.* **2005**, *46*, 143–151.
272. Ali, M.B.; Khatun, S.; Hahn, E.J.; Paek, K.Y. Enhancement of phenylpropanoid enzymes and lignin in *Phalaenopsis* orchid and their influence on plant acclimatisation at different levels of photosynthetic photon flux. *Plant Growth Regul.* **2005**, *49*, 137–146.
273. Kimura, M.; Yamamoto, Y.; Seki, M.; Sakurai, T.; Sato, M.; Abe, T.; Yoshida, S.; Manabe, K.; Shinozaki, K.; Matsui, M. Identification of *Arabidopsis* genes regulated by high light-stress using cDNA microarray. *Photochem. Photobiol.* **2003**, *77*, 226–233.
274. Ruhland, C.T.; Xiong, F.S.; Clark, W.D.; Day, T.A. The influence of ultraviolet-B radiation on growth, hydroxycinnamic acids and flavonoids of *Deschampsia antarctica* during Springtime ozone depletion in Antarctica. *Photochem. Photobiol.* **2005**, *5*, 1086–1093.
275. Lesniewska, E.; Adrian, M.; Klinguer, A.; Pugin, A. Cell wall modification in grapevine cells in response to UV stress investigated by atomic force microscopy. *Ultramicroscopy* **2004**, *100*, 171–178.
276. Pristov, J.B.; Spasić, M.; Spasojević, M. Converting low dose radiation to redox signaling. *Plant Signal Behav.* **2013**, *8*, e23151.
277. Messenger, D.J.; McLeod, A.R.; Fry, S.C. The role of ultraviolet radiation, photosensitizers, reactive oxygen species and ester groups in mechanisms of methane formation from pectin. *Plant Cell Environ.* **2009**, *32*, 1–9.
278. Pristov, J.B.; Jovanović, S.V.; Mitrović, A.; Spasojević, I. UV-irradiation provokes generation of superoxide on cell wall polygalacturonic acid. *Physiol. Plant.* **2013**, *148*, 574–581.
279. Barka, E.A.; Kalantari, S.; Makhlouf, J.; Arul, J. Impact of UV-C Irradiation on the cell wall-degrading enzymes during ripening of tomato (*Lycopersicon esculentum* L.) fruit. *J. Agric. Food Chem.* **2000**, *48*, 667–671.

280. Bu, J.; Yua, Y.; Aisikaera, G.; Yinga, T. Postharvest UV-C irradiation inhibits the production of ethylene andthe activity of cell wall-degrading enzymes during softening of tomato (*Lycopersicon esculentum* L.) fruit. *Postharvest Biol. Technol.* **2013**, *86*, 337–345.
281. Yamasaki, S.; Noguchi, N.; Mimaki, K. Continuous UV-B irradiation induces morphological changes and the accumulation of polyphenolic compounds on the surface of cucumber cotyledons. *J. Radiat. Res.* **2007**, *48*, 443–454.
282. Zagoskina, N.V.; Dubravina, G.A.; Alyavina, A.K.; Goncharuk, E.A. Effect of ultraviolet (UV-B) radiation on the formation and localization of phenolic compounds in tea plant callus cultures. *Russ. J. Plant Phys.* **2003**, *50*, 270–275.
283. Hilal, M.; Parrado, M.F.; Rosa, M.; Gallardo, M.; Orce, L.; Massa, E.D.; González, J.A.; Prado, F.E. Epidermal lignin deposition in quinoa cotyledons in response to UV-B radiation. *Photochem. Photobiol.* **2004**, *79*, 205–210.
284. Pontin, M.A.; Piccoli, P.N.; Francisco, R.; Bottini, R.; Martinez-Zapater, J.M.; Lijavetzky, D. Transcriptome changes in grapevine (*Vitis vinifera* L.) cv. Malbec leaves induced by ultraviolet-B radiation. *BMC Plant Biol.* **2010**, *10*, 224–237.
285. Yokawa, K.; Baluška, F. Pectins, ROS homeostasis and UV-B responses in plant roots. *Phytochemitry* **2014**, in press.
286. Lindroth, R.L. Impacts of elevated atmospheric CO_2 and O_3 on forests: Phytochemistry, trophic interactions, and ecosystem dynamics. *J. Chem. Ecol.* **2010**, *36*, 2–21.
287. Munné-Bosch, S.; Queval, G.; Foyer, C.H. The impact of global change factors on redox signaling underpinning stress tolerance. *Plant Physiol.* **2013**, *16*, 5–19.
288. Betzelberger, A.M.; Yendrek, C.R.; Sun, J.; Leisner, C.P.; Nelson, R.L.; Ort, D.R.; Ainsworth, E.A. Ozone exposure response for U.S. soybean cultivars: Linear reductions in photosynthetic potential, biomass and yield. *Plant Physiol.* **2012**, *160*, 1827–1839.
289. Emberson, L.D.; Bueker, P.; Ashmore, M.R.; Mills, G.; Jackson, L.S.; Agrawal, M.; Atikuzzaman, M.D.; Cinderby, S.; Engardt, M.; Jamir, C.; *et al.* A comparison of North American and Asian exposure-response data for ozone effects on crop yields. *Atmos. Environ.* **2009**, *43*, 1945–1953.
290. Wilkinson. S.; Davies, W.J. Drought, ozone, ABA and ethylene: New insights from cell to plant to community. *Plant Cell Environ.* **2010**, *33*, 510–525.
291. Racherla, P.N.; Adams, P.J. The response of surface ozone to climate change over the eastern United States. *Atmos. Chem. Phys.* **2008**, *8*, 871–885.
292. Günthardt-Goerg, M S.; Mc Quattie, C.J.; Scheidegger, C.; Rhiner, C.; Matyssek, R. Ozone-induced cytochemical and ultrastructural changes in leaf mesophyll cell walls. *Can. J. For. Res.* **1997**, *27*, 453–463.
293. Günthardt-Goerg, M.S. Different responses to ozone of tobacco, poplar, birch, and alder. *J. Plant Physiol.* **1996**, *148*, 207–214.
294. Paoletti, E.; de Marco, A.; Beddows, D.C.S.; Harrison, R.M.; Manning, W.J. Ozone levels in European and USA cities are increasing more than at rural sites, while peak values are decreasing. *Environ. Pollut.* **2014**, *192*, 295–299.

295. Paoletti, E.; Contran, N.; Bernasconi, P.; Günthardt-Goerg, M.S.; Wollenveider, P. Structural and physiological responses to ozone in Manna ash (*Fraxinus ornus* L.) leaves of seedlings and mature trees under controlled and ambient condition. *Sci. Total Environ.* **2010**, *408*, 2014–2024.
296. Richet, N.; Afif, D.; Huber, F.; Pollet, B.; Banvoy, J.; el Zein, R.; Lapierre, C.; Dizengremel, P.; Perré, P.; Cabané, M. Cellulose and lignin biosynthesis is altered by ozone in wood of hybrid poplar (*Populus tremula* × *alba*). *J. Exp. Bot.* **2011**, *62*, 3575–3586.
297. Bussotti, F.; Agati, G.; Desotgiu, R.; Matteini, P.; Tani, C. Ozone foliar symptoms in woody plant species assessed with ultrastructural and fluorescence analysis. *New Phytol.* **2005**, *166*, 941–955.
298. Bussotti, F.; Pancrazi, M.; Matteucci, G.; Gerosa, G. Leaf morphology and chemistry in *Fagus sylvatica* L. (beech) trees as affected by site factors and ozone: Results from Level II permanent monitoring plots in Italy. *Tree Physiol.* **2005**, *25*, 211–219.
299. Keutgen, A.J.; Pawelzik, E. Contribution of amino acids to strawberry fruit quality and their relevance as stress indicators under NaCl salinity. *Food Chem.* **2008**, *111*, 642–647.
300. Kanter, U; Heller, W.; Durner, J.; Winkler, J.B.; Engel, M.; Behrendt, H.; Holzinger, A.; Braun, P.; Hauser, H.; Ferreira, F.; *et al.* Molecular and immunological characterization of ragweed (*Ambrosia artemisiifolia* L.) pollen after exposure of the plants to elevated ozone over a whole growing season. *PLoS One* **2013**, *8*, e61518.
301. Minas, I.S.; Vicente, O.R.; Dhanapal, O.P.; Manganaris, G.A.; Goulas, V.; Vasilakakis, M.; Crisosto, H.; Molassiotis, A. Ozone-induced kiwifruit ripening delay is mediated by ethylene biosynthesis inhibition and cell wall dismantling regulation. *Plant Sci.* **2014**, *29*, 76–85.
302. Oksanen, E.; Häikiö, E.; Sober, J.; Karnosky, D.F. Ozone-induced H_2O_2 accumulation in field-grown aspen and birch is linked to foliar ultrastructure and peroxisomal activity. *New Phytol.* **2003**, *161*, 791–799.
303. Cho, K.; Shibato, J.; Kubo, A.; Kohno, Y.; Satoh, K.; Kikuchi, S.; Sarkar, A.; Agrawal, G.K.; Rakwal, R. Comparative analysis of seed transcriptomes of ambient ozone-fumigated two different rice cultivars. *Plant Signal Behav.* **2013**, *8*, e26300.
304. Cabané, M.; Pireaux, J.C.; Leger, E.; Weber, E.; Dizengremel, P.; Pollet, B.; Lapierre, C. Condensed lignins are synthesized in poplar leaves exposed to ozone. *Plant Physiol.* **2004**, *134*, 586–594.
305. Sharma, Y.K.; Davis, K.R. Ozone-induced expression of stress-related genes in *Arabidopsis thaliana*. *Plant Physiol.* **1994**, *105*, 1089–1096.
306. Eckey-Kaltenbach, H.; Ernst, D.; Heller, W.; Sandermann, H. Biochemical plant responses to ozone: IV. Cross-induction of defensive pathways in parsley (*Petroselinum crispum* L.) plants. *Plant Physiol.* **1994**, *104*, 67–74.
307. Wiese, C.B.; Pell, E.J. Oxidative modification of the cell wall in tomato plants exposed to ozone. *Plant Physiol. Biochem.* **2003**, *41*, 375–382.
308. Ranieri, A.; Castagna, A.; Pacini, J.; Baldan, B.; Mensuali Sodi, A.; Soldatini, G.F. Early production and scavenging of hydrogen peroxide in the apoplast of sunflowers plants exposed to ozone. *J. Exp. Bot.* **2003**, *54*, 2529–2540.
309. Ranieri, A.; Castagna, A.; Soldatini, G.F. Differential stimulation of ascorbate peroxidase isoforms by ozone exposure in sunflower plants. *J. Plant Physiol.* **2000**, *156*, 266–271.

310. Keutgen, A.J.; Pawelzik, E. Apoplastic antioxidative system responses to ozone stress in strawberry leaves. *J. Plant Physiol.* **2008**, *165*, 868–875.

311. Meehl, G.A.; Stocker, W.D.; Collins, P.; Friedlingstein, A.T.; Gaye, J.M.; Gregory, A.; Kitoh, R.; Knutti, J.M.; Murphy, A.; Noda, S.C.B.; *et al.* Global Climate Projections. In *Climate Change 2007: The Physical Science Basis. Contribution of Working Group I to the Fourth Assessment Report of the Intergovernmental Panel on Climate Change*; Solomon, S., Qin, D., Manning, M., Chen, Z., Marquis, M., Averyt, K.B., Tignor, M., Miller, H.L., Eds.; Cambridge University Press: Cambridge, UK, New York, NY, USA, 2007.

312. Leakey, A.D.B.; Bernacchi, C.J.; Dohleman, F.G.; Ort, D.R.; Long, S.P. Will photosynthesis of maize (*Zea mays*) in the US Corn Belt increase in future CO_2 rich atmospheres? An analysis of diurnal courses of CO_2 uptake under free-air concentration enrichment (FACE). *Glob. Chang. Biol.* **2004**, *10*, 951–962.

313. De Souza, A.P.; Arundale, R.A.; Dohleman, F.G.; Long, S.P.; Buckeridge, M.S. Will the exceptional productivity of *Miscanthus x giganteus* increase further under rising atmospheric CO2? *Agricult. For. Meteorol.* **2013**, *171*, 82–92.

314. King, J.S.; Pregitzer, K.S.; Zak, D.R.; Karnosky, D.F.; Isebrands, J.G.; Dickson, R.E.; Hendrey, G.R.; Sober, J. Fine root biomass and fluxes of soil carbon in young stands of paper birch and trembling aspen as affected by elevated atmospheric CO_2 and tropospheric O_3. *Oecologia* **2001**, *128*, 237–250.

315. Oksanen, E.; Sôber, J.; Karnosky, D.F. Interactions of elevated CO_2 and ozone in leaf morphology of aspen (*Populus tremuloides*) and birch (*Betula papyrifera*) in aspen FACE experiment. *Environ. Pollut.* **2001**, *115*, 437–446.

316. Kubiske, M.E.; Quinn, V.S.; Marquart, P.E.; Karnosky, D.F. Effects of elevated atmospheric CO_2 and/or O_3 on intra- and inter-specific competitive ability of aspen. *Plant Biol.* **2007**, *9*, 342–355.

317. Cseke, L.J.; Tsai, C.J.; Rogers, A.; Nelsen, M.P.; White, H.L.; Karnosky, D.F.; P Odila, G.K. Transcriptomic comparison in the leaves of two aspen genotypes having similar carbon assimilation rates but different partitioning patterns under elevated (CO_2). *New Phytol.* **2009**, *182*, 891–911.

318. Wei, H.; Gou, J.; Yordanov, Y.; Zhang, H.; Thakur, R.; Jones, W.; Burton. A. Global transcriptome profiling of aspen trees under elevated (CO_2) for identifying potential molecular mechanisms responsible for enhanced growth. *J. Plant Res.* **2013**, *126*, 305–320.

319. Kostiainen, K.; Saranpää, P.; Lundqvist, S.O.; Kubiske, M.E.; Vapaavuori, E. Wood properties of *Populus* and *Betula* in long-term exposure to elevated CO_2 and O_3. *Plant Cell Environ.* **2014**, *37*, 1452–1463.

320. Kim, K.; Labbé, N.; Warren, J.M.; Elder, T.; Rials, T.G. Chemical and anatomical changes in *Liquidambar styraciflua* L. xylem after long term exposure to elevated CO_2. *Environ. Pollut.* **2015**, *17*, 179–185.

321. Gibeaut, D.M.; Cramer, G.R.; Seemann, J.R. Growth, cell walls, and UDP-Glc dehydrogenase activity of *Arabidopsis thaliana* grown in elevated carbon dioxide. *J. Plant Physiol.* **2001**, *158*, 569–576.

322. Duan, Z.; Homma, A.; Kobayashi, M.; Nagata, M.; Kaneko, Y.; Fujiki, Y.; Nishida, I. Photoassimilation assimilate translocation and plasmodesmal biogenesis in the source leaves of *Arabidopsis thaliana* grown under an increased atmospheric CO_2 concentration. *Plant Cell. Physiol.* **2014**, *55*, 358–369.

323. Perez-Lopez, U.; Robredo, A.; Lacuesta, M.; Munoz-Rueda, A.; Mena-Petite, A. Atmospheric CO_2 concentration influences the contributions of osmolyte accumulation and cell wall elasticity to salt tolerance in barley cultivars. *J. Plant Physiol.* **2010**, *167*, 15–22.

324. Huyskens-Keil, S.; Herppich, W. High CO_2 effects on postharvest biochemical and textural properties of white asparagus (*Asparagus officinalis* L.) spears. *Postharvest Biol. Technol.* **2013**, *75*, 45–53.

325. Chen, S.; Olbrich, A.; Langenfeld-Heyser, R.; Fritz, E.; Polle. A. Quantitative X-ray microanalysis of hydrogen peroxide within plant cells. *Microsc. Res. Techniq*. **2009**, *72*, 49–60.

326. Ookawara, R.; Satoh, S.; Yoshioka, T.; Ishizawa, K. Expression of α-expansin and xyloglucan endotransglucosylase/hydrolase genes associated with shoot elongation enhanced by anoxia, ethylene and carbon dioxide in arrowhead (*Sagittaria pygmaea* Miq.) tubers. *Ann. Bot.* **2005**, *96*, 693–702.

327. May, P.; Liao, W.; Wu, Y.; Shuai, B.; Mc Combie, W.R.; Zhang, M.Q.; Liu, Q.A. The effects of carbon dioxide and temperature on microRNA expression in Arabidopsis development. *Nat. Commun.* **2013**, *4*, doi:10.1038/ncomms3145.

328. Cosgrove, D.J. Growth of the plant cell wall. *Nat. Rev. Mol. Cell Biol.* **2005**, *6*, 850–861.

329. Hijazi, M.; Velasquez, S.M.; Jamet, E.; Estevez, J.M.; Albenne, C. An update on post-translational modifications of hydroxyproline-rich glycoproteins: Toward a model highlighting their contribution to plant cell wall architecture. *Front. Plant Sci.* **2014**, *5*, 395.

330. Seifert, G.J.; Xue, H.; Acet, T. The Arabidopsis thaliana fasciclin like arabinogalactan protein 4 gene acts synergistically with abscisic acid signaling to control root growth. *Ann. Bot.* **2014**, *114*, 1125–1133.

10

The Utilization of Plant Facilities on the International Space Station—The Composition, Growth, and Development of Plant Cell Walls under Microgravity Conditions

Ann-Iren Kittang Jost [1,†], Takayuki Hoson [2,†] and Tor-Henning Iversen [3,†,*]

[1] Centre for Interdisciplinary Research in Space (CIRiS), NTNU Samfunnsforskning AS, Dragvoll Allé 38, Trondheim NO-7491, Norway; E-Mail: a.i.kittang.jost@ciris.no

[2] Department of Biology, Graduate School of Science, Osaka City University, Sumiyoshi-ku, Osaka 558-8585, Japan; E-Mail: hoson@sci.osaka-cu.ac.jp

[3] Department of Biology, The Norwegian University of Science and Technology (NTNU), Realfagbygget, Trondheim NO-7491, Norway

† These authors contributed equally to this work.

* Author to whom correspondence should be addressed; E-Mail: Tor-Henning.Iversen@bio.ntnu.no

Academic Editor: Philip J. Harris

Abstract: In the preparation for missions to Mars, basic knowledge of the mechanisms of growth and development of living plants under microgravity (micro-g) conditions is essential. Focus has centered on the g-effects on rigidity, including mechanisms of signal perception, transduction, and response in gravity resistance. These components of gravity resistance are linked to the evolution and acquisition of responses to various mechanical stresses. An overview is given both on the basic effect of hypergravity as well as of micro-g conditions in the cell wall changes. The review includes plant experiments in the US Space Shuttle and the effect of short space stays (8–14 days) on single cells (plant protoplasts). Regeneration of protoplasts is dependent on cortical microtubules to orient the nascent cellulose microfibrils in the cell wall. The space protoplast experiments demonstrated that the regeneration capacity of protoplasts was retarded. Two critical factors are the basis for longer space experiments: *a*. the effects of gravity on the molecular mechanisms for cell wall development, *b*. the availability of facilities and hardware for performing cell wall experiments in space and return of RNA/DNA back to the Earth. Linked to these aspects is a description of existing hardware functioning on the International Space Station.

Keywords: International Space Station (ISS); microgravity; *Arabidopsis thaliana*; European Modular Cultivation System (EMCS); BIOLAB; cell wall; gravity resistance; hypergravity; microtubules; protoplasts

1. Introduction

The plant cell wall in a growing plant is composed of a variety of components—cellulose, matrix polysaccharides, phenolics, and glycoproteins [1]. The wall, which surrounds the plant cell, is the major source of mechanical strength and form and despite its rigidity it permits and facilitates growth. The network of cell walls in a plant can be compared to both skin and bones in humans and animals. The cell walls surrounding the plasma membrane define the boundaries of the cell itself and represent the place where the plant cell meets a variety of environmental signals. In this way, the wall is the major location of plant responses to such signals [2,3]. Located at the surface and forming the interface between adjacent cells, the wall plays an important role in cell intercellular communication.

This review focuses on gravity as one of the environmental signals that is always present in a constant direction and magnitude on Earth. Throughout evolution, plants have utilized gravity as the most stable and reliable signal for their survival. The immediate effect of gravity enables plants to orient their leaves towards sunlight and to develop a root system to anchor and absorb water and minerals. Plant gravitropism is the most typical response when a plant is turned onto its side and the plant stem bends upwards towards the gravity direction and the root grows the opposite way. This occurs both in light and in darkness but the true gravitropic response requires darkness. It is also well recognized nowadays that the best environment for studying plant gravitropism in order to elucidate the mechanisms behind gravity effects is under microgravity conditions, *i.e.*, in space experiments.

Gravity resistance was introduced by one of the authors of the present article as a new term, and its nature and mechanism were initially studied in ground-based experiments using hypergravity conditions produced by centrifugation [4,5]. More recently, the same authors performed a space experiment on the International Space Station (ISS) to study the effect of microgravity on the gravity resistance phenomena [6]. These experiments were proposed and performed in order to explain that the initial curvature, the true gravitropism, and the following straight growth against gravity are quite different in origin and mechanisms. According to the results obtained, the mechanical resistance to the gravitational force may be a principal graviresponse in plants, independent of gravitropism. In the present review article, the results obtained, which have clarified the outline of the sequence of events in gravity resistance, will be described in more detail. The role of the cell wall in gravity resistance in plants will also be further elucidated.

Since the experiments in space are dependent on the availability of plant cultivation facilities, adequate hardware will be described in the present article. Historically, the review starts with a link back to the early plant experiments in the US Space Shuttle period. Initially, and due to the limited time periods in space (up to 15 days), freshly produced plant cell protoplasts were used to study the effect of short exposures to microgravity conditions and the effect on single cells. Plant protoplasts *i.e.*, cells from which the cell wall has been enzymatically removed, were produced immediately before the launch of the space shuttle, and the initial effects of the cell wall regeneration was followed for the first time under micro-g. Further on in the present article, a summary of these results obtained with cell walls is presented. After the introduction of a functional ISS in 2005, cultivation of intact plants originating from seeds was possible. The different types of facilities and their limitations and advantages for intact plant studies focusing on growth and development of the cell walls are presented.

The end of this review article deals with the molecular aspects of the cell wall development under micro-g conditions. In order to fully understand the cell wall composition, function, and regulation on the Earth, it is expected that results from space experiments will be useful. For the optimal construction of such experiments, sophisticated methods within molecular biology, molecular genetics, and genomics will have to be applied. A key question here, which has not yet been answered, is how to obtain and preserve RNA/DNA samples from the microgravity environment for further analysis on Earth.

2. Gravity Resistance

Plants show mechanical resistance to gravitational acceleration. Genetic and physiological evidence indicate that this response, termed gravity resistance, is distinct from gravitropism. Weeping, characterized by hanging branches, is caused by a genetic mutation. Weeping is not a mutant of gravitropism, because branches of most weeping trees show normal gravitropism [7]. However, weeping branches lose the capacity to maintain enough mechanical rigidity to resist gravitational acceleration [8]. On the other hand, stem organs of some aquatic or semi-aquatic plants such as rice coleoptiles elongate rapidly and grow more slender when underwater. This growth response is partly caused by reduction in the gravitational acceleration actually applied to their bodies due to buoyancy, and it is reversible [2,8]. The rigidity of rice coleoptiles when grown under submergence is lower than when grown in the air [2]. However, rice coleoptiles show normal gravitropic curvature under submergence, as in the air [9]. In the gravitropic curvature of stem organs and roots, the initial curvature, which represents true gravitropism, and the following straight growth against gravity are quite different in origin and mechanism. Gravity resistance may sustain the latter process [8]. The cell wall is the major source of mechanical rigidity for plant bodies, and so it may be responsible for gravity resistance. Evidence obtained by the ground-based studies and space experiments described below supports this hypothesis.

3. Gravity Resistance and Evolution

During evolution, a link has been created between the responses to various mechanical stresses and the establishment of land plants. When plant ancestors first emerged on land, probably about 450 million years ago in the mid-Ordovician, their bodies were surrounded by soft cell coverings, because it was not necessary for them to develop hard bodies. Plant ancestors were directly exposed to a gravitational acceleration of 1 g on land, and only plants that successfully evolved a tough body with rigid cell walls were able to resist the force and survive on land [10,11]. The development of the mechanisms of gravity resistance was indispensable for the transition of plant ancestors from an aquatic environment to a terrestrial environment and the consequent establishment of land plants. On land, plants also have been exposed to a variety of mechanical stresses, such as wind. An increase in cell wall rigidity is, as for the gravity signal, generally induced in response to mechanical stimuli [12–14], the same. In addition, the pathway leading to gravity resistance, described below, may be at least in part common with that of mechanical stresses. Thus, the development of the mechanisms of gravity resistance should be important for the acquisition of responses to mechanical stresses for plants [8].

4. Mechanisms Controlling Gravity Resistance

The mechanism of gravity resistance consists of signal perception, transformation and transduction of the perceived signal, and response to the transduced signal, as is seen in other environmental responses as well. The outline of the sequence of events in gravity resistance has already been described [5,15]. The gravity signal may be perceived by mechanoreceptors (mechanosensitive ion channels) on the plasma membrane, and it appears that amyloplast sedimentation in statocytes is not directly involved. The signal should be then transformed and transduced into the cells. Membrane sterols and cortical microtubules (CMTs) may be involved in this process. In the membranes, sterols consist of microdomains called rafts, and CMTs are linked to the rafts via microtubule-associated proteins (MAPs). Because the functions of CMTs, the plasma membrane, and the cell wall are mutually and deeply dependent, as shown by the results of studies using mutants and metabolic inhibitors, the structural continuum or physiological continuity of these cellular components may play a key role in signal transduction. The transduced signal then may induce the expression of diverse genes influencing the structure and function of various cellular components. These steps regulate the synthetic levels of cell wall constituents. At the same time, transformed signals may be transduced through the plasma membrane, and they might modify the activities of proton pumps or other ion channels, which regulates the cell wall environment. The signal transduced through the plasma membrane may also directly influence cellulose synthase activity or exocytosis of Golgi vesicles. In addition, the changes in orientation of CMTs may bring about reorientation of cellulose microfibrils. The modifications to the cell wall environment, in combination with the changes in the levels and structure of the cell wall constituents, determine the cell wall rigidity, leading to gravity resistance.

5. Cell Wall Changes under Hypergravity Conditions

Microgravity conditions can be produced by free fall or parabolic flight on Earth. However, the microgravity obtained by these methods is generally too brief to induce obvious changes in plant growth and morphogenesis. Therefore, hypergravity generated from centrifugation has been widely used to analyze plant responses to gravity. Since the pioneering study by Waldron and Brett [16] in pea seedlings, the effects of centrifugal acceleration on growth and development have been analyzed in various plant materials. As a result, hypergravity has been shown to suppress elongation growth but promote the lateral expansion of stem organs and roots [4,5,17]. The first analysis of changes in the mechanical properties of the cell wall under hypergravity conditions was carried out in garden cress seedlings [18]. It has been confirmed in various materials that hypergravity increases cell wall rigidity [4,5].

The structural aspects of the wall under different gravity conditions are the basis for understanding how the gravitational acceleration can influence the mechanical rigidity of the cell wall. The plant cell wall consists of two phases: a cellulosic microfibrillar phase and a matrix phase consisting of polysaccharides, phenolics, and glycoproteins (see e.g., Carpita and Gibeaut [1]). The rigidity of the cell wall is determined by the chemical nature of each wall constituent and the interactions among them, such as cross-link formation. In particular, the levels of wall constituents and the molecular mass of certain matrix polysaccharides are important for the regulation of the rigidity. Hypergravity has been shown to increase cell wall thickness in various organs [4,5,8]. In azuki bean epicotyls, hypergravity increased the level of cellulose microfibrils in the basal supporting regions, and those of matrix polysaccharides in the upper growing regions [19,20]. Cellulose accumulation was also induced by hypergravity in pollen tubes [21]. On the other hand, hypergravity caused the polymerization of certain matrix polysaccharides, the types of which differed between dicotyledonous plants and monocotyledonous *Gramineae* plants. In dicotyledons, hypergravity increased the molecular mass of xyloglucans, whereas hypergravity increased that of 1,3,1,4-β-glucans in *Gramineae* plants. Hypergravity also decreased the xyloglucan-degrading activity in dicotyledons and 1,3,1,4-β-glucanase activity in *Gramineae*. In addition, modifications to xyloglucan metabolism under hypergravity conditions were regulated by prompt and differential changes in the expression of xyloglucan endo-transglucosylase/hydrolase (XTH) genes [22]. In pollen tubes, hypergravity altered intracellular vesicle transport related to the synthesis of matrix polysaccharides [21]. Thus, modification of the metabolic turnover of matrix polysaccharides as well as their accumulation may be involved in making the cell wall mechanically rigid to resist the gravitational acceleration under hypergravity conditions.

Certain differentiated cell walls contain a significant amount of phenolic substances such as lignin and monophenols. The lignin content was increased in response to hypergravity treatment in various stem organs [8]. In wheat seedlings, the levels of monophenols, in particular of diferulic acid, were increased under hypergravity conditions [8]. Also, the activity of wall-bound peroxidase, which is responsible for polymerization of lignin and the coupling reaction that produces diferulic acid, was increased by hypergravity [20,23]. The stimulation of cross-link formation by phenolics may increase cell wall rigidity and thus enhance resistance to the gravitational acceleration.

6. Cell Wall Changes under Micro-g Conditions—Space Experiments

Experiments under true microgravity conditions in space are effective to study gravity responses in plants. Opportunities for space experiments have been limited, but the situation was greatly improved by the initiation of scientific operations on the ISS. Growth and development of various organs has been examined in space, but the results are controversial. Such a contradiction may be caused by differences in culture conditions other than gravity, as well as differences in materials, age, and parameters measured [4,5,17]. Careful consideration of these conditions has confirmed growth stimulation under microgravity conditions in space at least in seedlings of Arabidopsis and rice and in inflorescence stems of Arabidopsis [4,5,24,25]. The first analysis of changes in cell wall rigidity under microgravity conditions in space was carried out in RICE experiments on the Space Shuttle STS-95 mission, and it was confirmed that microgravity decreased cell wall rigidity in seedlings of Arabidopsis and rice and inflorescence stems of Arabidopsis [4,5,25].

A comprehensive analysis of the chemical properties of the cell wall constituents was also conducted in the RICE experiments. The levels of cellulose and matrix polysaccharides per unit length were lower in the space-grown rice coleoptiles and Arabidopsis hypocotyls than the controls, indicating that microgravity decreased the cell wall thickness. The space-grown materials also had a lower molecular mass of the matrix polysaccharides, which is mainly due to the decrease in the molecular size of 1,3,1,4-β-glucans in rice and xyloglucans in Arabidopsis. In Arabidopsis hypocotyls, the activity of xyloglucan-degrading enzymes was also increased in space. These changes in cell wall properties are opposite to those induced by hypergravity, as mentioned above. These results support the hypothesis that under microgravity in space the metabolic turnover of cell wall constituents of plant seedlings is activated, which leads to a decrease in cell wall rigidity [4,5,8].

As to phenolics, the changes in content and formation of lignin under microgravity conditions in space were first analyzed in hypocotyls of pine and mung bean [26]. The lignin content in space-grown materials was lower than in those grown at 1 g on Earth. Also, peroxidase activity in space-grown pine hypocotyls was lower than that of ground controls [26]. On the contrary, there were no significant effects of microgravity on lignin content in seedlings or leaves of wheat [27,28]. This discrepancy may be due to the differences in experimental conditions mentioned above.

7. Cell Walls and Gravity Resistance—Future Aspects

The results obtained by the ground-based studies and space experiments so far support the hypothesis that plants increase the rigidity of their cell walls via modifications to the metabolism of cell wall constituents to resist the gravitational acceleration. However, the type of plant materials used in space experiments is limited. Changes in cell wall properties need to be examined in further space experiments to confirm the universality of cell wall functions in gravity resistance. On the other hand, CMTs are involved in gravity responses in plants through the regulation of the orientation of cellulose microfibrils, as discussed below for the protoplasts. Under hypergravity conditions, the expression of most α- and β-tubulin genes was upregulated in Arabidopsis hypocotyls, depending on the magnitude of the gravitational force [29,30], and a reorientation of CMTs from transverse to longitudinal directions was induced [31]. To clarify the role of CMTs in gravity resistance, we are conducting Resist Tubule (PI, T. Hoson) and Aniso Tubule (PI, K. Soga) space experiments on the Japanese Experiment Module "Kibo" of the ISS.

The physiological functions of the cell wall are sustained by its intimate crosstalk with the symplast, and the plasma membrane plays an important role in regulating the crosstalk [15]. As mentioned above, mechanoreceptors located on the plasma membrane may be responsible for the perception of the gravity signal, and membrane rafts play a role in transformation and transduction of the perceived signal in gravity resistance. In addition, two other plasma membrane components, cellulose synthase complexes and H^+-ATPase (proton pump), are likely involved in gravity resistance [5,8,15]. The role of these membrane components and the interaction among them in gravity resistance remain to be clarified in future space experiments.

8. Micro-g and Cell Wall Free Plant Cells (Protoplasts)

Historically, the review of the plant space experiments starts with a link back to the US Space Shuttle period with focus on short stays in space and the effect on single cells. Important here is the use of plant protoplasts *i.e.*, cells from which the cell wall has been enzymatically removed. These cells represent a uniform population of true single cells where tissue specific variations are eliminated. The totipotency of the plant cell makes the protoplasts isolated from mature differentiated tissue competent to express genes that regulate mitotic activity. The lack of cell-to-cell interactions in newly isolated protoplasts makes them ideal objects for environmental changes and systematic modification of extracellular conditions. The effect of variations in gravity forces on cell wall regeneration can easily be detected in the initial stages of the protoplast wall reconstitution. In this context, it should be stressed that this is dependent on CMTs, which are believed to orient the nascent cellulose microfibrils in the regenerating cell wall.

The earliest experiments using protoplast cells started on a satellite launched from the Soviet Union in 1989 [32]. Rapeseed protoplasts were used and exposed for micro-g conditions for 14 days on Biokosmos 9 (Kosmos 2044). The major result of the space experiment was a significant decrease in the content of cellulose and hemicellulose. A callus culture was established from the protoplasts, and after retrieval a decreased number of peroxidase isoenzymes and a decreased total peroxidase activity were also detected. The callus cultures established during the two-week-long flight also showed a retarded growth and division compared to the ground control [32,33]. An attempt to get the retrieved tissue cultures to differentiate into mature plants was unsuccessful. Protoplasts exposed to simulated 0 g for 3 days on a 2-D clinostat and protoplasts, which were permitted to regenerate cell walls at ground conditions (1 g), were used as controls for the Biokosmos 9 experiment. A disadvantage of the Biokomos 9 experiment was, however, the lack of a 1 g centrifuge on board the space vehicle. This meant that it was difficult to differentiate between the effect of the weightlessness and the other environmental conditions (e.g., cosmic radiation) under space conditions. Later on, Skagen [34] attempted to clarify the effect of weightlessness compared to the other effects by using rapeseed protoplasts exposed to simulated 0 g.

The Biokosmos 9 experiment was followed by an experiment called "PROTO" which was flown on the Space Shuttle in 1992 with the IML-1 (International Microgravity Laboratory) mission [35–37]. Four years later, a new mission (S/MM-03; Shuttle Mission to Mir-03) was launched where a protoplast regeneration experiment was included. Both the IML-1 and the S/MM-03 confirmed some of the results obtained in Biokosmos 9; after 8–11 days of micro-g conditions the regeneration of the rapeseed protoplasts was retarded and mature plants did not develop from the retrieved calli obtained from the flight samples. In the space shuttle experiments, a centrifuge was placed on board the vehicle and functioned as a 1 g control. In addition, a 1 g ground control was used in the comparison to the micro-g results obtained.

The space flight protoplast experiments have shown that the critical step in the regeneration of the protoplast into an intact cell is the immediate reconstitution of the cell wall. In the period after the enzymatic removal of the cell walls from the mother hypocotyl plant tissue, the intention was that the cell-wall-free cells should be exposed to the micro-g conditions in space. For technical reasons, the protoplast samples in the experimental set-up were permitted to be placed in the shuttle cargo as a late access delivery. This meant that the freshly isolated protoplasts were only 17 hours old when they were exposed to the micro-g conditions in space for the first time. One of the astronauts on the shuttle had the special task to microscopically observe that the cells still were at the cell wall free stage when the Space Shuttle Biorack laboratory was opened in space. This made it possible to follow the regeneration of the walls throughout the complete space period from an early stage up until the small calli had developed.

9. Protoplasts and Cell Wall Regeneration

Already in the 1980s, one of the authors of the present review article knew that after the Space Shuttle period, whereby plant cells were exposed space conditions for short periods, a new era with longer duration cultivation of plants, and not only cells, would come in due time. Our space biology group had already been selected as one of the plant cell study groups with experiments for the Space Shuttle Challenger in late 1986. Challenger was launched for the first time in 1983 and after nine successful flights a disaster occurred, and the practical consequence for our group was that we were offered the Biokosmos 9 experiment as an alternative.

Our protoplast experiments and the results obtained and experience gained should eventually lead to preparation of an ISS experiment of longer growth duration. As a link to this work, Skagen [34], in her PhD-thesis and in several articles [38–40], had focused on CMTs, which were believed to orient the nascent cellulose microfibrils in the regenerating cell wall [41–44]. At that time, it was also known that the amount of CMTs was anticipated to affect both the cellulose incorporation [45] and the division capacity in cell cultures [46].

The importance of various g-forces and the orientation of CMTs in the protoplast cell wall regeneration have been described by [40]. As demonstrated in Figure 1 [40], the protoplasts exposed to 1, 7 and 10 g reoriented the CMTs in parallel arrays while protoplasts exposed to simulated 0 g conditions showed CMTs aligned in a random pattern after 48 hours. Based on the experience from the protoplast experiment and the various g-effects on the cell wall regeneration, Skagen [34] selected a number of key elements which should therefore be included in an ISS single cell experiment; the effects of various g-forces and naphtaleneacetic acid (NAA) on the CMT organization, density, wall reconstitution (cellulose synthesis and peroxidase activity), and evaluation of the significance of a dense parallel CMT array for cellulose incorporation and cell division.

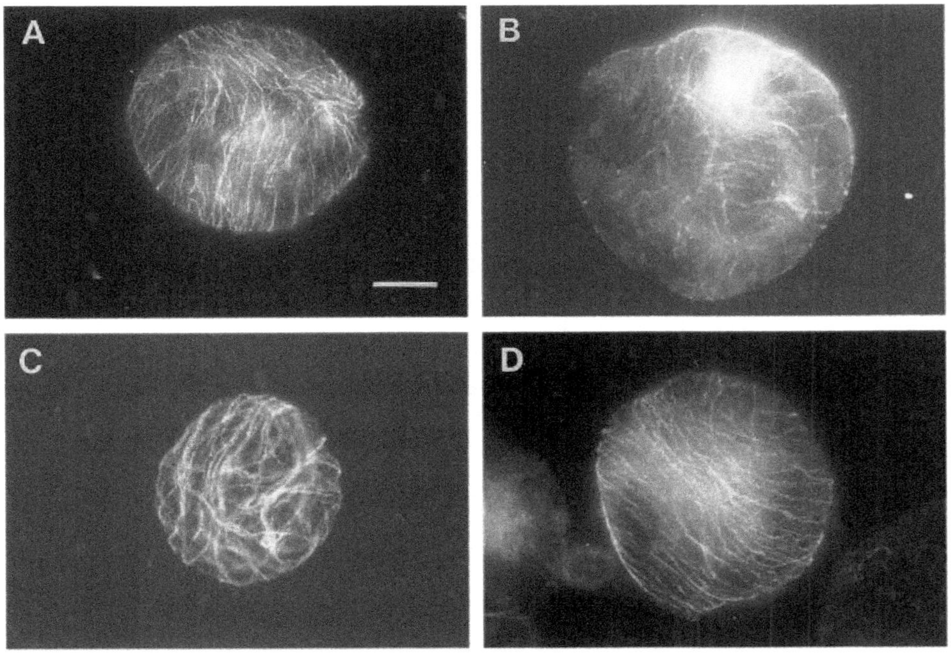

Figure 1. Overview of pattern of corticular microtubules (CMT) in protoplasts exposed to 48 hours of various g-forces. In contrast to most of the protoplasts exposed to 1, 7 or 10 g, which reoriented the CMT in parallel arrays (illustrated by the 1 g- (**A**) and 10 g-protoplast (**D**)), the 0 g-protoplasts showed CMT aligned in a random pattern after 48 h of g-treatment (**B**). Protoplasts with partly parallel arrays were observed in all groups, here illustrated by the 7 g-protoplast (**C**). An increased size among the 0 g-protoplasts compared to the 1 g- and hyper-g-protoplasts was observed (bar: 20 mm)—(original picture by Skagen and Iversen [40]).

A well prepared experimental set-up using rapeseed single cells based on our basic protoplast knowledge was submitted to ESA for tests on the ISS. However, in the final selection of our different proposals submitted to ESA for experiments in the European Modular Cultivation System (EMCS) the use of *Arabidopsis thaliana* as a model plant was given priority. Following is a summary of the adequateness of the EMCS for plant growth and development studies in general and the potential for future plant cell wall studies.

10. The EMCS and Plant Space Studies

The EMCS (Figure 2) and also the parallel space laboratory called BIOLAB were the result of a request to the ESA (the European Space Agency) from plant space scientists, and they were developed by EADS Astrium (Friedrichshafen, Germany) on behalf of ESA. The idea was to design two dedicated facilities to improve and standardize plant growth in the ISS. The EMCS Flight Model (EMCS FM) was brought to the ISS on the Space Shuttle ULF 1.1 in March 2006 after bio-compatibility and functionality tests [47]. The EMCS, which is the facility specially equipped for plant studies, has been equipped with two centrifuges to perform experiments in microgravity with variable *g*-levels up to 2.0 g. Detailed descriptions of the EMCS before being launched have been given by various authors [48–52]. In a more recent review article [53], the utilization of the EMCS on the ISS is presented.

The experience gained after 8 years of the EMCS in function on the ISS is that cultivation of higher plants can be performed under fully controlled environmental conditions and for long growth periods. So far, ten plant experiments have been performed (Table 1). One of these, the MULTIGEN 1 experiment, lasted for 85 days and *Arabidopsis thaliana* were grown from seeds to new plants producing embryos. In this specific experiment, circumnutations were in focus and, using the video technique (Figure 3) developed for that experiment, detailed representation of their growth and development could be followed over the total growth period as illustrated in Figure 4.

Table 1. Experiments performed in the EMCS.

Experiment	Principal investigator	Execution year	Status
TROPI-1	John Z. Kiss	2006-7	[54,55]
GRAVI-1	Dominique Driss-Ecole	2007	[56]
MULTIGEN-1	Tor-Henning Iversen	2007	[57,58]
Cell Wall/Resist Wall	Takayuki Hoson/ Kazuhiko Nishitani	2008	[6]
GENARA-A	Eugenie Carnero-Diaz/F. Javier Medina	2010	Experiment performed
TROPI-2	John Z. Kiss	2010	[59,60]
Plant Signalling	Imara Perera	2011	Experiment performed
Seedling Growth-1	John Z. Kiss	2013	[61]
GRAVI-2	Valérie Legué	2014	Experiment performed
Seedling Growth-2	John Z. Kiss/F. Javier Medina	2014	Experiment performed

A.

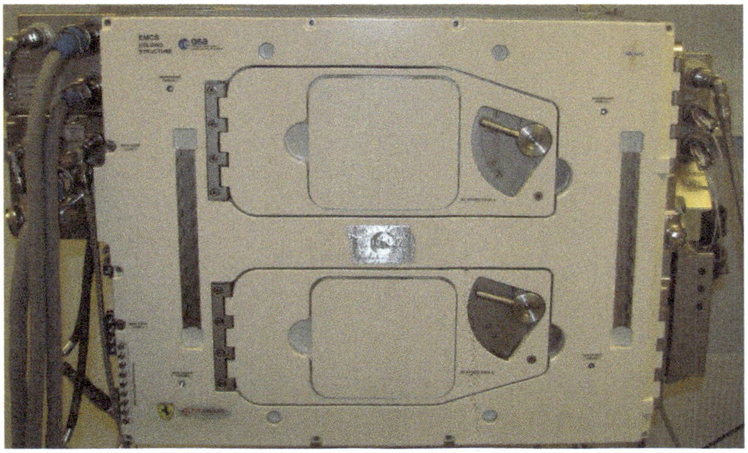

B.

Figure 2. The EMCS Flight Module seen from the front side (**A**) with two doors behind which there are two identical rotors (centrifuges). In (**B**), where the doors are open, a detailed view of the inside of the EMCS is shown. For more detailed descriptions of the EMCS please see [48–52].

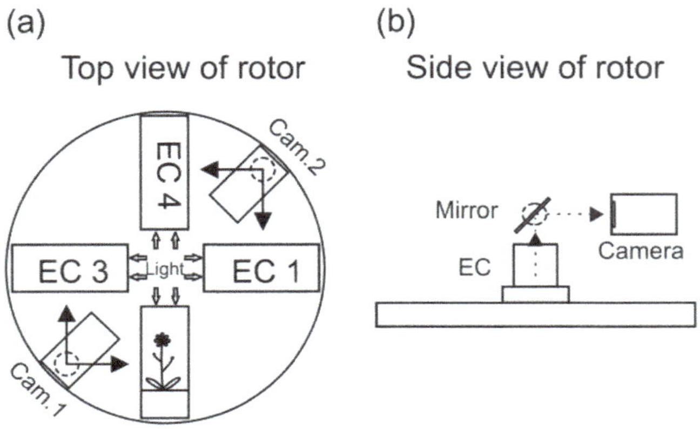

Figure 3. A schematic presentation of the rotor arrangement European Modular Cultivation System (EMCS) (**a**). The ECs are the plant cultivation chambers, and the plants can be photographed using the movable video cameras and with the help of mirrors (**b**) (see [53]).

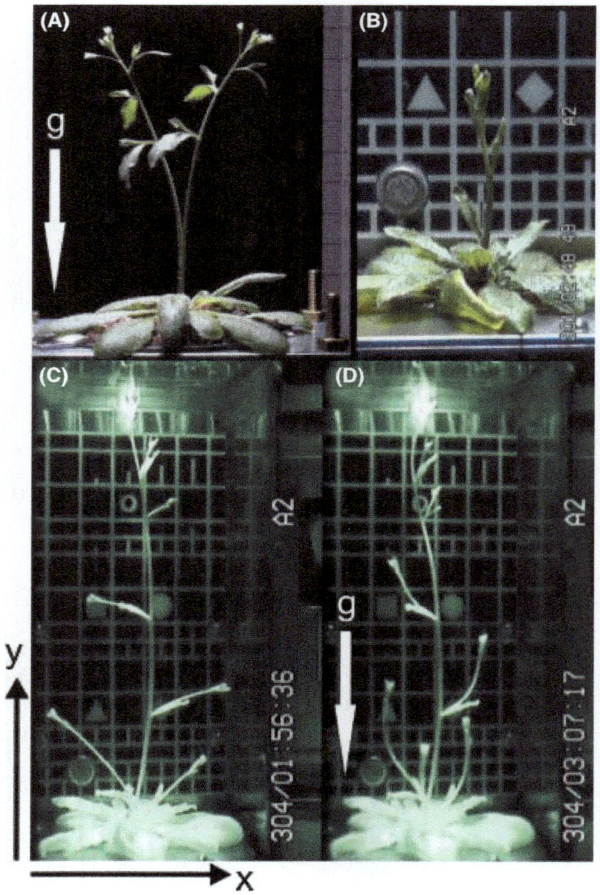

Figure 4. The picture shows an *Arabidopsis thaliana* plant growing in weightlessness on board the ISS in the plant cultivation chamber in the EMCS. The pictures show the different developmental stages and the effect of variable g-forces the plant can be exposed to (micro-g to 0.8 g): (**A**) 1 g on Earth with the force direction as illustrated by the arrow; (**B**) micro-g on the ISS; (**C**) micro-g on the ISS just before the centrifuges are started; (**D**) the same plant as in picture C, 70 min after the centrifuges are started, giving 0.8 g in the middle of the growth chamber with directions as illustrated by the arrow. For further details, see [58].

11. Space Experiments and Cell Wall Behavior

Despite the fact that the ISS has been in function since 2006 and quite a few plant experiments have been performed, the majority of these are already linked to the effect of micro-g on higher plant development. However, principally, samples could have been taken to observe the behavior of the cell wall characteristics, but so far this has not been implemented in any of the ten experiments (Table 1) performed in the EMCS.

The Cell Biology Experiment Facility (CBEF) installed in the Kibo Module also has been used for plant cultivation up to 62 days [62]. As described previously, in the present review article, the growth and cell wall mechanical properties of *Arabidopsis* grown for 33 days were followed in the Space Seed experiment on the Kibo Module on board the ISS [8]—for more detail on the results obtained see Section 6 "Cell Wall Changes under Micro-g Conditions—Space Experiments" in the present review. For the scientific community interested in the effect of various gravity signals on the cell wall behavior, there are possibilities for using both the EMCS and the CBEF on the ISS. However, due to the selection procedure for potential cell wall studies and the long preparation period for ISS experiments, the present authors would advise potential cell wall scientists to contact us in an attempt to implement new relevant experiments in already accepted space experiments using the EMCS (e.g., Plant Development).

12. The Effects of Gravity on the Molecular Mechanisms for Cell Wall Development

Global gene expression analysis is a powerful tool to evaluate how plants adapt to microgravity. There have been a few space experiments where the transcriptome of higher plants and cell cultures have been compared with ground controls [60,63,64]. Stutte and co-authors [28] reported that 820 of the analyzed genes were found to be expressed both under microgravity and in the ground control samples, but none of these were found to be above the two-fold cut-off. Etiolated Arabidopsis seedlings and undifferentiated cultured Arabidopsis cells have been exposed to microgravity during the BRIC-16 flight on STS-131. The post-flight analyses revealed transcription profiles that were quite similar for the cells and tissues. The genes regulated a five-fold or less in microgravity, were mainly associated with pathogen response and wounding. The response was moderate in the seedling, but more dramatic for the cell culture. Genes associated with cell wall metabolism and cell elongation were among the genes encoding transcription factors that were down regulated five-fold. As part of the TROPI-2 experiment [59] in EMCS, Arabidopsis seedlings were used to perform transcriptome analyses [60]. The experiment was primarily designed to study tropisms of plants using image analysis, and the preservation to fix the seedlings for RNA extraction on the rotor immediately following the experiment was not feasible due to hardware limitations. Still, by comparing gene expression profiles from a range of gravity treatments, they were able to identify genes involved in e.g., cell-wall development. Among the genes identified when comparing spaceflight and ground were e.g., expansin-related genes (ATEXPA4, ATEXPA7), three extensin-related genes (At4g08380, At5g06630 and LEUCIN-RICH REPEAT/EXTENSIN 1-LRX1) and pectinases and chitinases. When comparing the transcriptome of all the g-treated spaceflight samples with the microgravity samples, only 27 genes were at least two-fold differently regulated and among them was a cell wall kinase (WAK3, At1g48100).

The gene expression studies in space experiments are challenging because the time period from the end of an experiment (*i.e.*, after the stop of the 1× g rotor) until the point in time when the samples are preserved is critical. Studies have shown that genes are induced by gravity and mechanical stimulation less than 2 min in the Arabidopsis root apex [65]. The time period from the end of an experiment (*i.e.*, after the stop of the 1 × g rotor) until the samples are preserved is critical for gene expression studies. The BPS growth system on ISS allows for a sample collection and preservation within 2 min [28]. For samples in the EMCS, the preservation will vary depending of what kind of unique experiment equipment is used. When using equipment like MULTIGEN-1 and TROPI (see Table 1), approximately 1–1.5 h must be anticipated from the point in time when the rotor stops until the last sample is preserved. This depends on how quickly the samples can be removed from the EMCS Experimental Container and be inserted in e.g., MELFI (Minus Eighty Laboratory Freezer for ISS; [66]). When using GRAVI-2 equipment (optimized for experiments on etiolated lentil seedlings), the preservation can take place automatically when the samples are still under a controlled environment and a selected gravity level. A modified version of the GRAVI-2 hardware are being developed to fly the Plant Development experiment that will allow optimal growth of *Arabidopsis thaliana* seedlings under light conditions with preservation in e.g., RNAlater under a controlled environment [53]. When testing the preservation protocol for the MULTIGEN-2 experiment (finally included as part of the Plant Development experiment), the RNA quality and quantity was satisfactory for microarray analysis [66]. However, among the genes induced in the RNAlater preserved samples, three salt inducible transcription factors (ZAT10, SZF1, and SZF2) were identified, suggesting that the high salt concentration in RNAlater causes salt stress before the transcription has stopped. The results do indicate that not all the biological processes are stopped instantly by the RNAlater, and further optimization of the preservation protocol will be performed.

The Random Positioning Machines (RPMs) and clinostats allow studies of simulated microgravity effects on plants on the ground. A RPM consists of two planar frames that can be rotated independently with random speeds and directions [67–69]. In several cases, it has been demonstrated that higher plants grown on RPMs perform a similar growth and cellular response as under true microgravity conditions for the parameters studied [24,68,70]. When exposing *Arabidopsis thaliana* seedlings on a RPM under light conditions for 16 h the results demonstrated a moderate to low regulation of 55 genes (<0.2% of the analyzed genes; [71]). Most of the differentially expressed genes could be linked to either a stress or light response. The AtEXP8 was among the significantly repressed genes, and this gene encodes an expansin with a known cell-wall loosening function in plants [72] linked to auxin and both photo- and gravitropic growth responses [73]. Gene expression analyses on a DNA chip have been carried out on 6 day old plants using a 3-D clinostat in two experiments, the first lasting for 1 h and the second for 6 days [74]. Both the 1 h and the 6-day-long exposures showed a differential expression of a relatively large number of genes (400 and 500 genes, respectively). Among the induced genes were Ext (At3g244890) that has been linked to cell wall modifications and plant tip growth [75]. This gene was shown to be induced after a 1 h exposure to simulated microgravity, but not after a 24, 96 or 144 h exposure. This indicated an early response for redistribution/reformation of the composition in the cell wall structures [74]. In the same study, the XTR6 (At4g25810) gene was found to be significantly induced, especially after 24 h. The XTR6 correlates with plant cell expansion/elongation by decorating cellulose micro fibrils and forming cross-links between them [76].

13. Conclusions

The results obtained from the ground-based studies and space experiments have supported the hypothesis that plants increase the rigidity of their cell walls via modifications to the metabolism of cell wall constituents in gravity resistance. Because the type of plant materials used in space experiments is limited, the universality of this hypothesis needs to be confirmed in further space experiments. In addition, the roles of the symplast and the plasma membrane sustaining the functions of the cell wall remain to be clarified.

The earliest space experiments using rapeseed protoplasts exposed for micro-g conditions for 14 days on Biokosmos 9 demonstrated a significant decrease in the content of cellulose and hemicellulose. From the protoplasts retrieved after flight, the callus cultures established showed a retarded growth and division compared to the ground control, and the tissue cultures were not able to differentiate into mature plants. Both the IML-1 and the S/MM-03 experiments also using protoplasts confirmed some of the results obtained in Biokosmos 9; after 8–11 days of micro-g conditions the regeneration of the rapeseed protoplasts was retarded and mature plants did not develop from the retrieved calli obtained from the flight samples. The space flight protoplast experiments have shown that the critical step in the regeneration of the protoplast into an intact cell is the immediate reconstitution of the cell wall. CMTs are believed to orient the nascent cellulose microfibrils in the regenerating cell, and apparently the amount of the CMTs is anticipated to affect both the cellulose incorporation and the division capacity in cell cultures.

The results from the performed space experiments, including the few focusing on cell walls, are difficult to interpret and often inconclusive due to the use of various flight hardware, growth conditions, preservations, the small number of experiments, and few replicates. Future space experiments should therefore have standardized conditions e.g., for plant growth [77,78]. Whenever possible, the preservation for gene expression analyses should take place under environmentally controlled conditions. For the "omic data" (e.g., genomics and proteomics), a general database should be developed [77] giving the science community access to these limited and exclusive space experiment datasets.

Cell wall molecular mechanisms that are well established on Earth (e.g., cell wall loosening and selective cell wall rearrangement mechanism) are important candidates for space experiments.

Acknowledgements

We wish to thank our coworkers in the respective laboratories for their contributions to this article. We would also like to thank the European Space Center (ESA) and the Norwegian Space Center for their financial support to parts of the experimental basis for the results presented in this article. The ground-based studies and space experiments by one of the authors and his coworkers behind this article were supported by the Japan Aerospace Exploration Agency, Japan Space Forum, and Grants-in-Aid for Scientific Research from the Japan Society for the Promotion of Science.

Author Contributions

In this review article the three authors have contributed equally in the selection of articles and writing the manuscript.

Conflicts of Interest

The authors declare no conflict of interest. The funding sponsors had no role in the design of the study; in the collection, analyses, or interpretation of data; in the writing of the manuscript, and in the decision to publish the results.

References

1. Carpita, N.C.; Gibeaut, D.M. Structural models of primary cell walls in flowering plants: Consistency of molecular structure with the physical properties of the walls during growth. *Plant J.* **1993**, *3*, 1–30.
2. Hoson, T. Apoplast as the site of response to environmental signals. *J. Plant. Res.* **1998**, *111*, 167–177.
3. Hoson, T. Physiological functions of plant cell coverings. *J. Plant Res.* **2002**, *115*, 277–282.

4. Hoson, T.; Soga, K. New aspects of gravity responses in plant cells. *Int. Rev. Cytol.* **2003**, *229*, 209–244.
5. Hoson, T.; Saito, Y.; Soga, K.; Wakabayashi, K. Signal perception, transduction, and response in gravity resistance. Another graviresponse in plants. *Adv. Space Res.* **2005**, *36*, 1196–1202.
6. Hoson, T.; Matsumoto, S.; Soga, K.; Wakabayashi, K.; Hashimoto, T.; Sonobe, S.; Muranaka, T.; Kamisaka, S.; Kamada, M.; Omori, K.; *et al.* Growth and cell wall properties in hypocotyls of *Arabidopsis tua6* mutant under microgravity conditions in space. *Biol. Sci. Space* **2009**, *23*, 71–76.
7. Nakamura, T.; Saotome, M.; Ishiguro, Y.; Itoh, R.; Higurashi, S.; Hosono, M.; Ishii, Y. The effects of GA_3 on weeping of growing choots of the Japanese cherry, *Prunus spachiana. Plant Cell Physiol.* **1994**, *35*, 523–527.
8. Hoson, T.; Wakabayashi, K. Role of the plant cell wall in gravity resistance. *Phytochemistry* **2014**, doi:10.1016/j.phytochem.2014.08.022.
9. Kutschera, U.; Siebert, C.; Masuda, Y.; Sievers, A. Effects of submergence on development and gravitropism in the coleoptile of *Oryza sativa* L. *Planta* **1991**, *183*, 112–119.
10. Hoson, T. The mechanism and significance of gravity resistance in plants. *J. Gravit. Physiol.* **2006**, *13*, 97–100.
11. Volkmann, D.; Baluška, F. Gravity: One of the driving forces for evolution. *Protoplasma* **2006**, *229*, 143–148.
12. Braam, J. In touch: plant responses to mechanical stimuli. *New Phytol.* **2005**, *165*, 373–389.
13. Hoson, T. Cellulose as the anti-gravitational polysaccharide. In *Cellulose: Structure and Properties, Derivatives and Industrial Uses*; Lejeune, A., Deprez, T., Eds.; Nova Science Publishers: New York, NY, USA, 2010; pp. 293–307.
14. Verhertbruggen, Y.; Marcus, S.E.; Chen, J.; Knox, J.P. Cell wall pectic arabinans influence the mechanical properties of *Arabidopsis thaliana* inflorescence stems and their response to mechanical stress. *Plant Cell Physiol.* **2013**, *54*, 1278–1288.
15. Hoson, T.; Soga, K.; Wakabayashi, K. Role of the cell wall-sustaining system in gravity resistance in plants. *Biol. Sci. Space* **2009**, *23*, 131–136.
16. Waldron, K.W.; Brett, C.T. Effects of extreme acceleration on the germination growth and cell wall composition of pea epicotyls. *J. Exp. Bot.* **1990**, *41*, 71–77.
17. Hoson, T. Plant growth and morphogenesis under different gravity conditions: Relevance to plant life in space. *Life* **2014**, *4*, 205–216.
18. Hoson, T.; Nishitani, K.; Miyamoto, K.; Ueda, J.; Kamisaka, S.; Yamamoto, R.; Masuda, Y. Effects of hypergravity on growth and cell wall properties of cress hypocotyls. *J. Exp. Bot.* **1996**, *47*, 513–517.
19. Nakano, S.; Soga, K.; Wakabayashi, K.; Hoson, T. Different cell wall polysaccharides are responsible for gravity resistance in the upper and the basal regions of azuki bean epicotyls. *Biol. Sci. Space* **2007**, *21*, 113–116.
20. Wakabayashi, K.; Nakano, S.; Soga, K.; Hoson, T. Cell wall-bound peroxidase activity and lignin formation in azuki bean epicotyls grown under hypergravity conditions. *J. Plant Physiol.* **2009**, *166*, 947–954.
21. Chebli, Y.; Pujol, L.; Shojaeifard, A.; Brouwer, I.; van Loon, J.J.; Geitmann, A. Cell wall assembly and intracellular trafficking in plant cells are directly affected by changes in the magnitude of gravitational acceleration. *PLOS ONE* **2013**, *8*, e58246.

22. Soga, K.; Wakabayashi, K.; Kamisaka, S.; Hoson, T. Effects of hypergravity on expression of XTH genes in azuki bean epicotyls. *Physiol. Plant.* **2007**, *131*, 332–340.
23. Wakabayashi, K.; Soga, K.; Hoson, T. Phenylalanine ammonia-lyase and cell wall peroxidase are cooperatively involved in the extensive formation of ferulate network in cell walls of developing rice shoots. *J. Plant Physiol.* **2012**, *169*, 262–267.
24. Matía, I.; González-Camacho, F.; Herranz, R.; Kiss, J.Z.; Gasset, G.; van Loon, J.J.; Marco, R.; Medina, F.J. Plant cell proliferation and growth are altered by microgravity conditions in spaceflight. *J. Plant Physiol.* **2010**, *167*, 184–193.
25. Hoson, T.; Soga, K.; Wakabayashi, K.; Hashimoto, T.; Karahara, I.; Yano, S.; Tanigaki, F.; Shimazu, T.; Kasahara, H.; Masuda, D.; *et al*. Growth stimulation in inflorescences of an *Arabidopsis* tubulin mutant under microgravity conditions in space. *Plant Biol.* **2014**, *16*, S91–S96.
26. Cowles, J.R.; LeMay, R.; Jahns, G.; Scheld, W.H.; Peterson, C. Lignification in young plant seedlings grown on earth and aboard the space shuttle. In *Plant Cell Wall Polymers. Biogenesis and Biodegradation*; Lewis, N.G., Paice, M.G., Eds.; ACS Symposium Series 399. American Chemical Society: Washington, DC, USA, 1989; pp. 203–213.
27. Levine, L.H.; Heyenga, A.G.; Levine, H.G.; Choi, J.-W.; Davin, L.B.; Krikorian, A.D.; Lewis, N.G. Cell-wall architecture and lignin composition of wheat developed in a microgravity environment. *Phytochemistry* **2001**, *57*, 835–846.
28. Stutte, G.W.; Monje, O.; Hatfield, R.D.; Paul, A.-L.; Ferl, R.J.; Simone, C.G. Microgravity effects on leaf morphology, cell structure, carbon metabolism and mRNA expression of dwarf wheat. *Planta* **2006**, *224*, 1038–1049.
29. Yoshioka, R.; Soga, K.; Wakabayashi, K.; Takeba, G.; Hoson, T. Hypergravity-induced changes in gene expression in *Arabidopsis* hypocotyls. *Adv. Space Res.* **2003**, *31*, 2187–2193.
30. Matsumoto, S.; Saito, Y.; Kumasaki, S.; Soga, K.; Wakabayashi, K.; Hoson, T. Up-regulation of tubulin genes and roles of microtubules in hypergravity-induced growth modifications in *Arabidopsis* hypocotyls. *Adv. Space Res*. **2007**, *39*, 1176–1181.
31. Soga, K.; Wakabayashi, K.; Kamisaka, S.; Hoson, T. Hypergravity induces reorientation of cortical microtubules and modifies growth anisotropy in azuki bean epicotyls. *Planta* **2006**, *224*, 1485–1494.
32. Iversen, T.-H.; Rasmussen, O.; Gmunder, F.; Baggerud, C.; Kordyum, E.L.; Lozovaya, V.V.; Tairbekov, M. The effect of microgravity on the development of plant protoplasts flown on Bikosmos 9. *Adv. Space Res*. **1992**, *12*, 121–131.
33. Rasmussen, O.S.; Klimchuk, D.A.; Kordyum, E.L.; Danevich, L.A.; Tarnavskaya, E.B.; Lozovaya, V.V.; Tairbekov, M.G.; Baggerud, C.; Iversen, T.-H. The effect of exposure to microgravity on the development and structural organisation of plant protoplasts flown on Biokosmos 9. *Physiol. Plant.* **1992**, *84*, 162–170.
34. Skagen, E.B. The Early Regeneration Process in Protoplasts from *Brassica napus* Hypocotyls Cultivated under Various Gravity Forces. Ph.D. Thesis, Norwegian University of Science and Technology: Trondheim, Norway, 1999.
35. Rasmussen, O.S.; Baggerud, C.; Larssen, H.C.; Evjen, K.; Iversen T.-H. The effect of 8 days of microgravity on regeneration of intact plants from protoplasts. *Physiol. Plant*. **1994**, *92*, 404–411.
36. Rasmussen, O.; Bondar, R.; Baggerud, C.; Iversen, T.-H. Development of plant protoplasts during the IML-1 mission. *Adv. Space Res*. **1994**, *14*, 189–196.

37. Iversen, T.-H.; Johnsson, A.; Skagen, E.B.; Beisvåg, T.; Chinga, G.; Andreassen, P.; Kittang, A.-I.; Wold, A.B.; Kristensen, A.; Rasmussen, O.S. Immobilized protoplasts under microgravity conditions. Preliminary results from the S/MM-03 mission. Available online: http://adsabs.harvard.edu/full/1996ESASP.390..29I (accessed on 3 August 2014).
38. Skagen, E.B. Cortical microtubule reorganization in protoplasts isolated from *Brassica napus* hypocotyl is affected by gravity. *J. Gravit. Physiol.* **1998**, *5*, 117–120.
39. Skagen, E.-B.; Iversen, T.-H. Simulated weightlessness and hyper-g results in opposite effects on the regeneration of the cortical microtubule array in protoplasts from *Brassica napus* hypocotyls. *Physiol. Plant.* **1999**, *106*, 318–325.
40. Skagen, E.-B.; Iversen, T.-H. Effect of simulated and real weightlessness on early regeneration stages of *Brassica napus* protoplasts. *In Vitro Cell. Dev. Biol.* **2000,** *36*, 312–318.
41. Ledbetter, M.C.; Porter, K.R. A "microtubule" in plant cell fine structure. *J. Cell Biol.* **1963**, *19*, 239–250.
42. Giddings, T.H., Jr.; Staehelin, L.A. Microtubule-mediated control of microfibril deposition: A re-examination of the hypothesis. In *The Cytoskeletal Basis of Plant Growth and Form*; Lloyd, C.W., Ed.; Academic Press: London, UK, 1998; pp. 85–99.
43. Williamson, R.E. Orientation of cortical microtubules in interphase plant cells. *Int. Rev. Cytol.* **1991**, *129*, 135–206.
44. Fisher, D.D.; Cyr, R.J. Extending the microtubule/microfibril paradigm. *Plant Physiol.* **1998**, *116*, 1043–1051.
45. Mejer, E.G.M.; Simmonds, D.H. Interphase microtubule arrays in cultured mesophyll protoplasts of higher plants. *Physiol. Plant.* **1988**, *72*, 511–517.
46. Hahne, G.; Hoffmann, F. Dimethyl sulfoxide can initiate cell divisions of arrested callus protoplast by promoting cortical microtubule assembly. *Proc. Natl. Acad. Sci. USA* **1984**, *81*, 5449–5453.
47. Fossum, K.R.; Kittang, A.-I.; Iversen, T.-H.; Brinckmann, E.; Schiller, P. Testing the European Modular Cultivation System (EMCS) for ISS Plant and Cell Research. *SAE Technical Paper* **2005**, doi:10.4271/2005-01-2841.
48. Brinckmann, E. Spaceflight opportunities on the ISS for plant research—the ESA perspective. *Adv. Space Res.* **1999**, *24*, 779–788.
49. Brinckmann, E. New facilities and instruments for developmental biology research in space. *Adv. Space Biol. Med.* **2003**, *9*, 253–280.
50. Brinckmann, E. ESA hardware for plant research on the International Space Station. *Adv. Space Res.* **2005**, *36*, 1162–1166.
51. Brinckmann, E.; Brillouet, C. Space plant research on the ISS with the European Modular Cultivation System and with BIOLAB. *SAE Technical Paper* **2000**, doi:10.4271/2000-01-2472.
52. Brinckmann, E.; Schiller, P. Experiments with small animals in BIOLAB and EMCS on the International Space Station. *Adv. Space Res.* **2002**, *30*, 809–814.
53. Kittang, A.-I.; Iversen, T.-H.; Fossum, K.R.; Mazars, C.; Carnero-Diaz, E.; Boucheron-Dubuisson, E.; le Disquet, I.; Legue, V.; Herranz, R.; Pereda-Loth, V.; *et al.* Exploration of plant growth and development using the European Modular Cultivation System facility on the International Space Station. *Plant Biol.* **2014**, *16*, 528–538.

54. Kiss, J.Z.; Kumar, P.; Bowman, R.N.; Steele, M.K.; Eodice, M.T.; Correll, M.J.; Edelmann, R.E. Biocompatibility studies in preparation for a spaceflight experiment on plant tropisms (TROPI). *Adv. Space Res.* **2007**, *39*, 1154–1160.
55. Millar, K.D.L.; Kumar, P.; Correll, M.J.; Mullen, J.L.; Hangarter, R.P.; Edelmann, R.E.; Kiss, J.Z. A novel phototropic response to red light is revealed in microgravity. *New Phytol.* **2010**, *186*, 648–656.
56. Driss-Ecole, D.; Legu, V.; Carnero-Diaz, E.; Perbal, G. Gravisensitivity and automorphogenesis of lentil seedling roots grown on board the International Space Station. *Physiologia Plantarum* **2008**, *134*, 191–201.
57. Solheim, B.G.B.; Johnsson, A.; Iversen, T.-H. Ultradian rhythms in *Arabidopsis thaliana* leaves in microgravity. *New Phytol.* **2009**, *183*, 1043–1052.
58. Johnsson, A.; Solheim, B.G.B.; Iversen, T.-H. Gravity amplifies and microgravity decreases circumnutations in *Arabidopsis thaliana* stems: Results from a space experiment. *New Phytol.* **2009**, *182*, 621–629.
59. Kiss, J.Z.; Millar, K.D.; Edelmann, R.E. Phototropism of *Arabidopsis thaliana* in microgravity and fractional gravity on the International Space Station. *Planta* **2012**, *236*, 635–645.
60. Correll, M.J.; Pyle, T.P.; Millar, K.D.L.; Sun, Y.; Yao, J.; Edelmann, R.E.; Kiss, J.Z. Transcriptome analyses of *Arabidopsis thaliana* seedlings grown in space: Implications for gravity-responsive genes. *Planta* **2013**, *238*, 519–533.
61. Kiss, J.Z.; Aanes, G.; Schiefloe, M.; Coelho, L.H.F.; Millar, K.D.L.; Edelmann, R.E. Changes in operational procedures to improve spaceflight experiments in plant biology in the European Modular Cultivation System. *Adv. Space Res.* **2014**, *53*, 818–827.
62. Yano, S.; Kasahara, H.; Masuda, D.; Tanigaki, F.; Shimazu, T.; Suzuki, H.; Karahara, I.; Soga, K.; Hoson, T.; Tayama, I.; *et al.* Improvements in and actual performance of the Plant Experiment Unit onboard Kibo, the Japanese Experiment Module on the International Space Station. *Adv. Space Res.* **2013**, *51*, 780–788.
63. Paul, A.L.; Popp, M.P.; Gurley, W.B.; Guy, C.; Norwood, K.L.; Ferl, R.J. Arabidopsis gene expression patterns are altered during spaceflight. *Adv. Space Res.* **2005**, *36*, 1175–1181.
64. Paul, A.L.; Zupanska, A.K.; Ostrow, D.T.; Zhang, Y.; Sun, Y.; Li, J.L.; Shanker, S.; Farmerie, W.G.; Amalfitano, C.E.; Ferl, R.J. Spaceflight transcriptomes: unique responses to a novel environment. *Astrobiology* **2012**, *12*, 40–56.
65. Kimbrough, J.M.; Salinas-Mondragon, R.; Boss, W.F.; Brown, C.S.; Sederoff, H.W. The fast and transient transcriptional network of gravity and mechanical stimulation in the *Arabidopsis* root apex. *Plant Physiol.* **2004**, *136*, 2790–2805.
66. Kittang, A.-I.; Kvaløy, B.; Winge, P.; Iversen, T.-H. Ground testing of *Arabidopsis* preservation protocol for the microarray analysis to be used in the ISS EMCS Multigen-2 experiment. *Adv. Space Res.* **2010**, *46*, 1249–1256.
67. Borst, A.G.; van Loon, J.J.W.A. Technology and developments for the random positioning machine. RPM. *Microgravity Sci. Technol.* **2009**, *21*, 287–292.
68. Hoson, T.; Kamisaka, S.; Masuda, Y.; Yamashita, M.; Buchen, B. Evaluation of the three-dimensional clinostat as a simulator of weightlessness. *Planta* **1997**, *203*, S187–S197.

69. Van Loon, J.J.W.A. Some history and the use of random positioning machine, RPM, in gravity related research. *Adv. Space Res.* **2007**, *39*, 1161–1165.
70. Kraft, T.F.; van Loon, J.J.; Kiss, J.Z. Plastid position in *Arabidopsis* columella cells is similar in microgravity and on a random-positioning machine. *Planta* **2000**, *211*, 415–422.
71. Kittang, A.-I.; Winge, P.; van Loon, J.J.W.A.; Bones, A.M.; Iversen, T.-H. Adaptation response of *Arabidopsis thaliana* to random positioning. *Adv. Space Res.* **2013**, *52*, 1320–1331.
72. Lee, Y.; Choi, D.; Kende, H. Expansins: ever-expanding numbers and functions. *Curr. Opin. Plant Biol.* **2001**, *4*, 527–532.
73. Esmon, C.A.; Tinsley, A.G.; Ljung, K.; Sandberg, G.; Hearne, L.B.; Liscum, E. A gradient of auxin and auxin-dependent transcription precedes tropic growth responses. *Proc. Natl. Acad. Sci. USA* **2006**, *103*, 236–241.
74. Soh, H.; Auh, C.; Soh, W.Y.; Han, K.; Kim, D.; Lee, S.; Rhee, Y. Gene expression changes in *Arabidopsis* seedlings during short- to long-term exposure to 3-D clinorotation. *Planta* **2011**, *234*, 255–270.
75. Bucher, M.; Brunner, S.; Zimmermann, P.; Zardi, G.I.; Amrhein, N.; Willmitzer, L.; Riesmeier, J.W. The expression of an extensin-like protein correlates with cellular tip growth in tomato. *Plant Physiol.* **2002**, *128*, 911–923.
76. Claisse, G.; Charrier, B.; Kreis, M. The *Arabidopsis thaliana* GSK3/Shaggy like kinase AtSK3-2 modulates floral cell expansion. *Plant Mol. Biol.* **2007**, *64*, 113–124.
77. Ruyters, G.; Braun, M. Plant biology in space: recent accomplishments and recommendations for future research. *Plant Biol.* **2014**, *16*, 4–11.
78. Wolff, S.A.; Coelho, L.; Karoliussen, I.; Brinckmann, E.; Jost, A.-I.K. Effects of the extraterrestrial environment on plants: Recommendations for future space experiments for the MELiSSA Higher Plant Compartment. *Life* **2014**, *4*, 189–204.

11

Vacuolar Sorting Receptor-Mediated Trafficking of Soluble Vacuolar Proteins in Plant Cells

Hyangju Kang [1] and Inhwan Hwang [1,2,*]

[1] Department of Life Sciences, Pohang University of Science and Technology, Pohang 790-784, Korea
[2] Division of Integrative Biosciences and Biotechnology, Pohang University of Science and Technology, Pohang 790-784, Korea

* Author to whom correspondence should be addressed; E-Mail: ihhwang@postech.ac.kr

---

Abstract: Vacuoles are one of the most prominent organelles in plant cells, and they play various important roles, such as degradation of waste materials, storage of ions and metabolites, and maintaining turgor. During the past two decades, numerous advances have been made in understanding how proteins are specifically delivered to the vacuole. One of the most crucial steps in this process is specific sorting of soluble vacuolar proteins. Vacuolar sorting receptors (VSRs), which are type I membrane proteins, are involved in the sorting and packaging of soluble vacuolar proteins into transport vesicles with the help of various accessory proteins. To date, large amounts of data have led to the development of two different models describing VSR-mediated vacuolar trafficking that are radically different in multiple ways, particularly regarding the location of cargo binding to, and release from, the VSR and the types of carriers utilized. In this review, we summarize current literature aimed at elucidating VSR-mediated vacuolar trafficking and compare the two models with respect to the sorting signals of vacuolar proteins, as well as the molecular machinery involved in VSR-mediated vacuolar trafficking and its action mechanisms.

Keywords: vacuolar sorting receptors; protein trafficking to vacuoles; sorting signals; molecular machinery; soluble vacuolar proteins

---

1. Introduction

Plant cells contain a variety of endomembrane compartments. Of these, the vacuole is the most prominent organelle, occupying up to 90% of the cellular volume in mesophyll cells. Two different types of vacuoles exist in plant cells: the lytic vacuole in vegetative cells and the protein storage vacuole (PSV) in seed cells [1,2]. These vacuoles play many important roles, such as the degradation of waste materials, ion and metabolite storage, and the maintenance of turgor pressure. To perform these activities, a large number of proteins localized to both the lumen and tonoplast are required. In plant cells, nascent vacuolar proteins are initially targeted to the endoplasmic reticulum (ER) and are subsequently transported to the vacuole via multiple routes depending on the individual proteins.

Of these multiple routes, the most prominent employs multiple intermediate organelles, *i.e.*, the Golgi apparatus, trans-Golgi network (TGN), and prevacuolar compartment (PVC) [3–9]. In this route, the process of protein trafficking from the ER to the vacuole comprises many distinct steps, each of which requires a large number of protein factors. In fact, numerous molecular factors involved in this route have been characterized at the molecular and cellular levels. One of these factors is the vacuolar sorting receptor (VSR), which plays a crucial role as a sorting receptor for soluble vacuolar proteins during their trafficking to the vacuole. VSRs are highly conserved among plant species, including BP80 (binding protein 80 kD) in pea, PV72 in pumpkin, and seven VSRs in Arabidopsis [8,10–14]. VSRs are type I membrane proteins, which consist of a large N-terminal luminal domain, a transmembrane domain (TMD), and a short C-terminal cytosolic tail [10,15]. The luminal domain binds to soluble vacuolar proteins, while the cytosolic tail interacts with various trafficking components for vesicle formation and for its own recycling [8,16–20].

Alternatively, certain vacuolar proteins are transported directly from the ER to the vacuole via a Golgi-independent route [21–23]. In particular, multiple tonoplast proteins including α-tonoplast intrinsic protein, calcineurin B-like protein 6, the two-pore potassium channel TPKb, the vacuolar H^+-ATPase VHA-a3, and the H+-pyrophosphatase AVP1 are transported via a Golgi-independent route [21–25]. In pumpkin seed cells, precursor-accumulating (PAC) vesicle-mediated trafficking transports proteins from the ER to the PSV via a Golgi-independent route [26]. Tonoplast proteins can also be transported by another route; the sucrose transporter SUC4 is transported from the ER to the TGN through the Golgi apparatus and is then directly targeted from the TGN to the tonoplast without passing through the PVC in an adaptor protein complex (AP)-3-dependent manner [27].

In this review, we focus on the VSR-mediated trafficking of soluble proteins to the vacuole. For other routes of vacuolar trafficking, we recommend other recent reviews [28–31]. Here, we summarize recent advances in elucidating VSR-mediated trafficking of soluble proteins to two types of vacuoles in plants, *i.e.*, lytic vacuoles and PSVs, and we discuss the two models describing this process.

2. Involvement of VSRs in Trafficking of Soluble Vacuolar Proteins to Lytic Vacuoles and PSVs

VSRs were originally identified as abundant proteins in the CCV fraction purified from plant extract [10]. Subsequently, many VSR homologs have been identified in association with the vacuolar trafficking process in various plant species, and they were thus proposed to function as sorting receptors of soluble vacuolar proteins [13,32,33]. The biological role of VSRs has been tested by examining protein trafficking to the vacuole. The protoplast system has been widely used to investigate the role of VSRs in lytic vacuolar trafficking. Transient expression of dominant negative mutant forms of AtVSR1 and AtVSR2 causes secretion and/or inhibition of various coexpressed vacuole-destined proteins, such as Spo:GFP (a fusion protein consisting of a sorting signal of sporamin protein from sweet potato (*Ipomoea batatas*) and green fluorescent protein (GFP)), AALP:GFP (a fusion of Arabidopsis aleurain-like protein and GFP), and Spo:amylase (a fusion of sporamin and amylase, destined for the lytic vacuole) [18,19,34,35]. These studies have provided strong evidence that VSRs are sorting receptors of soluble lytic vacuolar proteins, which was confirmed by genetic studies; in *atvsr1atvsr4* double-mutant plants, a small amount of AALP is secreted into the apoplast in leaf tissues [13]. In addition, in various *atvsr* single and double mutants (*atvsr1*, *arvsr1atvsr3* and *atvsr1atvsr4*), vacuolar trafficking capacity is reduced to varying degrees depending on the genotype, although these mutants do not show any noticeable defects in vacuolar trafficking in intact leaf tissues [35]. The defect in lytic vacuolar trafficking in *atvsr1atvsr4* double-mutant plants was complemented by transient expression of AtVSR1 or AtVSR4, further confirming that these proteins are involved in lytic vacuolar trafficking. Their role in lytic vacuolar protein sorting has also been directly supported by numerous ways; ER-retained soluble PV72 (PV72-HDEL) caused accumulation of AtALEU in the ER in Arabidopsis transgenic plants [36]. Amy-spo, a chimeric vacuolar cargo consisting of the N-terminal region of sporamin and amylase, was secreted into the apoplast when a mutant form of BP80, full-length BP80-Y612A, which was mistargeted to the plasma membrane was coexpressed in tobacco protoplasts [18]. In cultured cell lines of Arabidopsis, expression of the luminal domain of AtVSR1 caused co-secretion of various vacuolar proteins into the medium [37].

In addition, the role of VSRs in PSV trafficking has been confirmed by genetic studies. In a knock-out mutant of *AtVSR1*, significant amounts of 12S globulins and 2S albumins accumulate as precursors and are partially secreted into the extracellular matrix [33]. Zouhar *et al.* [13] subsequently confirmed this observation and demonstrated that of the seven AtVSR isoforms, two isoforms, AtVSR3 and AtVSR4, are also involved in PSV trafficking and are functionally redundant to AtVSR1. Interestingly, these *atvsr* mutants have smaller PSVs than the wild type, which may be due to the reduced levels of PSV proteins in the PSVs resulting from defects in PSV trafficking. However, single and double mutants of *atvrs1*, *atvsr3*, and *atvsr4* do not show any obvious defective phenotype in their vegetative tissues.

While the physiological roles of these AtVSR isoforms have been elucidated, the roles of AtVSR5 and AtVSR6 are not yet known. These two proteins also localize primarily to the PVC in protoplasts, as does AtVSR1 [35,38]. However, in contrast to other *atvsr* mutants, *atvsr5atvsr6* double-mutant plants did not exhibit defective trafficking of protein to the two vacuoles (lytic vacuole and PSV) when two lytic vacuolar cargoes, sporamin:GFP and AALP:GFP, and two PSV proteins, 12S globulins and 2S albumins, were examined [13,35]. The difference between these two VSR isoforms (VSR5 and VSR6) and other VSR isoforms in terms of protein trafficking to the vacuoles stems from the difference in their luminal domains; when the luminal domains of AtVSR1 and AtVSR5 were swapped, the vacuolar trafficking activity of the resulting mutants was determined by the luminal domain [35]. These results indicate that the luminal domain is involved in the specificity determination of AtVSR isoforms. However, we cannot exclude the possibility that AtVSR5 and AtVSR6 may also play a role in sorting vacuolar cargoes other than those examined. Further studies are necessary to determine the exact role of these two isoforms in plant cells.

3. Vacuolar Sorting Signals and Their Interactions with VSRs

When VSRs function as sorting receptors, one of their most important activities is the specific recognition of vacuolar proteins among the numerous organellar proteins that are simultaneously transported through the endomembrane compartments. Vacuolar proteins contain a specific sequence motif, the sorting signal, which is required for specific recognition by VSRs. The sorting signals of various vacuolar proteins are classified into two groups, sequence-specific vacuolar sorting signal (ssVSS) and C-terminal vacuolar sorting signal (ctVSS). The ssVSSs show a consensus sequence while the ctVSSs are poorly defined but generally composed of hydrophobic amino acids [39]. ssVSSs, which have been identified from lytic vacuolar proteins, such as barley proaleurain and sweet potato sporamin, include NPIR or similar sequences [6,40–42]. Indeed, peptides containing ssVSSs specifically bind to VSRs *in vitro* [10,16]. Consistent with the role of VSRs in the sorting of PSV proteins, the ctVSSs of Brazil nut 2S albumin and Arabidopsis 12S globulin strongly bind to BP80 and AtVSR1, respectively, while the C-terminal sorting signal of barley lectin shows weak binding [10,16,33]. These observations raise the intriguing question of how VSRs bind to two different types of vacuolar sorting signals, the ssVSS and ctVSS. One possibility is that they are recognized by different cargo binding sites in the luminal domain.

The mechanism underlying the specific recognition of sorting signals by VSRs has been elucidated at the molecular level. VSRs must bind to cargoes at the donor compartment and release them at the acceptor compartment. Ca^{2+} ion plays a crucial role in cargo binding to VSRs *in vitro*; cargo binding to VSRs is enhanced by high concentrations of Ca^{2+} [10,43–45]. Consistent with this finding, VSRs contain a Ca^{2+}-binding motif, the epidermal growth factor (EGF)-like motif, at the luminal domain. Moreover, Ca^{2+}-binding to the EGF-like motif is crucial for its interaction with other proteins [46]. Indeed, a luminal domain lacking the EGF-like motif exhibited a low level of cargo binding. These results support the notion that Ca^{2+} is involved in the interaction between cargoes and the luminal domains of VSRs [44,47]. However, it is still not fully understood how Ca^{2+} binding to the EGF-like motif contributes to cargo binding to the luminal domain.

Another important factor for the interaction between cargoes and their receptors is pH. In animal cells, the lysosomal cargo receptor mannose-6-phosphate (M6P) receptor (MPR) binds to M6P (the sorting signal of lysosomal cargoes) at the TGN and releases these cargoes at the late endosome [48]. In this process, the pH of these compartments is crucial for their interaction and dissociation; the pH of the TGN and the late endosomes is 6.0 and 5.5, respectively [49]. Therefore, the organelle in which cargo binding takes place has a higher pH than that for cargo release, indicating that a higher pH is favorable for cargo binding and a lower pH is favorable for cargo release. For VSRs, *in vitro* experiments have shown that the optimal pH for cargo binding to VSRs is pH 6–7, while they dissociate at pH 4 [10,47]. Currently, the pH levels of the plant organelles are not yet clearly defined; in fact, the results of two recent studies examining the pH of endomembrane compartments differ [50,51]. However, neither of these studies supports the *in vitro* result showing that cargo release from VSRs occurs at pH 4.0; thus other conditions, including the Ca^{2+} concentrations in the lumens of the compartments, may contribute to cargo binding to and release from VSRs during vacuolar trafficking.

4. Molecular Mechanisms of VSR-Mediated Vacuolar Trafficking

Extensive studies have been carried out investigating the molecular mechanisms of VSR-mediated vacuolar trafficking in various plant species. These studies have resulted in the proposal of two different models describing VSR-mediated vacuolar trafficking: VSR-mediated trafficking from the TGN to the PVC (Model I) and VSR-mediated trafficking from the ER to the TGN (Model II). These models are significantly different, particularly with respect to the locations of cargo binding to and release from the VSRs, as well as the use of carrier vesicles.

4.1. Model I: VSR-Mediated Trafficking from the TGN to the PVC

Model I closely resembles the models describing lysosomal and vacuolar trafficking in animal cells and yeast, respectively, whereas Model II appears to be specific to plant cells. One of the key differences in the two models is the localization of VSRs. In Model I, VSRs primarily localize to the PVC, while a significant proportion localize to the TGN. Indeed, immunogold labeling, immunostaining using anti-VSR antibody, and live cell imaging of GFP- or RFP-tagged VSRs have revealed that VSRs localize primarily to the PVC, with a significant proportion localizing to the TGN or trans-face of the Golgi [52–55]. The TMD and cytosolic tail of VSRs are sufficient for their localization to the PVC, because GFP-VSRs (with the luminal domain replaced with GFP) localize to the PVC [34,38,54]; thus, in Model I, VSRs bind to vacuolar cargoes at the TGN and release them at the PVC, thereby cycling between the TGN and PVC (Figure 1). In this model, the behavior of VSRs is conceptually similar to that of the sorting receptors MPRs in animal cells and Vps10p in yeast [56–59]. Indeed, the recycling of VSRs from the PVC to the TGN has been confirmed; The significant amounts of Spo:GFP and AALP:GFP were accumulated to the TGN in *vps29* mutant. Moreover, in the *vps29* mutant plants, VSRs accumulate to the PVC even when anterograde trafficking is inhibited by LatB, an inhibitor of actin filaments [60].

A large number of proteins function either directly or indirectly in VSR-mediated vacuolar trafficking. These protein factors include AtVTI11, EPSIN1 (renamed EpsinR1), AP-1, clathrin, actin filaments, and VPS29. Their physiological roles in VSR-mediated vacuolar trafficking have been confirmed genetically; like the *vsr1vsr4* double mutant, *atvti11*, *vps29*, *epsinr1*, and *ap1m2* mutant plants exhibit a defect in protein trafficking to the PSV and/or lytic vacuole. Of these proteins, EpsinR1 and AP-1 are monomeric and heterotetrameric adaptors of CCVs, respectively, thus raising the possibility that CCVs function as carriers during VSR-mediated vacuolar trafficking. Indeed, AtVSR1 interacts with EpsinR1. AP-1 may also have a direct interaction with VSRs since VSRs have an AP-1 binding motif, the YXXΦ motif, at their cytosolic tail. In fact, these proteins together with clathrin appear to form an interaction network at the TGN that is crucial for CCV formation [17,52,61]. In addition, μ-adaptin of AP-4 interacts with the cytosolic tail of VSR2 [20]. Another important interaction is the homomeric interaction between VSRs via a motif in the cytosolic tail [19], which is similar to oligomerization of MPRs in animal cells [62]. Therefore, overall, VSR-mediated vacuolar trafficking is conceptually similar to CCV-mediated trafficking of lysosomal cargoes in animal cells. In the interaction of VSRs with these proteins, the YXXΦ motif in the cytosolic tail is crucial; substitution of the Y residue of the YMPL motif with alanine causes mislocalization to the plasma membrane in addition to accumulation at the TGN [8,18,63,64]. This observation again supports the idea that VSRs are loaded into CCVs via the YXXΦ binding factor AP-1.

Figure 1. Vacuolar sorting receptor (VSR)-mediated soluble cargo transport from the *trans*-Golgi network (TGN) to the prevacuolar compartment (PVC): Model I. Nascent soluble vacuolar proteins are initially targeted to the endoplasmic reticulum (ER). Subsequently, they are transported to the Golgi apparatus in a COPII-dependent manner. Traveling from *cis*-Golgi to the TGN occurs via cisternal maturation. At the TGN, VSRs recognize their cargoes and receptor-cargo complexes are packaged into clathrin-coated vesicles with the help of AP-1 and/or EpsinR1. Clathrins and adaptor proteins dissociate from the clathrin-coated vesicles (CCVs) after vesicle release. After fusion of vesicles harboring vacuolar cargoes with the PVC, VSRs release their cargoes. VSRs are recognized by VPS29-containing retromers and recycle back to the TGN for the next round of cargo sorting. The exact mechanism of how retromers recognizes VSRs at the PVC remains elusive. It is also unknown whether retromers dissociate from the recycling vesicles in plants; thus retromers on the recycling vesicle are indicated with dotted lines. VSRs are also involved in trafficking of proteins to the PSV at the TGN via CCVs. However, it is not known which adaptors are involved in this pathway. It is also not clear whether the PVCs for the lytic vacuole and PSV are the same compartment or two different organelles. m, process occurring through maturation.

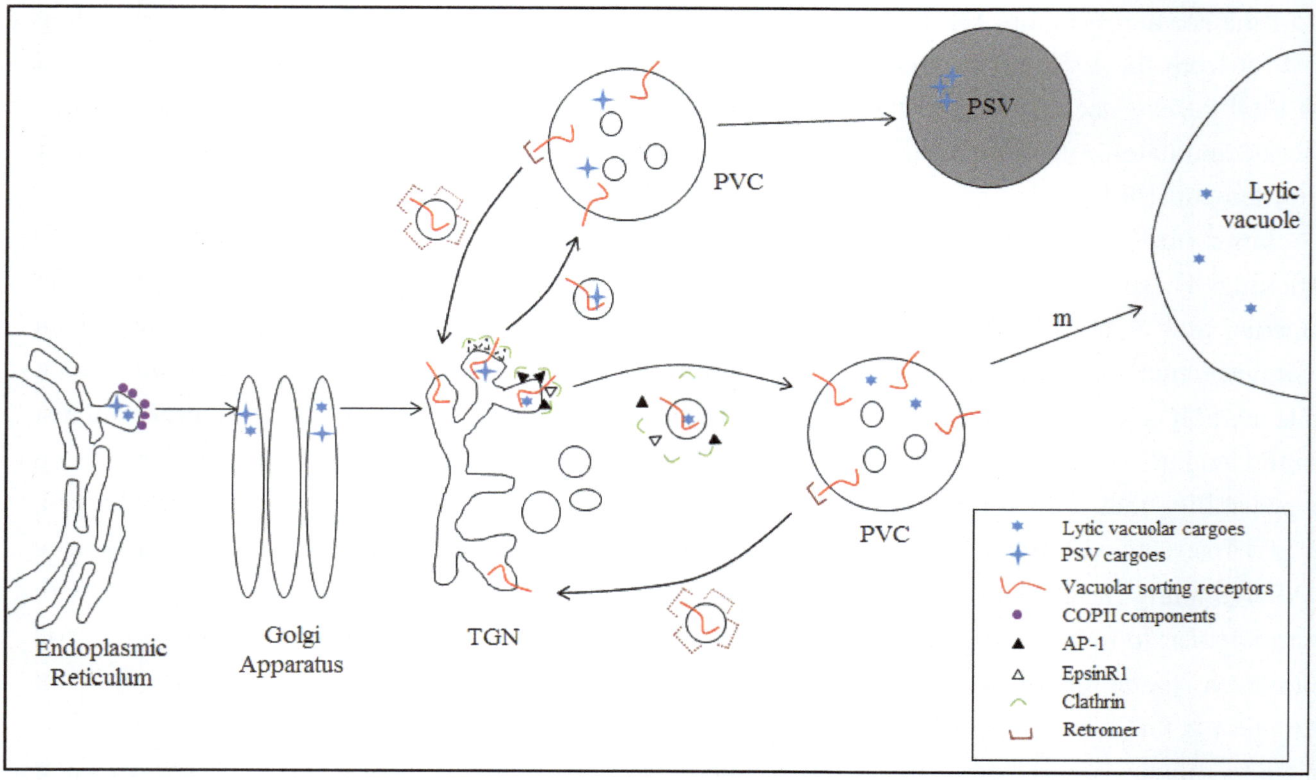

VSRs require additional molecules for their action. The majority of VSRs localize to the PVC. These VSRs must recycle to the TGN for cargo binding. In animal cells and yeast, retromer (a pentameric complex composed of two sorting nexins, VPS26, VPS29, and VPS35) is involved in the recycling of MPRs and Vps10p from the late endosome/PVC to the TGN [58,65–67]. All of the retromer components have been identified in plant cells. Moreover, VPS35, the receptor-binding subunit of retromer in animal cells and yeast [68], coimmunoprecipitates with Arabidopsis VSRs, and the VPS35/29/26 complex (cargo-selective subcomplex) localizes to the PVC [69]. Immunostaining experiments have suggested that SNXs and VPS29 localize to the PVC as well as the TGN [70–73]. Recently, however, both immunogold labeling and immunostaining experiments have demonstrated that SNX1 and SNX2a, as well as VPS29, localize to the TGN and not the PVC [74,75]. Currently, it is not easy to explain this inconsistency. Biologically active SNX1-XFP and VPS29-GFP are sensitive to both brefeldin A and wortmannin [70,71]. In addition, VPS29-GFP showed membrane localization in *snx1-1* and *snx2a-2snx2b-1* mutants. These results raise the possibility that retromer components localize to both the TGN and PVC, and they may not always function together [73]. Indeed, in mammalian cells, retromer localizes to both donor and acceptor membranes [76].

In addition, *vps29* mutant plants show defects in both PSV and lytic vacuolar trafficking [60,77]. A recent study has shown that VPS29 is crucial for recycling from the PVC to the TGN [60]. These results raise the possibility that retromer is involved in the recycling of VSRs from the PVC to the TGN. If this recycling indeed occurs, VSRs should have a specific sequence motif at the C-terminal cytosolic tail that can be recognized by retromer. Indeed, GFP-VSR2 (L615A), in which alanine is substituted for leucine in the YXXΦ motif, trafficks to the vacuole [8,18], raising the possibility that the YXXΦ motif also plays a role in retrograde trafficking. Similarly, yeast Vps10p, which also functions as a vacuolar sorting receptor, contains the YXXΦ motif, which is involved in recycling from the PVC to the TGN [78].

According to Model I, VSRs should bind to their cargoes at the TGN. Indeed, a large number of proteins involved in VSR-mediated trafficking localize to the TGN, including AtVTI11, EpsinR1, AP-1, and clathrin [17,79]. In addition, in protoplasts of *mag1-1/vps29* mutant plants, which have a defect in retrograde trafficking of VSRs from the PVC to the TGN, vacuolar cargoes accumulate to the TGN or are secreted into the medium [60]. However, a few critical questions should be answered before this model is conclusively accepted. As mentioned above, Ca^{2+} is an important cofactor for cargo binding to VSRs [10,45,47]. When the Ca^{2+} concentration was reduced from 1 mM to 10 μM, binding of the luminal domains of BP80 and AtVSR4 to a peptide containing the aleurain sorting signal dropped to 20% and 50%, respectively [45], indicating the importance of Ca^{2+} concentration for cargo binding to VSRs. The ER and Golgi contain 0.05–0.5 mM and 0.7–3 μM Ca^{2+}, respectively [80,81]. If the Ca^{2+} concentration of the TGN is similar to that of its neighboring organelle, the Golgi, it is too low to enable a high level of binding of cargoes to VSRs. However, the exact concentration of Ca^{2+} at the TGN is unknown. In addition, pH also plays a critical role in the binding of cargoes to VSRs. Currently, the exact pH levels of the TGN and PVC in plants are not well established. Two recent studies reported conflicting data on the pH of these organelles. One study reported that the pH of the TGN and PVC are 6.3 and 6.2, respectively [51], whereas the other study reported that the pH of the TGN and PVC are 6.1 and 7, respectively [50]. The underlying reason for the differences observed in the pH levels of these organelles is not fully understood. In fact, both studies employed a similar pH sensor, ratiometric pHluorin. It is likely that a combination of two factors, pH and Ca^{2+} concentration, determines the binding of cargoes to VSRs. In addition, the homomeric interaction of VSRs may also contribute to the binding of cargoes because the effective concentration of the luminal domain involved in cargo binding is increased by this homomeric interaction [19].

Another important question is how cargoes are released from VSRs at the PVC in Model I. In fact, there is no direct evidence for vacuolar cargo dissociation from VSRs in the PVC. A peptide containing an N-terminal sorting motif binds to VSRs optimally at pH 6–7 and dissociates at pH 4. However, according to two recent studies showing that the pH level of the PVC is 6.2 or 7 [50,51], cargoes may not be easily released from VSRs at the PVC. It is likely that the dissociation of cargoes from VSRs may also be determined by a combination of two factors, pH and Ca^{2+} concentration. However, additional studies are necessary to confirm the dissociation of cargoes from VSRs in the PVC.

4.2. VSR-Mediated Transport from the ER to the TGN: Model II

Recently, a new model has been proposed for VSR-mediated vacuolar trafficking in plant cells. This model is based on several observations that cannot be explained by Model I. Niemes *et al.* [82] showed that coexpression of mutant forms of SNXs or RNAi knockdown of SNXs cause accumulation of GFP-BP80 (a chimeric protein in which the luminal domain of BP80 is replaced by GFP) at the ER together with soluble vacuolar cargoes, while the secretion of α-amylase and the targeting of the Golgi protein Man1-RFP are not affected. In addition, ER-retained VSR mutants and ER-localized soluble PV72 accumulate soluble vacuolar proteins, but not Golgi proteins, in the ER [36,82]; thus, VSRs may bind to their cargoes at the ER/Golgi but not at the TGN. High concentrations of free Ca^{2+} in the ER could be favorable for binding of VSRs to cargo proteins in the ER [80,81]. In another study, Niemes *et al.* [74] found that SNX1, SNX2a, and VPS29 of retromer (involved in the recycling of VSRs) localize to the TGN in Arabidopsis protoplasts and in root cells of Arabidopsis and tobacco, as revealed by immunogold labeling and immunostaining. Stierhof *et al.* [75] also provided evidence for the TGN localization of SNX1 and SNX2a in Arabidopsis root cells by immunogold labeling. In addition, transient expression of SNX1 or SNX2a mutants in protoplasts, as well as RNAi knockdown of SNXs caused mislocalization of BP80 to the TGN. These results support Model II, in which VSRs cycle between the ER and TGN (Figure 2).

Another important difference between the two models is the trafficking of vacuolar proteins from the TGN to the PVC. Model II proposes that the trafficking of vacuolar proteins from the TGN to the PVC occurs in a receptor-independent manner without any carriers. It has been proposed that the TGN is converted to the PVC via organelle maturation [74]. Indeed, maturation of the TGN to the PVC has been suggested by electron microscopy [83]. Subsequently, the PVC may fuse with the vacuole. Rab5-to-Rab7 replacement is crucial for the PVC-vacuole fusion in Arabidopsis. The MON1/SAND-CCZ1 complex was shown to be involved in this Rab replacement process [84–86].

Figure 2. Vacuolar sorting receptor (VSR)-mediated cargo trafficking from the endoplasmic reticulum (ER) to the *trans*-Golgi network (TGN): Model II. After targeting of nascent vacuolar proteins to the ER, the folded vacuolar proteins bind to VSRs in the ER. Ca^{2+} plays a critical role in this binding. VSR-cargo complexes are transported to the cis-Golgi via COPII vesicles. The complexes are maintained until they reach the TGN, where the cargoes are released from the VSR due to the low concentrations of Ca^{2+}. VSRs are selectively recycled back to the ER by retromer. Vacuolar cargo-enriched domains of the TGN mature into the prevacuolar compartment (PVC). The endosomal sorting complexes required for transport (ESCRT) machinery might be involved in this maturation step. m, process occurring through maturation.

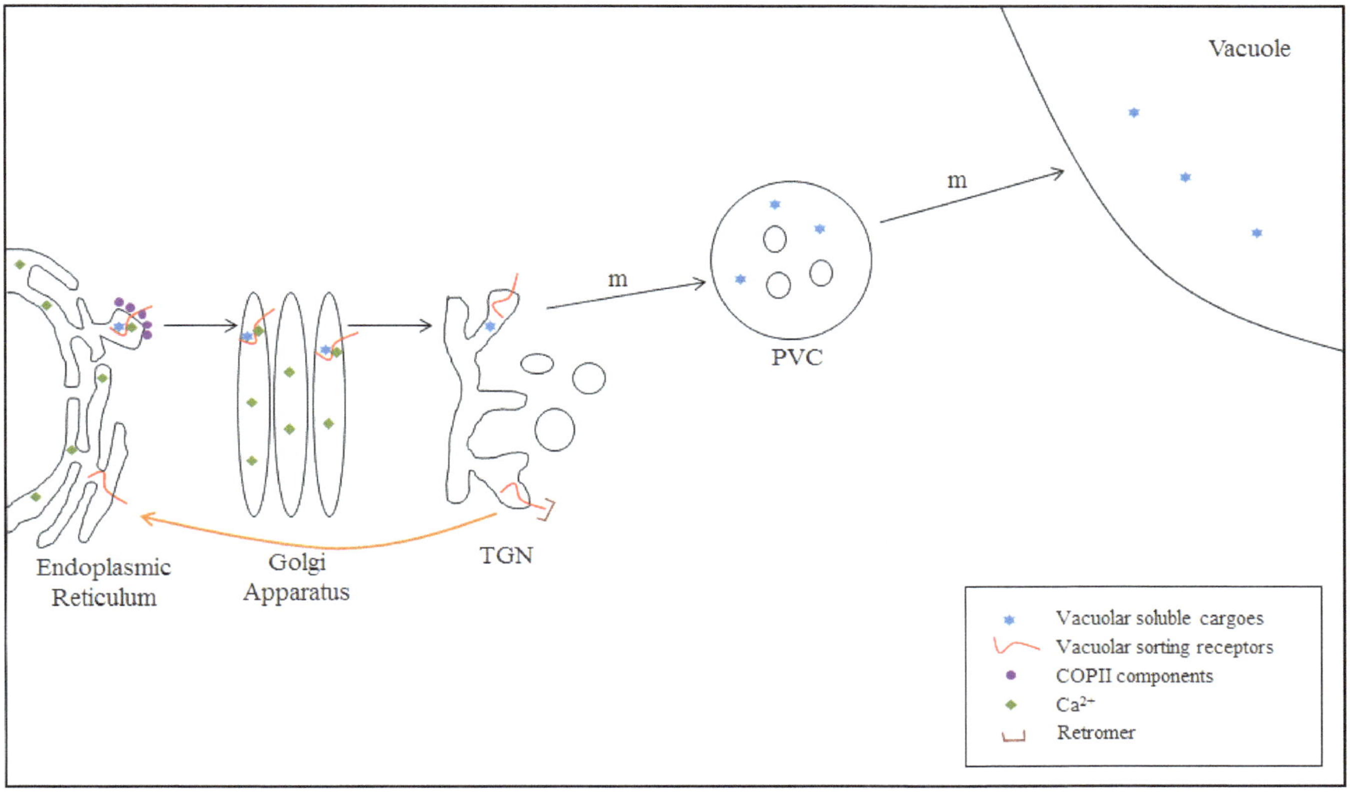

Model II raises the question of how vacuolar proteins are sorted from secreted cargo proteins at the TGN if the cargoes are dissociated from VSRs at the TGN and the vacuolar cargoes are transported from the TGN to the PVC via maturation of the TGN into the PVC. In animal cells, it has recently been demonstrated that secretory proteins are actively sorted at the TGN in a Ca^{2+}-dependent manner [87]. This active sorting involves Ca^{2+}-ATPase SPCA1 and Ca^{2+}-binding protein Cab45 at the membrane and in the lumen of the TGN, respectively [88–91]. In Arabidopsis, ECA3 (endoplasmic reticulum-type calcium ATPase 3) localizes to the Golgi/TGN/endosomes, and *eca3* mutant plants have a defect in transporting apoplastic peroxidases [92,93]. In addition, secretory vesicles form at the trans-side of the Golgi [94], which serves as a mechanism for discriminating secretory cargoes from vacuolar cargoes at the TGN. Moreover, sorting of secretory cargoes via a complex consisting of Cab45, Ca^{2+}, and Ca^{2+}-ATPase at the trans-Golgi may lead to a reduction in Ca^{2+} concentration at the TGN, thereby resulting in the release of vacuolar cargoes from VSRs. Based on these possibilities, Robinson and

Pimpl [95] suggested that Ca^{2+}-based vacuolar cargo sorting occurs at the TGN; thus, according to this model, soluble vacuolar proteins interact with VSRs at the ER and the cargo-VSR complexes are transported to the cis-Golgi via COPII vesicles together with secretory cargo proteins. At the trans-Golgi, secretory cargoes bind to the Ca^{2+}-binding protein together with Ca^{2+}-ATPase, while vacuolar cargoes remain in association with VSRs. Secretory cargoes are packaged into secretory vesicles at the "early" TGN, whereas vacuolar cargoes dissociate from VSRs at the "late" TGN due to the low concentration of Ca^{2+} and low pH achieved by the action of TGN-localized H^{+}-ATPase. At the "late" TGN, ligand-free VSRs recycle back to the ER via retromer, and vacuolar cargoes travel to the PVC via maturation of the TGN into the PVC and, finally, to the vacuole [95]. Currently, however, there is no experimental evidence for supporting the Ca^{2+}-based model in plants. On the contrary, there were several reports supporting the bulk flow secretion. For example, sporamin lacking the ssVSS and sec:GFP, a chimeric construct consisting of the leader sequence of binding protein (BiP) and GFP, were secreted into the apoplast [96,97]. sec:GFP may not have any Ca^{2+}-binding property which is required by this model. Thus, the Ca^{2+}-based model needs to be further tested in terms of both secretion and vacuolar transport in the future.

5. Conclusions and Perspectives

VSR-mediated trafficking of soluble vacuolar proteins through the endomembrane compartments has been studied extensively in plant cells. However, many questions remain unanswered. In fact, there are two different models for VSR-mediated trafficking of soluble vacuolar proteins in plant cells. One key issue is where cargoes are sorted by the receptors, VSRs, and then released from the VSRs. Another key issue is how cargoes are transported from the TGN to the PVC. Currently, it appears that more lines of evidence support Model I than Model II; however, further studies are necessary to exclude either of these two models or to support both of them. More direct evidence for cargo binding to, and release from, VSRs at their respective locations will be important for supporting both models. In addition, for Model I, more direct evidence is necessary for the involvement of clathrin at the TGN. For Model II, little information is currently available about the molecular machinery involved in VSR-mediated trafficking of soluble vacuolar proteins; thus the identification of these molecular factors will provide additional support for this model. Information about the exact pH and Ca^{2+} concentration at the Golgi, TGN, and PVC may also be crucial for explaining experimental results and designing future experiments.

Another intriguing question about the role of VSRs in vacuolar trafficking is how they are involved in trafficking proteins to the lytic vacuole and PSV in both leaf and seed cells. Lytic vacuolar proteins and PSV proteins have the sorting signals ssVSSs and ctVSSs, respectively. Thus, further studies are necessary to elucidate how VSRs function in both pathways with two different types of sorting signals.

Acknowledgments

This work was supported by a grant from the National Research Foundation of Korea (NRF) funded by the Korea government (MEST) (No. 2013070270).

Conflicts of Interest

The authors declare no conflict of interest.

References

1. Paris, N.; Stanley, C.M.; Jones, R.L.; Rogers, J.C. Plant cells contain two functionally distinct vacuolar compartments. *Cell* **1996**, *85*, 563–572.
2. Müntz, K. Deposition of storage proteins. *Plant Mol. Biol.* **1998**, *38*, 77–99.
3. Jürgens, G. Membrane trafficking in plants. *Annu. Rev. Cell Dev. Biol.* **2004**, *20*, 481–504.
4. Tang, B.L.; Wang, Y.; Ong, Y.S.; Hong, W. COPII and exit from the endoplasmic reticulum. *Biochim. Biophys. Acta* **2005**, *1744*, 293–303.
5. Traub, L.M. Common principles in clathrin-mediated sorting at the Golgi and the plasma membrane. *Biochim. Biophys. Acta* **2005**, *1744*, 415–437.
6. Hwang, I. Sorting and anterograde trafficking at the Golgi apparatus. *Plant Physiol.* **2008**, *148*, 673–683.
7. Richter, S.; Voß, U.; Jürgens, G. Post-Golgi traffic in plants. *Traffic* **2009**, *10*, 819–828.
8. Foresti, O.; Gershlick, D.C.; Bottanelli, F.; Hummel, E.; Hawes, C.; Denecke, J. A recycling-defective vacuolar sorting receptor reveals an intermediate compartment situated between prevacuoles and vacuoles in tobacco. *Plant Cell* **2010**, *22*, 3992–4008.
9. Jung, C.; Lee, G.J.; Jang, M.; Lee, M.; Lee, J.; Kang, H.; Sohn, E.J.; Hwang, I. Identification of sorting motifs of AtβFruct4 for trafficking from the ER to the vacuole through the Golgi and PVC. *Traffic* **2011**, *12*, 1774–1792.
10. Kirsch, T.; Paris, N.; Butler, J.M.; Beevers, L.; Rogers, J.C. Purification and initial characterization of a potential plant vacuolar targeting receptor. *Proc. Natl. Acad. Sci. USA* **1994**, *91*, 3403–3407.
11. Shimada, T.; Kuroyanagi, M.; Nishimura, M.; Hara-Nishimura, I. A pumpkin 72-kDa membrane protein of precursor-accumulating vesicles has characteristics of vacuolar sorting receptor. *Plant Cell Physiol.* **1997**, *38*, 1414–1420.
12. Paris, N.; Rogers, S.W.; Jiang, L.; Kirsch, T.; Beevers, L.; Phillips, T.E.; Rogers, J.C. Molecular cloning and further characterization of a probable plant vacuolar sorting receptor. *Plant Physiol.* **1997**, *115*, 29–39.
13. Zouhar, J.; Muñoz, A.; Rojo, E. Functional specialization within the vacuolar sorting receptor family: VSR1, VSR3 and VSR4 sort vacuolar storage cargo in seeds and vegetative tissues. *Plant J.* **2010**, *64*, 577–588.
14. De Marcos Lousa, C.; Gershlick, D.C.; Denecke, J. Mechanisms and concepts paving the way towards a complete transport cycle of plant vacuolar sorting receptors. *Plant Cell* **2012**, *24*, 1714–1732.
15. Paris, N.; Neuhaus, J.M. BP-80 as a vacuolar sorting receptor. *Plant Mol. Biol.* **2002**, *50*, 903–914.
16. Kirsch, T.; Saalbach, G.; Raikhel, N.V.; Beevers, L. Interaction of a potential vacuolar targeting receptor with amino- and carboxyl-terminal targeting determinants. *Plant Physiol.* **1996**, *111*, 469–474.
17. Song, J.; Lee, M.H.; Lee, G.J.; Yoo, C.M.; Hwang, I. Arabidopsis EPSIN1 plays an important role in vacuolar trafficking of soluble cargo proteins in plant cells via interactions with clathrin, AP-1, VTI11, and VSR1. *Plant Cell* **2006**, *18*, 2258–2274.

18. DaSilva, L.L.; Foresti, O.; Denecke, J. Targeting of the plant vacuolar sorting receptor BP80 is dependent on multiple sorting signals in the cytosolic tail. *Plant Cell* **2006**, *18*, 1477–1497.
19. Kim, H.; Kang, H.; Jang, M.; Chang, J.H.; Miao, Y.; Jiang, L.; Hwang, I. Homomeric interaction of AtVSR1 is essential for its function as a vacuolar sorting receptor. *Plant Physiol.* **2010**, *154*, 134–148.
20. Gershlick, D.C.; de Marcos Lousa, C.; Foresti, O.; Lee, A.J.; Pereira, E.A.; DaSilva, L.L.; Bottanelli, F.; Denecke, J. Golgi-dependent transport of vacuolar sorting receptors is regulated by COPII, AP1, and AP4 protein complexes in tobacco. *Plant Cell* **2014**, *26*, 1308–1329.
21. Jiang, L.; Rogers, J.C. Integral membrane protein sorting to vacuoles in plant cells: Evidence for two pathways. *J. Cell Biol.* **1998**, *143*, 1183–1199.
22. Bottanelli, F.; Foresti, O.; Hanton, S.; Denecke, J. Vacuolar transport in tobacco leaf epidermis cells involves a single route for soluble cargo and multiple routes for membrane cargo. *Plant Cell* **2011**, *23*, 3007–3025.
23. Viotti, C.; Krüger, F.; Krebs, M.; Neubert, C.; Fink, F.; Lupanga, U.; Scheuring, D.; Bouttè, Y.; Frescatada-Rosa, M.; Wolfenstetter, S.; *et al*. The endoplasmic reticulum is the main membrane source for biogenesis of the lytic vacuole in Arabidopsis. *Plant Cell* **2013**, *25*, 3434–3449.
24. Park, M.; Kim, S.J.; Vitale, A.; Hwang, I. Identification of the protein storage vacuole and protein targeting to the vacuole in leaf cells of three plant species. *Plant Physiol.* **2004**, *134*, 625–639.
25. Isayenkov, S.; Isner, J.C.; Maathuis, F.J. Rice two-pore K^+ channels are expressed in different types of vacuoles. *Plant Cell* **2011**, *23*, 756–768.
26. Ikuko, H.N.; Shimada, T.; Hatano, K.; Takeuchi, Y.; Nishimura, M. Transport of storage proteins to protein storage vacuoles is mediated by large precursor-accumulating vesicles. *Plant Cell* **1998**, *10*, 825–836.
27. Wolfenstetter, S.; Wirsching, P.; Dotzauer, D.; Schneider, S.; Sauer, N. Routes to the tonoplast: The sorting of tonoplast transporters in Arabidopsis mesophyll protoplasts. *Plant Cell* **2012**, *24*, 215–232.
28. Marty, F. Plant vacuoles. *Plant Cell* **1999**, *11*, 587–599.
29. Pedrazzini, E.; Komarova, N.Y.; Rentsch, D.; Vitale, A. Traffic routes and signals for the tonoplast. *Traffic* **2013**, *14*, 622–628.
30. Rojas-Pierce, M. Targeting of tonoplast proteins to the vacuole. *Plant Sci.* **2013**, *211*, 132–136.
31. Viotti, C. ER and vacuoles: Never been closer. *Front. Plant Sci.* **2014**, *5*, e20.
32. Laval, V.; Masclaux, F.; Serin, A.; Carrière, M.; Roldan, C.; Devic, M.; Pont-Lezica, R.F.; Galaud, J.P. Seed germination is blocked in Arabidopsis putative vacuolar sorting receptor (atbp80) antisense transformants. *J. Exp. Bot.* **2003**, *54*, 213–221.
33. Shimada, T.; Fuji, K.; Tamura, K.; Kondo, M.; Nishimura, M.; Hara-Nishimura, I. Vacuolar sorting receptor for seed storage proteins in *Arabidopsis thaliana*. *Proc. Natl. Acad. Sci. USA* **2003**, *100*, 16095–16100.
34. DaSilva, L.L.; Taylor, J.P.; Hadlington, J.L.; Hanton, S.L.; Snowden, C.J.; Fox, S.J.; Foresti, O.; Brandizzi, F.; Denecke, J. Receptor salvage from the prevacuolar compartment is essential for efficient vacuolar protein targeting. *Plant Cell* **2005**, *17*, 132–148.

35. Lee, Y.; Jang, M.; Song, K.; Kang, H.; Lee, M.H.; Lee, D.W.; Zouhar, J.; Rojo, E.; Sohn, E.J.; Hwang, I. Functional identification of sorting receptors involved in trafficking of soluble lytic vacuolar proteins in vegetative cells of Arabidopsis. *Plant Physiol.* **2013**, *161*, 121–133.
36. Watanabe, E.; Shimada, T.; Tamura, K.; Matsushima, R.; Koumoto, Y.; Nishimura, M.; Hara-Nishimura, I. An ER-localized form of PV72, a seed-specific vacuolar sorting receptor, interferes the transport of an NPIR-containing proteinase in Arabidopsis leaves. *Plant Cell Physiol.* **2004**, *45*, 9–17.
37. Shen, J.; Suen, P.K.; Wang, X.; Lin, Y.; Lo, S.W.; Rojo, E.; Jiang, L. An *in vivo* expression system for the identification of cargo proteins of vacuolar sorting receptors in Arabidopsis culture cells. *Plant J.* **2013**, *75*, 1003–1017.
38. Miao, Y.; Yan, P.K.; Kim, H.; Hwang, I.; Jiang, L. Localization of green fluorescent protein fusions with the seven Arabidopsis vacuolar sorting receptors to prevacuolar compartments in tobacco BY-2 cells. *Plant Physiol.* **2006**, *142*, 945–962.
39. Vitale, A.; Hinz, G. Sorting of proteins to storage vacuoles: How many mechanisms? *Trends Plant Sci.* **2005**, *10*, 316–323.
40. Holwerda, B.C.; Padgett, H.S.; Rogers, J.C. Proaleurain vacuolar targeting is mediated by short contiguous peptide interactions. *Plant Cell* **1992**, *4*, 307–318.
41. Koide, Y.; Matsuoka, K.; Ohto, M.; Nakamura, K. The N-terminal propeptide and the C-terminus of the precursor to 20-kilo-dalton potato tuber protein can function as different types of vacuolar sorting signals. *Plant Cell Physiol.* **1999**, *40*, 1152–1159.
42. Xu, X.Y.; Lee, K.H.; Dong, T.; Jeong, J.C.; Jin, J.B.; Kanno, Y.; Kim, D.H.; Kim, S.Y.; Seo, M.; Bressan, R.A.; *et al.* A vacuolar β-glucosidase homolog that possesses glucose-conjugated abscisic acid hydrolyzing activity plays an important role in osmotic stress responses in Arabidopsis. *Plant Cell* **2012**, *24*, 2184–2199.
43. Shimada, T.; Watanabe, E.; Tamura, K.; Hayashi, Y.; Nishimura, M.; Hara-Nishimura, I. A vacuolar sorting receptor PV72 on the membrane of vesicles that accumulate precursors of seed storage proteins (PAC vesicles). *Plant Cell Physiol.* **2002**, *43*, 1086–1095.
44. Watanabe, E.; Shimada, T.; Kuroyanagi, M.; Nishimura, M.; Hara-Nishimura, I. Calcium-mediated association of a putative vacuolar sorting receptor PV72 with a propeptide of 2S albumin. *J. Biol. Chem.* **2002**, *277*, 8708–8715.
45. Suen, P.K.; Shen, J.; Sun, S.S.M.; Jiang, L. Expression and characterization of two functional vacuolar sorting receptor (VSR) proteins, BP-80 and AtVSR4 from culture media of transgenic tobacco BY-2 cells. *Plant Sci.* **2010**, *179*, 68–76.
46. Selander-Sunnerhagen, M.; Ullner, M.; Persson, E.; Eleman, O.; Stenflo, J.; Drakenberg, T. How an epidermal growth factor (EGF)-like domain binds calcium. High resuolution NMR structure of the calcium form of the NH2-terminal EGF-like domain in coagulation factor X. *J. Biol. Chem.* **1992**, *267*, 19642–19649.
47. Cao, X.; Rogers, S.W.; Butler, J.; Beevers, L.; Rogers, J.C. Structural requirements for ligand binding by a probable plant vacuolar sorting receptor. *Plant Cell* **2000**, *12*, 493–506.
48. Kornfeld, S. Structure and function of the mannose 6-phosphate/insulinlike growth factor II receptors. *Annu. Rev. Biochem.* **1992**, *61*, 307–330.

49. Casey, J.R.; Grinstein, S.; Orlowski, J. Sensors and regulators of intracellular pH. *Nat. Rev. Mol. Cell Biol.* **2010**, *11*, 50–61.
50. Martinière, A.; Bassil, E.; Jublanc, E.; Alcon, C.; Reguera, M.; Sentenac, H.; Blumwald, E.; Paris, N. *In vivo* intracellular pH measurements in tobacco and Arabidopsis reveals an unexpected pH gradient in the endomembrane system. *Plant Cell* **2013**, *25*, 4028–4043.
51. Shen, J.; Zeng, Y.; Zhuang, X.; Sun, L.; Yao, X.; Pimpl, P.; Jiang, L. Organelle pH in the Arabidopsis endomembrane system. *Mol. Plant* **2013**, *6*, 1419–1437.
52. Sanderfoot, A.A.; Ahmed, S.U.; Marty-Mazars, D.; Rapoport, I.; Kirchhausen, T.; Marty, F.; Raikhel, N.V. A putative vacuolar cargo receptor partially colocalizes with AtPEP12p on a prevacuolar compartment in Arabidopsis roots. *Proc. Natl. Acad. Sci. USA* **1998**, *95*, 9920–9925.
53. Li, Y.B.; Rogers, S.W.; Tse, Y.C.; Lo, S.W.; Sun, S.S.; Jauh, G.Y.; Jiang, L. BP-80 and homologs are concentrated on post-Golgi: Probable lytic prevacuolar compartments. *Plant Cell Physiol.* **2002**, *43*, 726–742.
54. Tse, Y.C.; Mo, B.; Hillmer, S.; Zhao, M.; Lo, S.W.; Robinson, D.G.; Jiang, L. Identification of multivesicular bodies as prevacuolar compartments in *Nicotiana tabacum* BY-2 cells. *Plant Cell* **2004**, *16*, 672–693.
55. Kim, H.; Park, M.; Kim, S.J.; Hwang, I. Actin filaments play a critical role in vacuolar trafficking at the Golgi complex in plant cells. *Plant Cell* **2005**, *17*, 888–902.
56. Gabel, C.A.; Goldberg, D.E.; Kornfeld, S. Lysosomal enzyme oligosaccharide phosphorylation in mouse lymphoma cells: Specificity and kinetics of binding to the mannose 6-phosphate receptor *in vivo*. *J. Cell Biol.* **1982**, *95*, 536–542.
57. Marcusson, E.G.; Horazdovsky, B.F.; Cereghino, J.L.; Gharakhanian, E.; Emr, S.D. The sorting receptor for yeast vacuolar carboxypeptidase Y is encoded by the VPS10 gene. *Cell* **1994**, *77*, 579–586.
58. Arighi, C.N.; Hartnell, L.M.; Aguilar, R.C.; Haft, C.R.; Bonifacino, J.S. Role of the mammalian retromer in sorting of the cation-independent mannose 6-phosphate receptor. *J. Cell Biol.* **2004**, *165*, 123–133.
59. Bonifacino, J.S.; Rojas, R. Retrograde transport from endosomes to the trans-Golgi network. *Nat. Rev. Mol. Cell Biol.* **2006**, *7*, 568–579.
60. Kang, H.; Kim, S.Y.; Song, K.; Sohn, E.J.; Lee, Y.; Lee, D.W.; Hara-Nishimura, I.; Hwang, I. Trafficking of vacuolar proteins: The crucial role of Arabidopsis vacuolar protein sorting 29 in recycling vacuolar sorting receptor. *Plant Cell* **2012**, *24*, 5058–5073.
61. Park, M.; Song, K.; Reichardt, I.; Kim, H.; Mayer, U.; Stierhof, Y.D.; Hwang, I.; Jürgens, G. Arabidopsis μ-adaptin subunit AP1M of adaptor protein complex 1 mediates late secretory and vacuolar traffic and is required for growth. *Proc. Natl. Acad. Sci. USA* **2013**, *110*, 10318–10323.
62. Ghosh, P.; Dahms, N.M.; Kornfeld, S. Mannose 6-phosphate receptors: New twists in the tale. *Nat. Rev. Mol. Cell Biol.* **2003**, *4*, 202–212.
63. Happel, N.; Höning, S.; Neuhaus, J.M.; Paris, N.; Robinson, D.G.; Holstein, S.E. Arabidopsis mu A-adaptin interacts with the tyrosine motif of the vacuolar sorting receptor VSR-PS1. *Plant J.* **2004**, *37*, 678–693.

64. Saint-Jean, B.; Seveno-Carpentier, E.; Alcon, C.; Neuhaus, J.M.; Paris, N. The cytosolic tail dipeptide Ile-Met of the pea receptor BP80 is required for recycling from the prevacuole and for endocytosis. *Plant Cell* **2010**, *22*, 2825–2837.
65. Seaman, M.N. Cargo-selective endosomal sorting for retrieval to the Golgi requires retromer. *J. Cell Biol.* **2004**, *165*, 111–122.
66. Seaman, M.N.; Marcusson, E.G.; Cereghino, J.L.; Emr, S.D. Endosome to Golgi retrieval of the vacuolar protein sorting receptor, Vps10p, requires the function of the VPS29, VPS30, and VPS35 gene products. *J. Cell Biol.* **1997**, *137*, 79–92.
67. Seaman, M.N.; McCaffery, J.M.; Emr, S.D. A membrane coat complex essential for endosome-to-Golgi retrograde transport in yeast. *J. Cell Biol.* **1998**, *142*, 665–681.
68. Nothwehr, S.F.; Ha, S.A.; Bruinsma, P. Sorting of yeast membrane proteins into an endosome-to-Golgi pathway involves direct interaction of their cytosolic domains with Vps35p. *J. Cell Biol.* **2000**, *151*, 297–310.
69. Oliviusson, P.; Heinzerling, O.; Hillmer, S.; Hinz, G.; Tse, Y.C.; Jiang, L.; Robinson, D.G. Plant retromer, localized to the prevacuolar compartment and microvesicles in Arabidopsis, may interact with vacuolar sorting receptors. *Plant Cell* **2006**, *18*, 1239–1252.
70. Jaillais, Y.; Fobis-Loisy, I.; Miège, C.; Rollin, C.; Gaude, T. AtSNX1 defines an endosome for auxin-carrier trafficking in Arabidopsis. *Nature* **2006**, *443*, 106–109.
71. Jaillais, Y.; Santambrogio, M.; Rozier, F.; Fobis-Loisy, I.; Miège, C.; Gaude, T. The retromer protein VPS29 links cell polarity and organ initiation in plants. *Cell* **2007**, *130*, 1057–1070.
72. Phan, N.Q.; Kim, S.J.; Bassham, D.C. Overexpression of Arabidopsis sorting nexin AtSNX2b inhibits endocytic trafficking to the vacuole. *Mol. Plant* **2008**, *1*, 961–976.
73. Pourcher, M.; Santambrogio, M.; Thazar, N.; Thierry, A.M.; Fobis-Loisy, I.; Miège, C.; Jaillais, Y.; Gaude, T. Analyses of sorting nexins reveal distinct retromer-subcomplex functions in development and protein sorting in *Arabidopsis thaliana*. *Plant Cell* **2010**, *22*, 3980–3991.
74. Niemes, S.; Langhans, M.; Viotti, C.; Scheuring, D.; San Wan Yan, M.; Jiang, L.; Hillmer, S.; Robinson, D.G.; Pimpl, P. Retromer recycles vacuolar sorting receptors from the trans-Golgi network. *Plant J.* **2010**, *61*, 107–121.
75. Stierhof, Y.D.; Viotti, C.; Scheuring, D.; Sturm, S.; Robinson, D.G. Sorting nexins 1 and 2a locate mainly to the TGN. *Protoplasma* **2013**, *250*, 235–240.
76. McGough, I.J.; Cullen, P.J. Recent advances in retromer biology. *Traffic* **2011**, *12*, 963–971.
77. Shimada, T.; Koumoto, Y.; Li, L.; Yamazaki, M.; Kondo, M.; Nishimura, M.; Hara-Nishimura, I. AtVPS29, a putative component of a retromer complex, is required for the efficient sorting of seed storage proteins. *Plant Cell Physiol.* **2006**, *47*, 1187–1194.
78. Cooper, A.A.; Stevens, T.H. Vps10p cycles between the late-Golgi and prevacuolar compartments in its function as the sorting receptor for multiple yeast vacuolar hydrolases. *J. Cell Biol.* **1996**, *133*, 529–541.
79. Uemura, T.; Ueda, T.; Ohniwa, R.L.; Nakano, A.; Takeyasu, K.; Sato, M.H. Systematic analysis of SNARE molecules in Arabidopsis: Dissection of the post-Golgi network in plant cells. *Cell Struct. Funct.* **2004**, *29*, 49–65.

80. Ordenes, V.R.; Moreno, I.; Maturana, D.; Norambuena, L.; Trewavas, A.J.; Orellana, A. *In vivo* analysis of the calcium signature in the plant Golgi apparatus reveals unique dynamics. *Cell Calcium* **2012**, *52*, 397–404.
81. Stael, S.; Wurzinger, B.; Mair, A.; Mehlmer, N.; Vothknecht, U.C.; Teige, M. Plant organellar calcium signaling: An emerging field. *J. Exp. Bot.* **2012**, *63*, 1525–1542.
82. Niemes, S.; Labs, M.; Scheuring, D.; Krueger, F.; Langhans, M.; Jesenofsky, B.; Robinson, D.G.; Pimpl, P. Sorting of plant vacuolar proteins is initiated in the ER. *Plant J.* **2010**, *62*, 601–614.
83. Scheuring, D.; Viotti, C.; Krüger, F.; Künzl, F.; Sturm, S.; Bubeck, J.; Hillmer, S.; Frigerio, L.; Robinson, D.G.; Pimpl, P.; *et al.* Multivesicular bodies mature from the trans-Golgi network/early endosome in Arabidopsis. *Plant Cell* **2011**, *23*, 3463–3481.
84. Cui, Y.; Zhao, Q.; Gao, C.; Ding, Y.; Zeng, Y.; Ueda, T.; Nakano, A.; Jiang, L. Activation of the Rab7 GTPase by the MON1-CCZ1 complex is essential for PVC-to-vacuole trafficking and plant growth in Arabidopsis. *Plant Cell* **2014**, *26*, 2080–2097.
85. Ebine, K.; Inoue, T.; Ito, J.; Ito, E.; Uemura, T.; Goh, T.; Abe, H.; Sato, K.; Nakano, A.; Ueda, T. Plant vacuolar trafficking occurs through distinctly regulated pathways. *Curr. Biol.* **2014**, *24*, 1375–1382.
86. Singh, M.K.; Krüger, F.; Beckmann, H.; Brumm, S.; Vermeer, J.E.; Munnik, T.; Mayer, U.; Stierhof, Y.D.; Grefen, C.; Schumacher, K.; Jürgens, G. Protein delivery to vacuole requires SAND protein-dependent Rab GTPase conversion for MVB-vacuole fusion. *Curr. Biol.* **2014**, *24*, 1383–1389.
87. Von Blume, J.; Duran, J.M.; Forlanelli, E.; Alleaume, A.M.; Egorov, M.; Polishchuk, R.; Molina, H.; Malhotra, V. Actin remodeling by ADF/cofilin is required for cargo sorting at the trans-Golgi network. *J. Cell Biol.* **2009**, *187*, 1055–1069.
88. Scherer, P.E.; Lederkremer, G.Z.; Williams, S.; Fogliano, M.; Baldini, G.; Lodish, H.F. Cab45, a novel (Ca^{2+})-binding protein localized to the Golgi lumen. *J. Cell Biol.* **1996**, *133*, 257–268.
89. Von Blume, J.; Alleaume, A.M.; Cantero-Recasens, G.; Curwin, A.; Carreras-Sureda, A.; Zimmermann, T.; van Galen, J.; Wakana, Y.; Valverde, M.A.; Malhotra, V. ADF/cofilin regulates secretory cargo sorting at the TGN via the Ca^{2+} ATPase SPCA1. *Dev. Cell* **2011**, *20*, 652–662.
90. Von Blume, J.; Alleaume, A.M.; Kienzle, C.; Carreras-Sureda, A.; Valverde, M.; Malhotra, V. Cab45 is required for Ca($^{2+}$)-dependent secretory cargo sorting at the trans-Golgi network. *J. Cell Biol.* **2012**, *199*, 1057–1066.
91. Curwin, A.J.; von Blume, J.; Malhotra, V. Cofilin-mediated sorting and export of specific cargo from the Golgi apparatus in yeast. *Mol. Biol. Cell* **2012**, *23*, 2327–2338.
92. Li, X.; Chanroj, S.; Wu, Z.; Romanowsky, S.M.; Harper, J.F.; Sze, H. A distinct endosomal Ca^{2+}/Mn^{2+} pump affects root growth through the secretory process. *Plant Physiol.* **2008**, *147*, 1675–1689.
93. Mills, R.F.; Doherty, M.L.; Lopez-Marques, R.L.; Weimar, T.; Dupree, P.; Palmgren, M.G.; Pittman, J.K.; Williams, L.E. ECA3, a Golgi-localized P2A-type ATPase, plays a crucial role in manganese nutrition in Arabidopsis. *Plant Physiol.* **2008**, *146*, 116–128.
94. Kang, B.H.; Nielsen, E.; Preuss, M.L.; Mastronarde, D.; Staehelin, L.A. Electron tomography of RabA4b- and PI-4Kbeta1-labeled trans Golgi network compartments in Arabidopsis. *Traffic* **2011**, *12*, 313–329.

95. Robinson, D.G.; Pimpl, P. Receptor-mediated transport of vacuolar proteins: A critical analysis and a new model. *Protoplasma* **2014**, *251*, 247–264.
96. Matsuoka, K.; Bassham, D.C.; Raikhel, N.V.; Nakamura, K. Different sensitivity to wortmannin of two vacuolar sorting signals indicates the presence of distinct sorting machineries in tobacco cells. *J. Cell Biol.* **1995**, *130*, 1307–1318.
97. Batoko, H.; Zheng, H.Q.; Hawes, C.; Moore, I. A rab1 GTPase is required for transport between the endoplasmic reticulum and golgi apparatus and for normal golgi movement in plants. *Plant Cell* **2000**, *12*, 2201–2218.

12

Overexpression of a Gene Involved in Phytic Acid Biosynthesis Substantially Increases Phytic Acid and Total Phosphorus in Rice Seeds

Yusuke Tagashira, Tomoe Shimizu, Masanobu Miyamoto, Sho Nishida and Kaoru T. Yoshida *

Graduate School of Agricultural and Life Sciences, The University of Tokyo, Bunkyo-ku, Tokyo 113-8657, Japan; E-Mails: tagashira0812@gmail.com (Y.T.); 1176hours@gmail.com (T.S.); masanyobu@yahoo.co.jp (M.M.); asho3@mail.ecc.u-tokyo.ac.jp (S.N.)

* Author to whom correspondence should be addressed; E-Mail: ayosida@mail.ecc.u-tokyo.ac.jp

Academic Editors: Francesca Sparvoli, Roberto Pilu and Eleonora Cominelli

---

Abstract: The manipulation of seed phosphorus is important for seedling growth and environmental P sustainability in agriculture. The mechanism of regulating P content in seed, however, is poorly understood. To study regulation of total P, we focused on phytic acid (inositol hexakisphosphate; $InsP_6$) biosynthesis-related genes, as $InsP_6$ is a major storage form of P in seeds. The rice (*Oryza sativa* L.) low phytic acid mutant *lpa1-1* has been identified as a homolog of archael 2-phosphoglycerate kinase. The homolog might act as an inositol monophosphate kinase, which catalyzes a key step in $InsP_6$ biosynthesis. Overexpression of the homolog in transgenic rice resulted in a significant increase in total P content in seed, due to increases in $InsP_6$ and inorganic phosphates. On the other hand, overexpression of genes that catalyze the first and last steps of $InsP_6$ biosynthesis could not increase total P levels. From the experiments using developing seeds, it is suggested that the activation of $InsP_6$ biosynthesis in both very early and very late periods of seed development increases the influx of P from vegetative organs into seeds. This is the first report from a study attempting to elevate the P levels of seed through a transgenic approach.

Keywords: ectopic expression; mineral element; molecular breeding; *Oryza sativa* L.; phosphorus; phytic acid; seed; translocation

---

1. Introduction

Early development of seedlings is completely dependent on seed nutrient reserves. Seeds accumulate a large amount of phosphorus (P) to sustain seedling growth. Seeds store P mainly in the form of phytic acid (inositol hexakisphosphate; $InsP_6$), with approximately 70% to 80% of total P stored in the form of $InsP_6$ [1]. After imbibition, phytase hydrolyzes $InsP_6$ in seeds, and the resulting available P is remobilized into shoots and roots.

Initial seedling growth is supported by available P in the seeds. As the plant develops, it proceeds from P-heterotrophy (P supply from seed) to P-autotrophy (uptake of external P via roots). In maize (*Zea mays* L.), the P-heterotrophic growth phase continues for 4 d after sowing, and the P-autotrophic phase starts after 16 d after sowing [2]. From 5 to 15 d after sowing, seedling growth is supported by both seed P and external P. The supply of nutrients from seed reserves to support early seedling development is, therefore, substantial. Ros *et al.* (1997) investigated the effect of seed P levels on early growth of rice and confirmed the beneficial effects from an increase in seed P content on plant growth and, in particular, the growth of roots [3].

In *Arabidopsis thaliana* L., reduced total P content was observed in seeds of the *atpap26* mutant, in which acid phosphatase activity was markedly reduced in the leaves. There was also a significant decrease in remobilization of P from old, senesced leaves to new leaves and to seeds [4]. Seed germination of the *atpap26* mutant was delayed. These results suggest that a reduction in total P content has a negative effect on seed performance. On the contrary, higher seed P content has led to more rapid seedling emergence and larger biomass in several species [5,6].

Control of seed total P content is important not only for seed performance but also for environmental sustainability of P in agriculture [1]. However, the control mechanism of total P content in seed is poorly understood. The majority of seed P is stored in the form of $InsP_6$, so it is plausible that total P content in seed might increase if the $InsP_6$ content is increased. In fact, $InsP_6$ and total P contents are closely correlated in beans [7,8]. To elevate the $InsP_6$ level in a seed, it is important to activate the $InsP_6$ biosynthetic pathway by increasing the expression level of the rate-limiting enzymes in that pathway. *Myo*-inositol 3-phosphate synthase (MIPS), which catalyzes the first step of $InsP_6$ biosynthesis and inositol metabolism, has been considered as a key enzyme in these pathways [9] (Figure 1). There are two MIPS genes in the rice genome, *RINO1* (Os03g0192700) and *RINO2* (Os10g0369900). *RINO1* is responsible for $InsP_6$ biosynthesis in rice seeds, because *RINO1* transcript levels are extremely high in developing seeds and *RINO2* mRNA is scarcely detected [10]. Another important step is the last step of $InsP_6$ biosynthesis, from inositol pentaphosphate ($InsP_5$) to $InsP_6$, which is catalyzed by inositol 1,3,4,5,6-penta*kis*phosphate 2-kinase (IPK1) (Figure 1). The rice *IPK1* gene, *OsIPK1* (Os04g0661200), is highly expressed in developing seeds [10]. Activation of the first or last step of $InsP_6$ biosynthesis might lead to activation of $InsP_6$ biosynthesis and an increase in total P content in seeds.

The step from $InsP_1$ to inositol diphosphate ($InsP_2$) is another key step in the $InsP_6$ biosynthetic pathway, as $InsP_1$ lies at an important branch point between $InsP_6$ biosynthesis and inositol metabolism (Figure 1). Kim *et al.* (2008) identified a rice low phytic acid mutant (*lpa1*) as a homolog of 2-phosphoglycerate kinase (2-PGK), which catalyzes the step from 2-phosphoglycerate to 2,3-bisphosphoglycerate in archaea [11]. From structural analogy of substrates and products, the protein encoded by rice *lpa1* might function as an $InsP_1$ kinase, which catalyzes the step from $InsP_1$ to $InsP_2$ [1,11].

If the protein catalyzes the key branch point step of $InsP_6$ biosynthesis, overexpression of the protein might lead to an increase in $InsP_6$ accumulation in seeds.

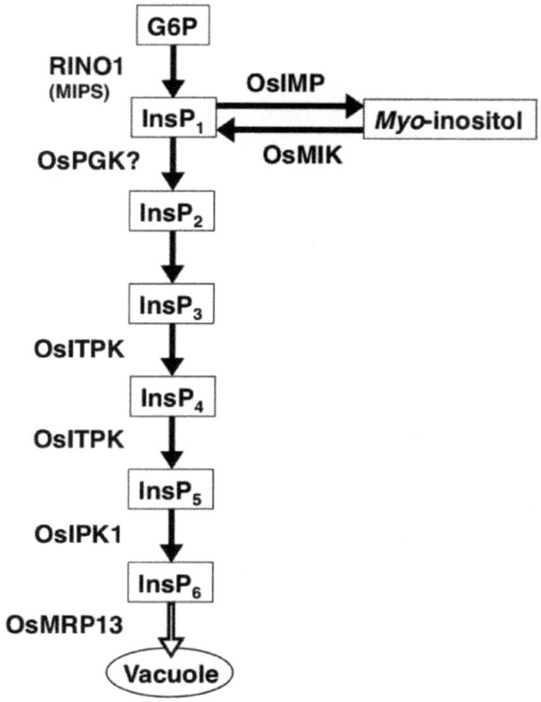

Figure 1. Scheme for biosynthesis of phytic acid (Ins P_6) and inositol in rice.

In the rice genome, there are two putative 2-PGK homologs, which were identified by a BLAST search. We designated the homologs as OsPGK1 (Lpa1; Os02g0819400) and OsPGK2 (Os09g057220). Alignment of the deduced amino acid sequence of OsPGK1 indicated an approximately 60% similarity to OsPGK2. OsPGK1 might play a major role in phytic acid biosynthesis in developing seeds, because the mutation of this gene resulted in a severe lpa phenotype [11,12]. The expression pattern of *OsPGK1* is immature seed-specific, which is apparent from the microarray data in RiceXPro (http://ricexpro.dna.affrc.go.jp).

In this study, to increase $InsP_6$ content in rice seeds, we generated transgenic rice plants that overexpressed the rice genes *RINO1*, *OsIPK1*, and *OsPGK1* under the control of an actin promoter. We designated the *RINO1*, *OsIPK1*, and *OsPGK1* overexpressors as RINO1ox, OsIPK1ox, and OsPGK1ox, respectively. We investigated $InsP_6$ and total P content in seeds from these transgenic plants and discovered that only OsPGK1ox significantly increased $InsP_6$ and total P contents in the seed, compared to the non-transformant (NT) seed. We discuss the causal mechanisms of these increases.

2. Results

2.1. The Effect of RINO1 or OsIPK1 Overexpression on $InsP_6$ Accumulation in Seeds

We first observed the seed phenotypes of RINO1ox and OsIPK1ox, as these enzymes have been well characterized in $InsP_6$ biosynthesis [9,10]. From the three fixed progeny lines of each transformant, we selected a RINO1ox line and an OsIPK1ox line, based on their high expression in T_3 plants (Figure 2A). $InsP_6$

content in T_4 seeds was measured by ion chromatography. $InsP_6$ content in the RINO1ox and OsIPK1ox seeds was slightly higher than in NT seed, albeit not significantly so. (Figure 2B). Seed total P content in both transgenic plants was identical to that of NT seed (Figure 2C).

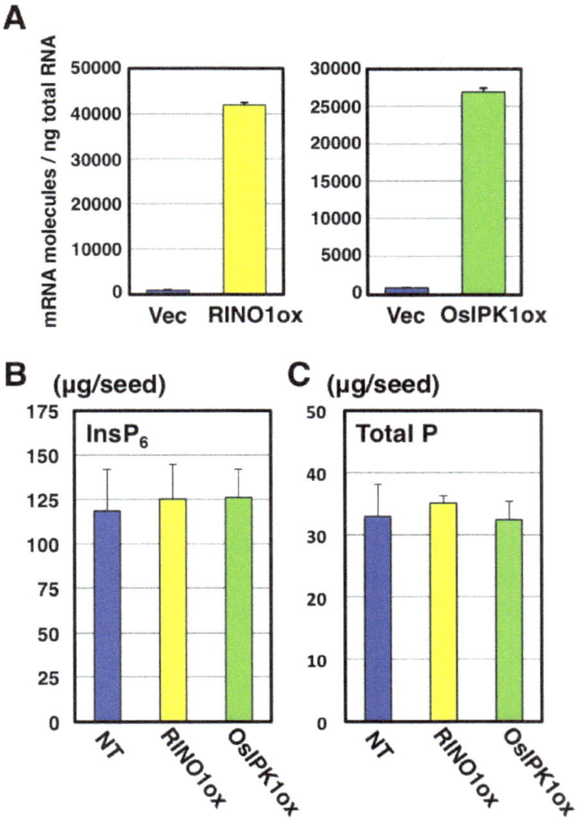

Figure 2. Gene expression and seed phenotype of RINO1ox and OsIPK1ox transgenic plants. (**A**) Quantitative RT-PCR analysis of *RINO1* (left) and *OsIPK1* (right) genes with cDNA templates from leaves of vector control (Vec) and RINO1ox or OsIPK1ox plants 7 d after germination ($n = 3$); (**B**) $InsP_6$ content in mature seeds obtained from non-transformant (NT) and two transgenic plants was determined by ion chromatography ($n \geq 7$); (**C**) Total P content in NT and two transgenic seeds was measured by colorimetric assay ($n \geq 5$). Each value (**A** to **C**) represents the mean ± SD.

2.2. Seed Phenotype of the OsPGK1 Overexpressing Transformants

As a next step, we generated transgenic rice plants carrying a rice *2-PGK* homolog gene (*OsPGK1*), driven by a rice actin promoter. Five primary transformants (OsPGK1ox) were obtained after regeneration from antibiotic-resistant calli. All plants produced viable T_1 seeds. After measuring $InsP_6$ content in T_1 seeds, we selected two transgenic lines (OsPGK1ox-1 and OsPGK1ox-2), which exhibited increased seed $InsP_6$ contents, and the fixed progeny lines were used in subsequent experiments. We confirmed overexpression of the *OsPGK1* gene in these transgenic lines (Figure 3A). We selected an azygous line (AZ), which was the transformed line without the transgene, derived from an OsPGK1ox-2 line.

We examined the phenotype of T_4 seeds. Although the dried mature seed weights of the OsPGK1ox lines were significantly less than weights of NT and AZ seeds (Figure 3B), there were no significant differences between the NT and OsPGK1ox lines in terms of germination rate, early seedling growth, and plant height. There was a significant increase in total $InsP_6$ content in the OsPGK1ox-1 seeds and a slight increase in total $InsP_6$ content in the OsPGK1ox-2 seeds compared with NT and AZ seeds (Figure 3C). Contrary to expectations, Pi content also increased in the seeds of OsPGK1ox lines (Figure 3D). Seed total P content was significantly higher in both transgenic lines than in NT and AZ (Figure 3E). Total P content increased by 1.29-fold for OsPGK1ox-2 seeds and 1.37-fold for OsPGK1ox-1 seeds, compared with NT seeds.

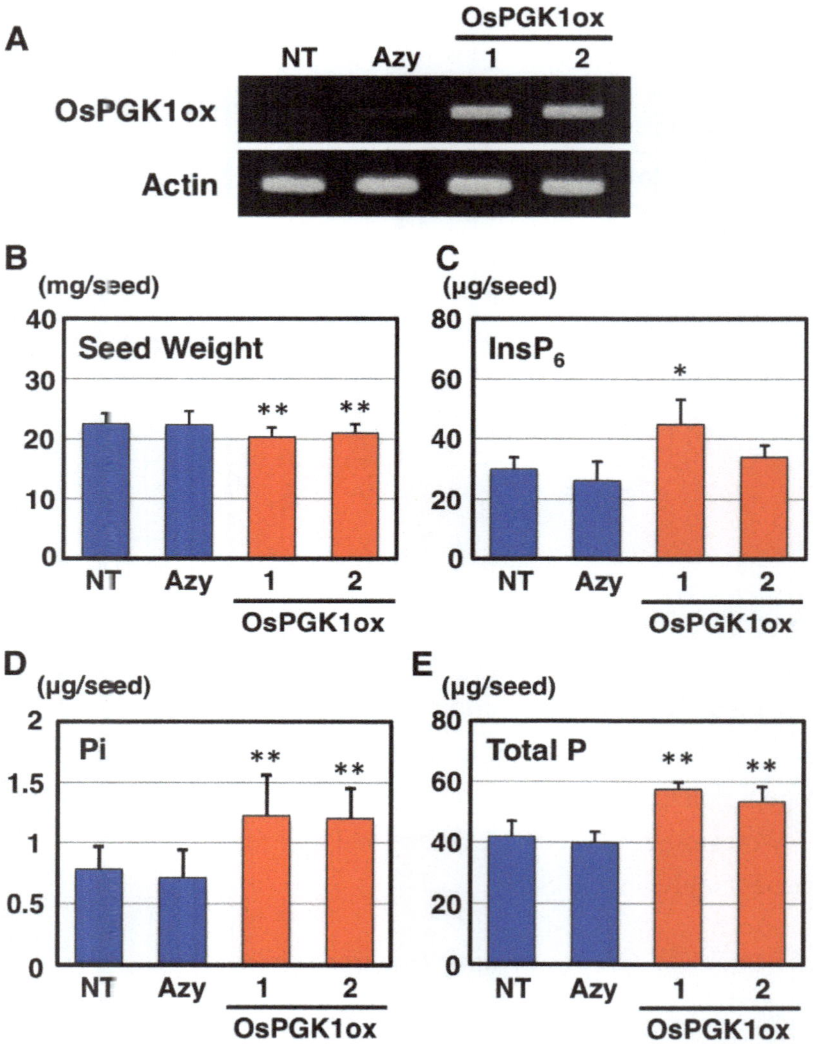

Figure 3. Gene expression and seed phenotype of OsPGK1ox transgenic plants. (**A**) Semi-quantitative RT-PCR analysis of the expression of *OsPGK1*. Total RNA was extracted from flowers of non-transformants (NT), azygous plants (Azy), and two independent OsPGK1ox lines (1 and 2) just before flowering. Actin was used as a reference; (**B**) Seed weights ($n = 20$); (**C**) $InsP_6$ content was determined by ion chromatography ($n = 4$); (**D**) Inorganic phosphate content was measured by colorimetric assay using molybdate staining ($n = 10$); (**E**) Total P content was measured by ICP-OES analysis ($n = 6$). Each value (**B** to **E**) represents the mean ± SD. * and ** indicate $p < 0.05$ and $p < 0.01$, respectively.

2.3. Accumulation of $InsP_6$ and Total Phosphorus in Developing Seeds of OsPGK1ox

To determine changes in $InsP_6$ content during seed development, seeds of OsPGK1ox-1 and NT were analyzed by ion chromatography at 7, 10, 15, 20, and 25 d after flowering (DAF; Figure 4A). $InsP_6$ content in NT seeds increased from 7 to 20 DAF and stopped at 20 DAF. In contrast, the increase in $InsP_6$ content in the OsPGK1ox-1 seeds continued from 7 to 25 DAF. $InsP_6$ content in the OsPGK1ox-1 seeds at 25 DAF was significantly higher than $InsP_6$ content in the NT seeds. Although $InsP_6$ content in the OsPGK1ox-1 seeds was slightly higher than $InsP_6$ content in the NT seeds from 7 to 20 DAF, the rates of $InsP_6$ content increase were similar for the OsPGK1ox-1 and NT seeds during this period (Figure 4A).

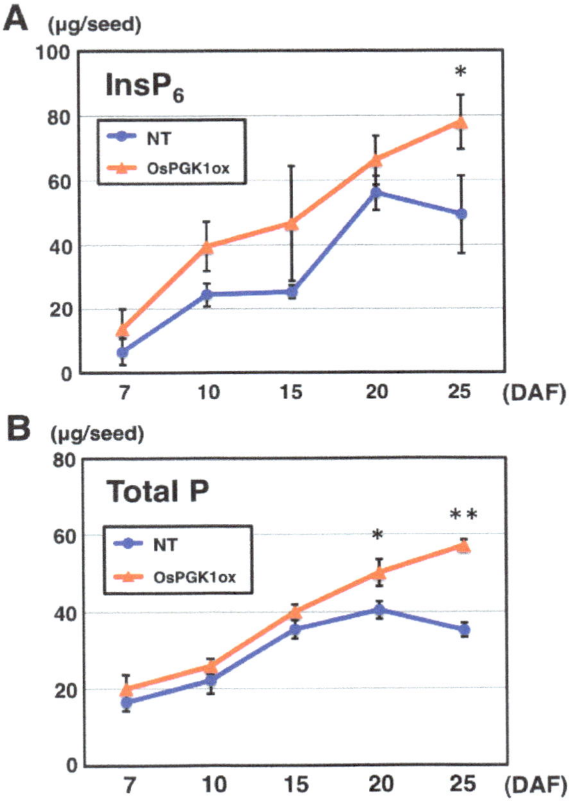

Figure 4. Changes in $InsP_6$ (**A**) and total P (**B**) content in immature seeds of OsPGK1ox-1 and NT during seed development from 7 to 25 d after flowering (DAF). $InsP_6$ and total P contents were determined by ion chromatography and ICP-OES analyses, respectively. Each value represents the mean ± SD of three replicates. * and ** indicate $p < 0.05$ and $p < 0.01$, respectively.

Similar results were obtained for total P content in developing seeds analyzed by inductively coupled plasma optical-emission spectrometry (ICP-OES; Figure 4B). Total P content in NT seeds rapidly increased until 15 DAF, and then gradually increased between 15 and 20 DAF. The increase in total P content stopped at 20 DAF in the NT seeds. In contrast, the increase in total P content in the OsPGK1ox-1 seeds continued from 7 to 25 DAF. Total P content in the OsPGK1ox-1 seeds at 20 and 25 DAF was significantly higher than total P content in the NT seeds. These results indicate that an influx of P from vegetative organs to seeds continued after 20 DAF only in the OsPGK1ox-1 plants. Although total P content in the OsPGK1ox-1 line, from 7 to 20 DAF, was slightly higher than total P content in the NT line, the rates of increase in total P were similar for OsPGK1ox-1 and NT seeds during this period (Figure 4B).

2.4. Mineral Contents in the Seeds of OsPGK1ox

In OsPGK1ox plants, a reduced number of grains per panicle was observed compared to NT panicles. The reductions were 40.4% for OsPGK1ox-1 and 27.1% for OsPGK1ox-2. We measured the contents of several mineral elements (Mg, K, Ca, Fe, and Zn) in OsPGK1ox-1 seeds, using ICP-OES analysis, to determine if the decrease in grains per panicle affected mineral influx from vegetative organs into seeds (Figure 5). Calcium was the negative control, because the influx of Ca into a seed via the phloem is limited during seed development [13]. The OsPGK1ox seeds tended to exhibit increased contents of mineral elements, except for Ca, as compared with NT seeds (Figure 5). Contents of mineral elements in the OsPGK1ox seeds increased by 1.07-fold for Mg and K, 1.05-fold for Fe, and 1.13-fold for Zn, compared to the NT seeds. However, total P content in the same OsPGK1ox-1 seeds was 1.37-fold higher than total P content in the NT seeds (Figure 3E).

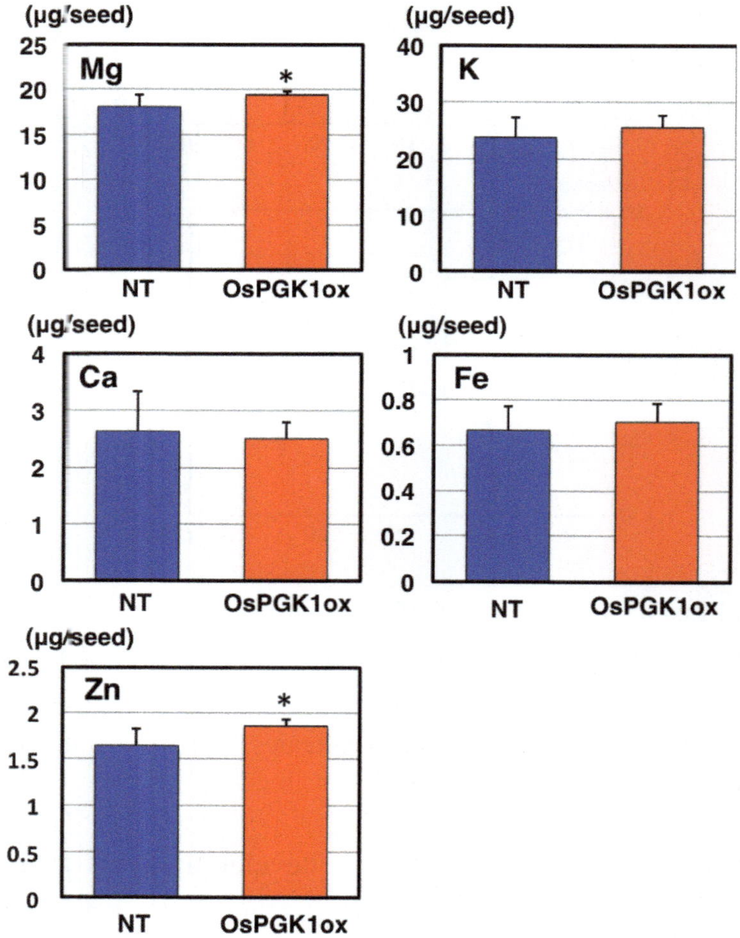

Figure 5. Contents of Mg, K, Ca, Fe, and Zn in mature seeds of OsPGK1ox and NT. The mineral contents were measured by ICP-OES analysis. The analysis in Figures 3E and 5 was performed simultaneously using the same seed samples. Each value represents the mean ± SD of six replicates. * indicates $p < 0.05$.

2.5. Expression of Phytic Acid Biosynthesis-related Genes in OsPGK1ox

To determine whether the overexpression of *OsPGK1* affects expression levels of other phytic acid-biosynthesis related genes, we analyzed the expression of several genes (*RINO1*, *OsIMP*, *OsMIK*, *OsITPK*, *OsIPK1*, *OsMRP13*, and *OsST*; Figure 1) by RT-PCR (Figure 6). OsIMP is a *myo*-inositol monophosphatase, OsMIK is a *myo*-inositol kinase, and OsITPK is an inositol 1,3,4-tris*kis*phosphate 5/6-kinase. OsMRP13/OsABCC13 is an ortholog of AtMRP5, which is a putative transporter of $InsP_6$ [14]. OsST is an ortholog of barley (*Hordeum vulgare* L.) HvST, which is a member of the sulfate transporter gene family and is the causal gene of a barley *lpa* mutant [15].

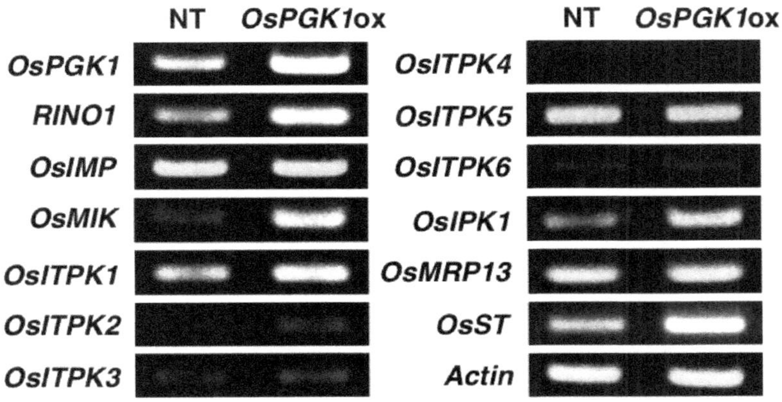

Figure 6. Semi-quantitative RT-PCR analysis of the expression of 13 phytic acid biosynthesis-related genes in the OsPGK1ox and NT plants. Total RNA was extracted from the roots of non-transformant (NT) and OsPGK1ox-1 3 d after germination. Actin was used as a reference.

In our previous studies, real-time RT-PCR analysis of mRNA levels revealed strong up-regulation of the phytic acid-biosynthesis related genes (e.g., *RINO1*, *OsIPK1*, and *OsITPK*) in immature rice seeds. The expression levels of these genes were more than tenfold greater in seeds than in vegetative tissues [10]. It is difficult to detect changes in the expression levels of the genes using analysis of total RNA extracted from immature seeds. To detect transcript level changes induced by overexpression of *OsPGK1*, we used total RNA extracted from roots as RT-PCR templates. $InsP_6$ in roots could be detected by ion chromatography even if the $InsP_6$ content was 1% of that of seeds (data not shown). Therefore, we concluded that the $InsP_6$ synthetic pathway is also active in roots, even though activity was lower in roots than in seeds.

To determine the effect of overexpression of *OsPGK1* on expression of the other genes, we used total RNA extracted from roots for RT-PCR analysis. Among phytic acid biosynthesis-related genes, expression of the *OsMIK*, *OsITPK2*, *OsITPK3*, *OsITPK4*, and *OsITPK6* genes was barely detected in the NT roots (Figure 6). However, expression of *OsMIK* gene was markedly induced, and expression of *OsITPK2* and *OsITPK3* genes was slightly induced in OsPGK1ox roots. Additionally, the expression levels of *RINO1*, *OsITPK1*, and *OsIPK1* in OsPGK1ox were higher than in the NT (Figure 6). The expression of *OsST* was also higher in OsPGK1ox than in NT, although we do not know its function in $InsP_6$ biosynthesis. The expression levels of transcripts of *OsIMP* and a putative phytic acid transporter, *OsMRP13*, displayed similar differences between OsPGK1ox and NT.

3. Discussion

There have been many reports on *lpa* mutants and on lpa transgenic plants [1,16]. In *lpa* seeds, the reduction in $InsP_6$ content is often accompanied by a molar-equivalent increase in Pi content, so total P content is almost unchanged. There are, however, two *lpa* mutants that have a reduction in total P content. Arabidopsis *atmrp5*, a putative transporter for $InsP_6$, reduced total P content by 35% [14], and barley *lpa1-1/hvst*, a homolog of a sulfate transporter, reduced total P content by 15% [15,17]. There have been no reports on mutants or transgenic plants with increases in $InsP_6$ or total P content. To our knowledge, this is the first report of significant increases in both $InsP_6$ and total P contents in seeds.

Overexpression of *RINO1* or *OsIPK1* was not effective in activating $InsP_6$ biosynthesis (Figure 2). In the RINO1ox plants, accumulation of $InsP_1$ and its dephosphorylated product, inositol, has been observed [18]. This suggests that $InsP_1$ is rapidly dephosphorylated into inositol in RINO1ox plants (Figure 1), which might explain the limited increase in $InsP_6$ in RINO1ox seeds (Figure 2). For OsIPK1ox, it is possible that there is a rate-limiting step before the last step of $InsP_6$ biosynthesis, which is catalyzed by OsIPK1, and, therefore, overexpression of *OsIPK1* has a very slight effect on activation of $InsP_6$ biosynthesis. We do not know if expression of $InsP_6$ biosynthesis-related genes is induced in both RINO1ox and OsIPK1ox lines. In any case, the fact that only the overexpression of *OsPGK1* led to activation of $InsP_6$ biosynthesis suggests that *OsPGK1* is a key gene for $InsP_6$ synthesis and that OsPGK1 is involved in the branch and, probably, the rate-limiting step from $InsP_1$ to $InsP_2$.

From the RT-PCR analysis results, we revealed that the expression of many phytic acid biosynthesis-related genes increased in OsPGK1ox plants (Figure 6). Among these genes, the expression of *OsMIK*, *myo*-inositol kinase, which catalyzes the step from *myo*-inositol to $InsP_1$, was markedly induced in OsPGK1ox plants. If OsPGK1 is involved in the branch step from $InsP_1$ to $InsP_2$, overexpression of *OsPGK1* might lead to a deficiency in $InsP_1$. It is plausible that the expression of *OsMIK* and *RINO1* might be raised in OsPGK1ox to compensate for the $InsP_1$ deficiency.

The rates of $InsP_6$ content increase, from 7 to 20 DAF, were similar for the OsPGK1ox and NT seeds (Figure 4A). This indicated that the effect of overexpression of *OsPGK1* was barely observed from 7 to 20 DAF. In contrast, after 20 DAF, synthesis of $InsP_6$ continued in the OsPGK1ox seed, but not in the NT seed (Figure 4A). Also, $InsP_6$ content in the OsPGK1ox seed was higher than that in NT seed at 7 DAF, although the difference was not significant.

Expression levels of the $InsP_6$ biosynthesis-related genes are usually highest between 7 and 21 DAF [10]. Microarray analysis revealed that the level of transcripts in *OsPGK1* seed is also high during that period, but not during the other periods (RiceXPro). It is plausible that the overexpression of *OsPGK1* was very effective in activating $InsP_6$ biosynthesis before 7 DAF and after 20 DAF, which explained the increase in $InsP_6$ content in OsPGK1ox seeds at very early and very late periods (Figure 4A). We plan to compare the gene expression levels in OsPGK1ox and NT seeds before 7 DAF and after 20 DAF in future studies.

In this study, the increase in $InsP_6$ content was accompanied by an increase in total P content. In the OsPGK1ox line, the increase in total P content continued after 20 DAF and total P content at 7 DAF was higher than total P content in the NT line, as was the case for $InsP_6$. This indicates that influx of P from vegetative organs into seeds is more active in OsPGK1ox, and the activity of $InsP_6$ biosynthesis affects the influx of P. The signal for determining the amount of P influx and involvement of OsPGK1 in P influx has not been elucidated. In Arabidopsis, AtPAP26 encoding purple acid phosphatase is up-regulated during leaf senescence and the *atpap26* mutant displayed delayed leaf senescence and reduced seed total P content [4]. This indicates that AtPAP26 plays a key role in remobilization of P from old leaves to seeds. Waters and Grusak (2008) investigated quantitative trait loci (QTL) that control seed P concentration in two Arabidopsis recombinant inbred line populations [19]. Some phosphate transporters belonging to the phosphate transporter 1 (Pht1) family were listed as candidate genes in the QTL regions [19]. The Pht1 Pi transporters are active in the uptake of Pi from the soil and its translocation within the plant, and some transporters might affect P remobilization from vegetative organs into seeds. Attention should be paid to both the phosphatases and the Pi transporters that are specific for P remobilization.

In this study, the number of grains per panicle was greatly reduced in the OsPGK1ox plant. The level of *myo*-inositol in the OsPGK1ox cells might be lower than NT cells, because the overexpression of *OsPGK1* induced the high expression of *OsMIK* gene (Figures 1 and 6). *myo*-Inositol is a central compound in diverse biochemical processes, including signal transduction, stress protection, cell wall biogenesis, growth regulation, and membrane trafficking [20]. Therefore, it is possible that a decrease of *myo*-inositol level in shoot apical meristems caused by ectopic expression of *OsPGK1* leads to an alteration in the development of inflorescences and floral organs in the OsPGK1ox plant. This might result in the reduced number of grains per panicle.

The reduced number of grains per panicle might influence the translocation of mineral elements. However, contents of mineral elements other than P increased only slightly in the OsPGK1ox line (Figure 5). Therefore, the large increase in P content in the seeds of OsPGK1ox was mainly due to the specific response to overexpression of *OsPGK1*, not to the reduced number of grains per panicle. We plan to attempt to generate transgenic plants that overexpress *OsPGK1* under the control of a seed specific promoter, such as the 18-kDa oleosin promoter, which promotes expression in the embryo and aleurone layer, where phytic acid is synthesized [16].

Many reports have suggested that seed P content has a beneficial effect on seed performance, in terms of germination rate and early seedling growth [3–6]. However, OsPGK1ox seed performance was similar to that of NT seed, although seed total P content in OsPGK1ox increased by 1.3 to 1.4-fold. Further study is needed to determine the reason for identical seed performance of the OsPKG1ox and NT plants.

Control of the total P content of seeds is important to enhance P sustainability and decrease the environmental impacts of agriculture [1]. We hope that this report provides the first step toward manipulating seed total P.

4. Experimental Section

4.1. Transformation of Rice Plants

OsIPK1 or *OsPGK1* cDNA were amplified by RT-PCR using total RNA prepared from immature seeds (cv. Nipponbare) at 10 DAF. The gene-specific primer pairs were 5'-CTGATTCTGTGTGGGGATGG-3' and 5'-AAATTCGGCCTACTGCTGAG-3' for *OsIPK1*, and 5'-GGGAGGCCTCTTCTTGATTC-3' and 5'-TTGACACCGGAGGCACTATG-3' for *OsPGK1*. Amplified cDNA with a length of 1578 bp for *OsIPK1* or 2523 bp for *OsPGK1* was cloned into the binary vector pBIAct/nos containing the rice *Actin1* promoter [21] and *nos* terminator. The method of plasmid construction was similar to that described previously [18]. Transgenic rice (*Oryza sativa* L., cv. Kitaake) was produced by the *Agrobacterium*-mediated method and grown in a glasshouse. The method to produce *RINO1*-overexpressing rice was described previously [18].

The presence of introduced genes was confirmed by PCR using specific primer sets for the *Act1* promoter (5'-TCCCTCAGCATTGTTCATCG-3') and the *RINO1* (5'-ACCAGCTCCGTCGTGTCGTA-3'), *OsIPK1* (5'-GGTGCCGGTTGTCCCTTGTC-3'), and *OsPGK1* (5'-GCCTTGCATCCCATGAGTTG-3') genes.

4.2. Measurements of Seed Components

$InsP_6$ content in T_4 seeds was measured by ion-chromatography and the Pi levels of T_4 seeds were measured by molybdate-staining assay. The detailed method of Pi and $InsP_6$ measurements was described previously [22]. Total P content was determined by either colorimetric assay or by inductively coupled plasma optical-emission spectrometry (ICP-OES) analysis. The detailed methods using the colorimetric assay were described in Kuwano *et al.* (2009) [16]. The methods of ICP-OES analysis of total P and other minerals (Mg, K, Ca, Fe, and Zn) were described previously [23]. Before analysis, the developing T_4 seeds at 7, 10, 15, 20, 25 DAF were vacuum-freeze dried for 2 d and mature seeds were dried for 2 d at 60 °C.

4.3. Expression Analysis

Quantitative RT-PCR analysis of *RINO1* and *OsIPK1* genes was performed with cDNA templates from leaves of vector control plants carrying the empty vector, RINO1ox (T_3), or OsIPK1ox (T_3) plants at 7 d after germination. The detailed methods were described in Suzuki *et al.* (2007) [10].

For RT-PCR analysis, total RNA was prepared from flowers just before flowering or roots at 3 d after germination of NT (cv. Kitaake) and OsPGK1ox plants. The detailed methods of RT-PCR analysis were described previously [23]. The *Actin* gene was used as the control. The gene-specific PCR primer sets are listed in Table S1. The experiment was repeated three times with three biologically independent RNA samples.

5. Conclusions

To regulate total P in seeds, we focused on $InsP_6$ biosynthesis-related genes because $InsP_6$ is a major storage form of P in seeds. We generated transgenic rice plants that overexpressed $InsP_6$-biosynthesis key genes, *RINO1*, *OsIPK1*, and *OsPGK1*. Only the overexpression of *OsPGK1* resulted in a significant increase in seed total P, due to the increases in $InsP_6$ and Pi. It is strongly suggested that the overexpression of *OsPGK1* may lead the increase in influx of P from vegetative organs into seeds and may activate $InsP_6$ biosynthesis. This is a first report to elevate the seed P levels through a transgenic approach.

Acknowledgments

The ICP-OES analysis was supported by Japan Advanced Plant Science Network. The authors are grateful to the members of Laboratory of Plant Molecular Genetics at the University of Tokyo for comments, and participation in discussions.

Author Contributions

Kaoru T. Yoshida and Yusuke Tagashira conceived and designed the experiments; Tomoe Shimizu, Masanobu Miyamoto and Kaoru T. Yoshida generated the transgenic rice and selected the progeny lines; Yusuke Tagashira performed the other experiments; Sho Nishida contributed the analysis of ICP-OES; Kaoru T. Yoshida and Yusuke Tagashira wrote the paper; Kaoru T. Yoshida, Yusuke Tagashira and Sho Nishida read and approved the final manuscript.

Conflicts of Interest

The authors declare no conflict of interest.

References

1. Raboy, V. Approaches and challenges to engineering seed phytate and total phosphorus. *Plant Sci.* **2009**, *177*, 281–296.
2. Nadeem, M.; Mollier, A.; Morel, C.; Prud'homme, L.; Vives, A.; Pellerin, S. Remobilization of seed phosphorus reserves and their role in attaining phosphorus autotrophy in maize (*Zea mays* L.) seedlings. *Seed Sci. Res.* **2014**, *24*, 187–194.
3. Ros, C.; Bell, R.W.; White, P.F. Effect of seed phosphorus and soil phosphorus applications on early growth of rice (*Oryza sativa* L.) cv. IR66. *Soil Sci. Plant Nutr.* **1997**, *43*, 499–509.
4. Robinson, W.D.; Carson, I.; Ying, S.; Ellis, K.; Plaxton, W.C. Eliminating the purple acid phosphatase AtPAP26 in *Arabidopsis thaliana* delays leaf senescence and impairs phosphate remobilization. *New Phytol.* **2012**, *196*, 1024–1029.
5. De Marco, D.G. Effect of seed weight, and seed phosphorus and nitrogen concentration on the early growth of wheat seedlings. *Aust. J. Exp. Agric.* **1990**, *30*, 545–549.
6. Thompson, B.D.; Bell, R.W.; Bolland, M.D.A. Low seed phosphorus concentration depresses early growth and nodulation of narrow-leafed lupin. *J. Plant Nutr.* **1992**, *15*, 1193–1214.
7. Lolas, G.M.; Markakis, P. Phytic acid and other phosphorus compounds of beans (*Phaseolus vulgaris* L.). *J. Agric. Food Chem.* **1975**, *23*, 13–15.
8. Cichy, K.A.; Caldas, G.V.; Snapp, S.S.; Blair, M.W. QTL analysis of seed iron, zinc, and phosphorus levels in an Andean bean population. *Crop Sci.* **2009**, *49*, 1742–1750.
9. Yoshida, K.T.; Wada, T.; Koyama, H.; Mizobuchi-Fukuoka, R.; Naito, S. Temporal and spatial patterns of accumulation of the transcript of *myo*-inositol-1-phosphate synthase and phytin-containing particles during seed development in rice. *Plant Physiol.* **1999**, *119*, 65–72.
10. Suzuki, M.; Tanaka, K.; Kuwano, M.; Yoshida, K.T. Expression pattern of inositol phosphate-related enzymes in rice (*Oryza sativa* L.): Implications for the phytic acid biosynthetic pathway. *Gene* **2007**, *405*, 55–64.

11. Kim, S.I.; Andaya, C.B.; Goyal, S.S.; Tai, T.H. The rice *OsLpa1* gene encodes a novel protein involved in phytic acid metabolism. *Theor. Appl. Genet.* **2008**, *117*, 769–779.
12. Zhao, H.-J.; Liu, Q.-L.; Ren, X.-L.; Wu, D.-X.; Shu, Q.-Y. Gene identification and allele-specific marker development for two low phytic acid mutations in rice (*Oryza sativa* L.). *Mol. Breed.* **2008**, *22*, 603–612.
13. Epstein, E.; Bloom, A.J. *Mineral Nutrition of Plants: Principles and Perspectives*, 2nd ed.; Sinauer Associates, Inc.: Sunderland, UK, 2004; p. 400.
14. Nagy, R.; Grob, H.; Weder, B.; Green, P.; Klein, M.; Frelet-Barrand, A.; Schjoerring, J.K.; Brearley, C.; Martinoia, E. The *Arabidopsis* ATP-binding cassette protein AtMRP5/AtABCC5 is a high affinity inositol hexa*kis*phosphate transporter involved in guard cell signaling and phytate storage. *J. Biol. Chem.* **2009**, *284*, 33614–33622.
15. Ye, H.; Zhang, X.Q.; Broughton, S.; Westcott, S.; Wu, D.; Lance, R.; Li, C. A nonsense mutation in a putative sulphate transporter gene results in low phytic acid in barley. *Funct. Integr. Genomics* **2011**, *11*, 103–110.
16. Kuwano, M.; Mimura, T.; Takaiwa, F.; Yoshida, K.T. Generation of stable "low phytic acid" transgenic rice through antisense repression of the 1D-*myo*-inositol 3-phosphate synthase gene (*RINO1*) using the 18-kDa oleosin promoter. *Plant Bioltechnol. J.* **2009**, *7*, 96–105.
17. Ockenden, I.; Dorsch, J A.; Reid, M.M.; Lin, L.; Grant, L.K.; Raboy, V.; Lott, J.N.A. Characterization of the storage of phosphorus, inositol phosphate and cations in grain tissues of four barley (*Hordeum vulgare* L.) low phytic acid genotypes. *Plant Sci.* **2004**, *167*, 1131–1142.
18. Kusuda, H.; Koga, W.; Kusano, M.; Oikawa, A.; Saito, K.; Hirai, M.Y.; Yoshida, K.T. Ectopic expression of *myo*-inositol 3-phosphate synthase induces a wide range of metabolic changes and confers salt tolerance in rice. *Plant Sci.* **2015**, *232*, 49–56.
19. Waters, B.M.; Grusak, M.A. Quantitative trait locus mapping for seed mineral concentrations in two *Arabidopsis thaliana* recombinant inbred populations. *New Phytol.* **2008**, *179*, 1033–1047.
20. Irvine, R.F.; Schell, M.J. Back in the water: The return of the inositol phosphates. *Nat. Rev. Mol. Cell Biol.* **2001**, *2*, 327–338.
21. Zhang, W.; McElroy, D.; Wu, R. Analysis of rice *Act1* 5'region activity in transgenic rice plants. *Plant Cell* **1991**, *3*, 1155–1165.
22. Kuwano, M.; Ohyama, A.; Tanaka, Y.; Mimura, T.; Takaiwa, F.; Yoshida, K.T. Molecular breeding for transgenic rice with low-phytic-acid phenotype through manipulating *myo*-inositol 3-phosphate synthase gene. *Mol. Breed.* **2006**, *18*, 263–272.
23. Iwai, T.; Takahashi, M.; Oda, K.; Terada, Y.; Yoshida, K.T. Dynamic changes in the distribution of minerals in relation to phytic acid accumulation during rice seed development. *Plant Physiol.* **2012**, *160*, 2007–2014.

13

Determining the Composition of Lignins in Different Tissues of Silver Birch

Kurt V. Fagerstedt [1,*], Pekka Saranpää [2], Tarja Tapanila [2], Juha Immanen [1,3], Juan Antonio Alonso Serra [1] and Kaisa Nieminen [2]

[1] Department of Biosciences, Division of Plant Biology, P.O. Box 65, University of Helsinki, FI-00014 Helsinki, Finland; E-Mails: juha.immanen@helsinki.fi (J.I.); juan.alonsoserra@helsinki.fi (J.A.A.S.)

[2] Natural Resources Institute Finland, Jokiniemenkuja 1, Fi-01300 Vantaa, Finland; E-Mails: pekka.saranpää@luke.fi (P.S.); tarja.tapanila@luke.fi (T.T.); kaisa.nieminen@luke.fi (K.N.)

[3] Institute of Biotechnology, P.O. Box 65, Helsinki University, FI-00014 Helsinki, Finland

* Author to whom correspondence should be addressed; E-Mail: kurt.fagerstedt@helsinki.fi

Academic Editor: Philip Harris

Abstract: Quantitative and qualitative lignin analyses were carried out on material from the trunks of silver birch (*Betula pendula* Roth) trees. Two types of material were analyzed. First, whole birch trunk pieces were cryosectioned into cork cambium, non-conductive phloem, the cambial zone (conductive phloem, cambium and differentiating xylem), lignified xylem and the previous year's xylem; material that would show differences in lignin amount and quality. Second, clonal material from one natural birch population was analyzed to show variations between individuals and between the lignin analysis methods. The different tissues showed marked differences in lignin amount and the syringyl:guaiacyl (S/G) ratio. In the non-conductive phloem tissue containing sclereids, the S/G ratio was very low, and typical for phloem fibers and in the newly-formed xylem, as well as in the previous year's xylem, the ratio lay between five and seven, typical for broadleaf tree xylem. Clonal material consisting of 88 stems was used to calculate the S/G ratios from the thioacidolysis and CuO methods, which correlated positively with an R^2 value of 0.43. Comparisons of the methods indicate clearly that the CuO method is a good alternative to study the monomeric composition and S/G ratio of wood lignins.

Keywords: acetyl bromide; *Betula pendula*; cupric oxide; lignin analysis methods; phloem; thioacidolysis; xylem

---

1. Introduction

Lignin is an aromatic, heterogenic cell wall heteropolymer that is essential for mechanical support, water transport and disease resistance in trees [1]. Lignin can be fully removed, but only under drastic chemical conditions using either strong acid (e.g., sulfite pulping process) conditions under elevated temperatures and pressures or with ionic solvents. As the removal of lignin is costly, there has been considerable interest in reducing lignin content by genetic engineering of trees as a means of improving the efficiency of the pulping process, reducing the use of chemicals and improving profitability [2]. Lignin renders wood xylem recalcitrant in the biorefinery processes when bioethanol is being produced. Lignin is also a valuable polymer and serves as a platform chemical for various compounds, replacing fossil raw materials, such as carbon fibers, dispersants, vanillin and new types of plastics [3]. Hence, it is important to know the lignin composition to understand how it can be processed further into new products.

Deciduous wood lignin is composed of syringyl (S) and/or guaiacyl (G) units linked by a series of ether and carbon-carbon bonds. The ether β-*O*-4 linkages are both frequent and labile, which makes them the target of the delignification process. In contrast, the carbon-carbon linkages are resistant, especially the biphenyl 5–5 bonds involving the aromatic C-5 position, which is available for interunit linkages only in G units. Thus, conifer wood lignin essentially made of G units is less susceptible to kraft delignification than deciduous wood lignin [4].

The most commonly-used methods for quantitative analysis of lignin are the Klason and AcBr methods. Klason is a gravimetric method and only measures insoluble material after hydrolysis with 72% H_2SO_4. It is often combined with spectrophotometric determination of dissolved lignin [5]. Klason is regarded as being generally reliable for woody stem material. Recently, this procedure has been modified to allow application to a sample size of 50 mg [6]. The accuracy of lignin determination depends also on tissue and cell wall type [7]. Various non-lignin components, such as tannins, non-extracted polysaccharides and proteins [7,8], can be present in the lignin residue and, thus, limit the applicability of the Klason method in tissues other than xylem. The acetyl bromide method has been reported to be the simplest and fastest among the methods evaluated, presenting similar or best recovery of lignin in all tissues assessed [9]. It is based on the solubilization of lignin and the determination of absorbance values at 280 nm. The acetyl bromide protocol is based on the formation of acetyl derivatives in non-substituted OH groups and bromide replacement of the Cα-OH groups to produce a complete solubilization of the cell wall material under acidic conditions. However, an overestimation of the lignin content can occur due to the oxidative degradation of structural polysaccharides (e.g., xylans) during the incubation of the cell wall with the acid solution [9,10]. One difficulty with the acetyl bromide method is the need for a well-defined lignin standard to calibrate the method [7].

Chemical degradation methods are worthwhile when compositional information on lignin is needed. Thioacidolysis has been applied to a wide variety of materials and isolated lignins [11,12]. Thioacidolysis is an acid-catalyzed method that results in β-*O*-4 cleavage. The detection of C_6C_3 products (*p*-hydroxyphenyl (H), guaiacyl (G) and syringyl (S) monomers) gives information on the composition of non-condensed alkyl aryl structures, and the results can be used for calculating the S/G-ratios.

Alkaline cupric (II) oxide oxidation (CuO) is a technique commonly used to analyze the composition of lignins in complex sample matrixes, such as soils and sediments [13,14]. Relatively mild oxidation with CuO induces cleavage of β-*O*-4 ether bonds in lignin. The lignin macromolecule is hydrolyzed, yielding phenolic CuO oxidation products with aldehydic, ketonic and acidic side chains [15–17]. The degradation products obtained by both thioacidolysis and CuO methods can be analyzed by a variety of means, including gas chromatography/mass spectrometry (GC/MS). Neither thioacidolysis nor the CuO method completely depolymerize lignin, and therefore, the result does not cover the total content of lignin, but gives valuable information about the lignin substructures.

The aim of the present piece of work is to study the compositions of birch lignins using both the CuO and thioacidolysis methods. For this purpose, we chose the wood material carefully: First, to use the methods on wood material of natural variability, we used 22 different birch clones created to represent different genotypes found in natural stands. Second, to estimate how large variations in the sample ranging from outer phloem to mature xylem will affect lignin composition, we used longitudinal cryosections carefully cut from all of the different tissues of the silver birch trunk. Hence, we had material that contained different types of lignin (phloem fibers *vs.* secondary xylem), high amounts of pectin, proteins and hemicelluloses (cambial region) and tissues with high concentrations of hardwood-type lignin with a high S/G-ratio. Our major aim was to test the reliability and pertinence of the CuO and thioacidolysis methods.

2. Results and Discussion

2.1. Lignin Composition of Different Silver Birch Tissue Fractions

Different tissues were cryosectioned from 13-year-old whole silver birch stems and divided into five fractions (Figure 1). The fractions were subjected to CuO treatment to cleave β-*O*-4 ether bonds in lignin. The phenolic compounds derived are shown in Figures 2–4. The four major groups include guaiacyl (G), syringyl (S), hydroxycinnamyl (C) and *p*-hydroxyphenyl (H) moieties.

The fractions showed distinctive differences in the products of CuO treatment (Figure 2; note the different scales in the graphs). Aldehydes (vanillin and syringaldehyde) were the major products, with most of the vanillin in the non-conductive phloem, while syringaldehyde was the major product in the lignified xylem. The acid content (vanillic and syringic acid) increased from cork cambium to the xylem. Large variations between the samples were noticeable in the acids.

4-Hydroxy compounds were the products detected in Fractions 1 to 3, but there were only traces in the xylem. Within the hydroxycinnamyl group, ferulic acid was the dominating compound with minor amounts in the xylem (Figure 3). The amount of *p*-coumaric increased from the cork cambium to the cambium and was detected only in minor amounts in the xylem. The amount of ferulic acids in all fractions was at about the same level, except in the cambial zone (Figure 3). Ferulic acid has been shown to take part in the lignin polymer, and it has been postulated as an initiation or nucleation site at the beginning of lignin polymerization [18].

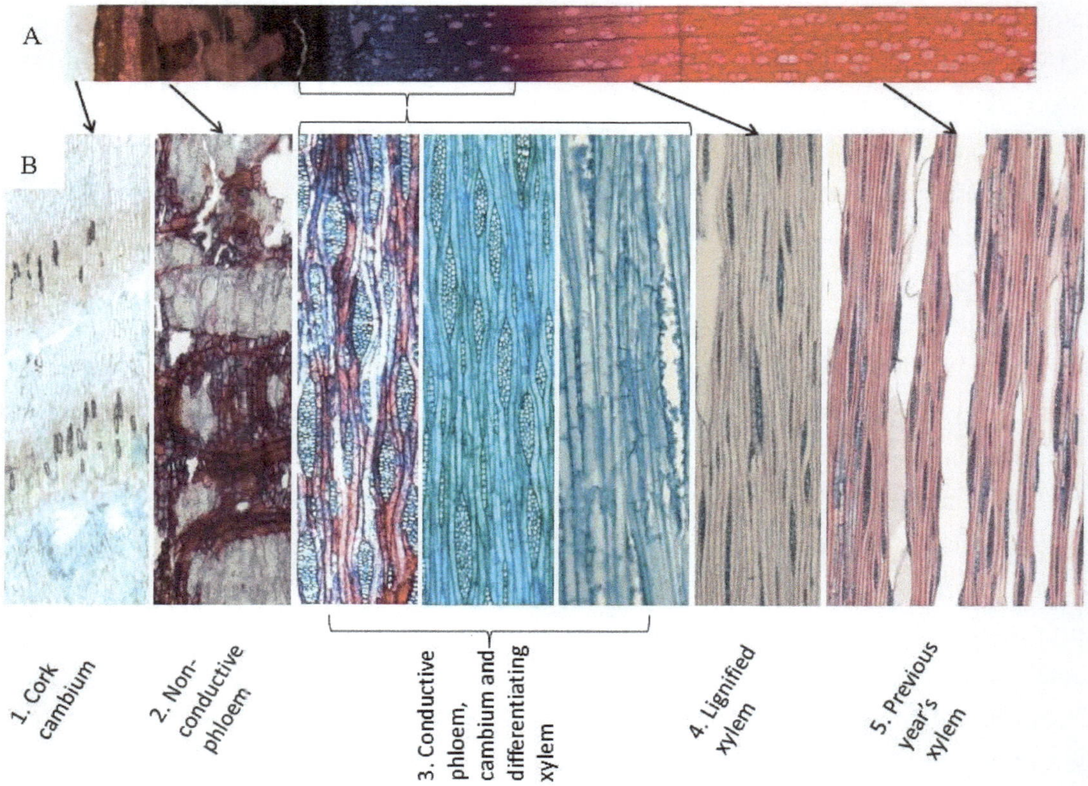

Figure 1. Photomicrographs of the silver birch sample sections with a cryomicrotome. (**A**) A cross-section through the stem. Mainly cellulosic cell walls are stained blue with Alcian blue and lignified tissues red with safranin. (**B**) Longitudinal sections presenting the actual samples for lignin analyses: 1, cork cambium; 2, non-conductive phloem; 3, combined conductive phloem, cambium and differentiating xylem; 4, lignified xylem; 5, previous year's xylem.

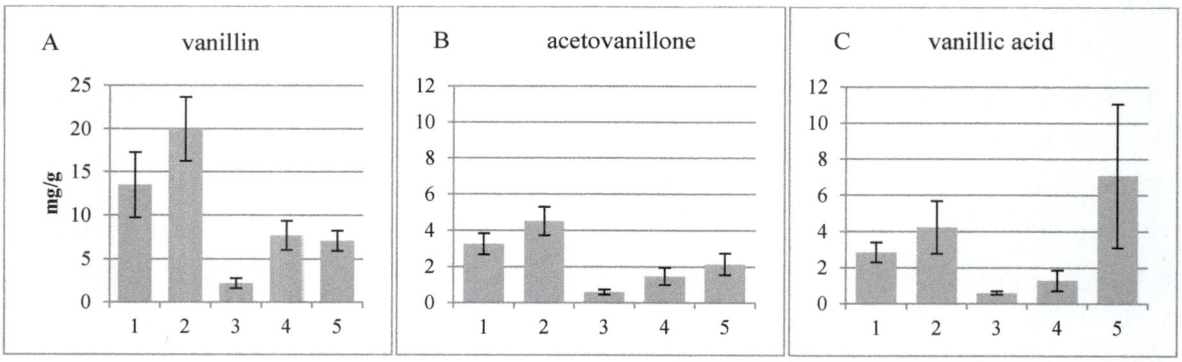

Figure 2. *Cont.*

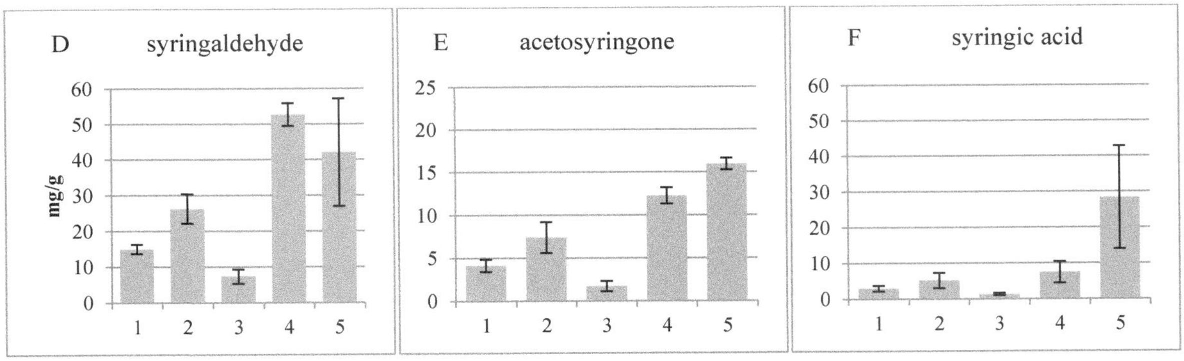

Figure 2. Guaiacyl (**A–C**) and syringyl (**D–F**) units released by the CuO treatment of the different tissues analyzed with the GC-MS. 1, Cork cambium; 2, non-conductive phloem; 3, combined conductive phloem, cambium and differentiating xylem; 4, lignified xylem; 5, previous year's xylem.

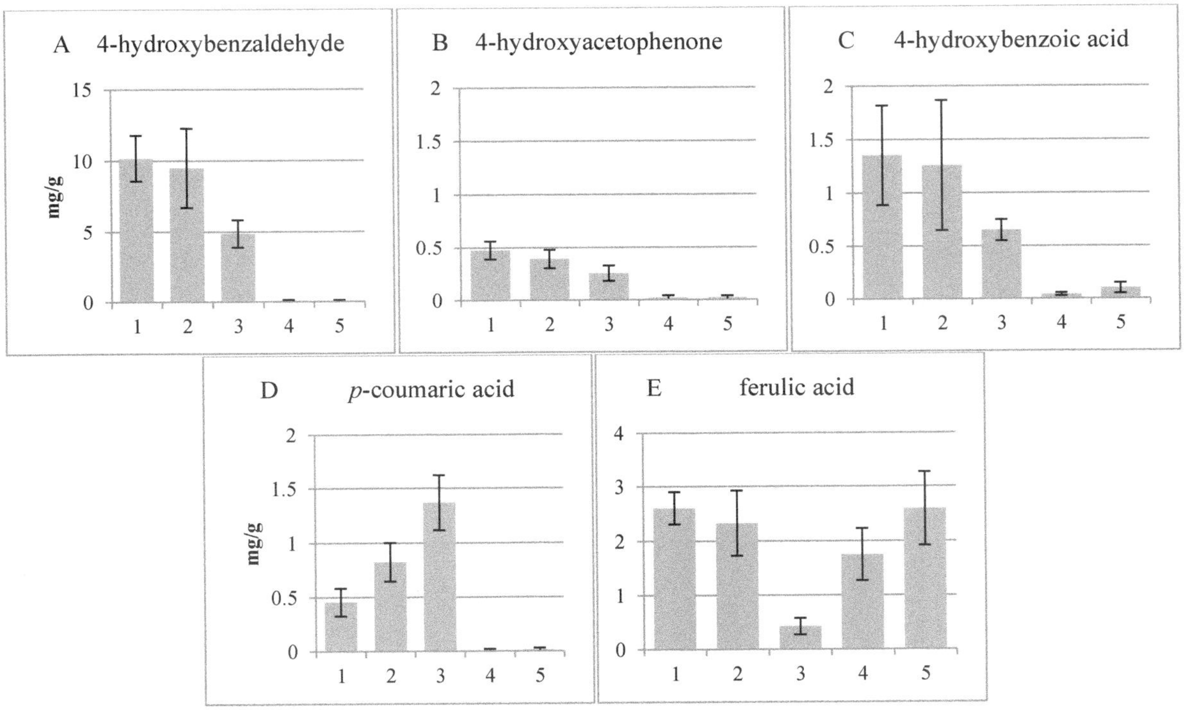

Figure 3. *p*-Hydroxyl phenols (**A–C**) and hydroxycinnamyls (**D,E**) released by the CuO treatment. 1, Cork cambium; 2, non-conductive phloem; 3, combined conductive phloem, cambium and differentiating xylem; 4, lignified xylem; 5, previous year's xylem.

Figures 4 and 5 show the amounts per g of dry weight of the S-, G-, P- and C-groups and the S/G ratios in the different fractions. The S-group contains syringaldehyde, acetosyringone and syringic acid, the G-group vanillin, acetovanillone and vanillic acid, the P-group 4-hydroxybenzaldehyde, 4-hydroxyacetophenone and 4-hydroxybenzoic acid and the C-group *p*-coumaric acid and ferulic acid. The S/G ratio was low in the cork cambium and non-conductive phloem, intermediate in the cambial region and high in the xylem. In the cambial region, the S- and G-group compounds were at a low level.

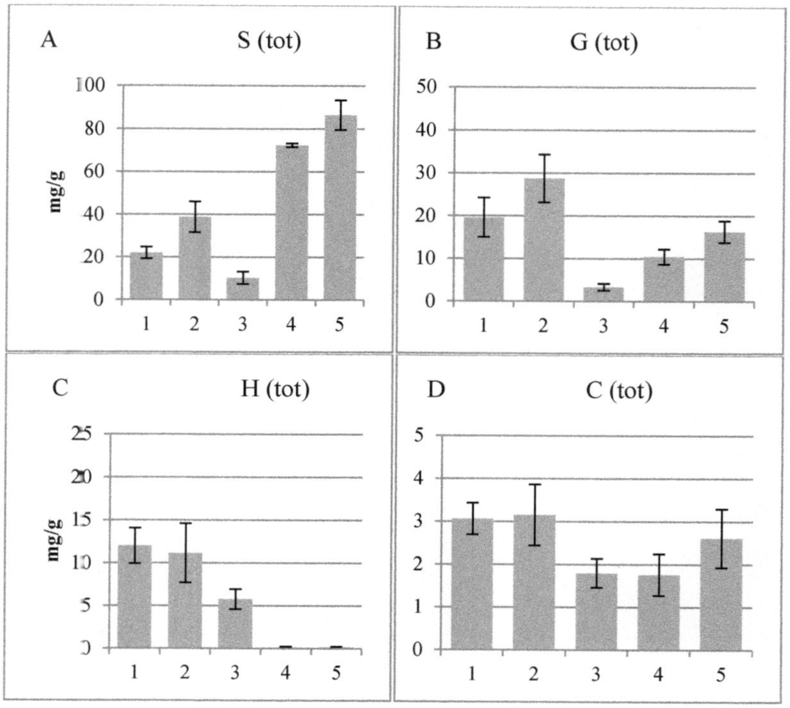

Figure 4. Total amounts per g of dry weight of the syringyl (S)-, the guaiacyl (G)-, *p*-hydroxyphenyls (H)- and hydroxycinnamyls (C)-groups in the different fractions obtained using the CuO method. (**A**) The S-group contains syringaldehyde, acetosyringone and syringic acid; (**B**) the G-group vanillin, acetovanillone and vanillic acid; (**C**) the P-group 4-hydroxybenzaldehyde, 4-hydroxyacetophenone and 4-hydroxybenzoic acid; and (**D**) the C-group *p*-coumaric acid and ferulic acid. 1, Cork cambium; 2, non-conductive phloem; 3, combined conductive phloem, cambium and differentiating xylem; 4, lignified xylem; 5, previous year's xylem.

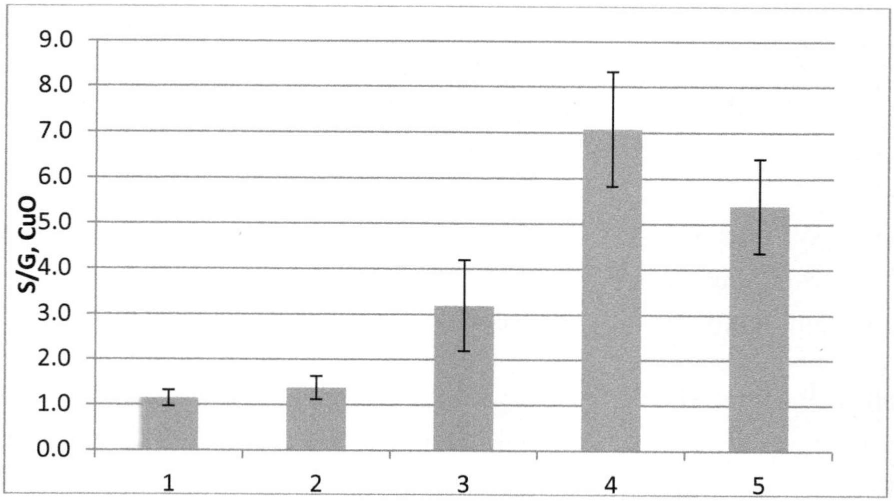

Figure 5. The S/G ratios as calculated from the CuO analysis results. 1, Cork cambium; 2, non-conductive phloem; 3, combined conductive phloem, cambium and differentiating xylem; 4, lignified xylem; 5, previous year's xylem.

Figure 6B shows that the total lignin contents, determined using the acetyl bromide method, of the different fractions were similar. This is in accordance with the sums of the monomer concentrations released by the CuO method (Figure 6A) for all of the other fractions, except Fraction 3. This fraction, which contained the cambial region where lignification is likely to be at the minimum, yielded the lowest sum of monomers. The absence of lignin in this tissue can also be seen in the staining in Figure 1. This region contains predominantly cellulose, hemicelluloses, pectin and proteins [19]. This discrepancy between Figure 6A,B could be partially explained by the fact that the *p*-hydroxycinnamic acids are bifunctional molecules with carboxylic and phenolic binding sites, and they can be involved in both ester and ether linkages to other cell wall components. It is also true that degraded xylans (due to perchloric acid) have strong absorbance at 280 nm, leading to overestimation of lignin with the AcBr method. The cambial region is high in protein content [19], but proteins precipitate during sample preparation and should not interfere with the AcBr method.

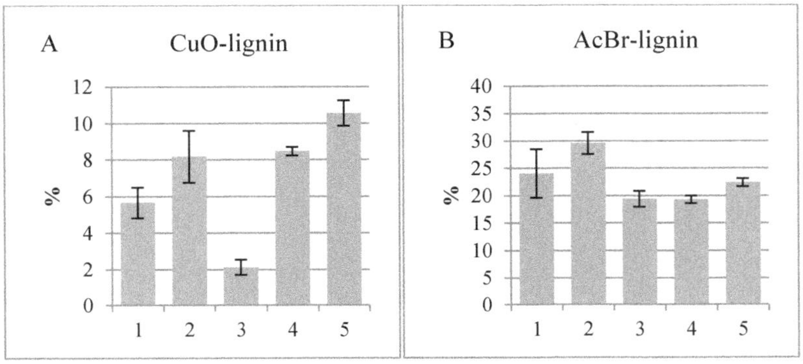

Figure 6. The same silver birch tissues analyzed by CuO and AcBr methods. (**A**) The sum of the monomers released by the CuO method; (**B**) the total lignin amount estimated by the AcBr method. 1, Cork cambium; 2, non-conductive phloem; 3, combined conductive phloem, cambium and differentiating xylem; 4, lignified xylem; 5, previous year's xylem.

2.2. Variation in Lignins in a Natural Silver Birch Population

Figure 7 shows a comparison of CuO and thioacidolysis in the estimation of the S/G ratio. The positive correlation was rather low with R^2 being 0.43 and the linear regression equation $y = 0.681x + 0.6635$. It was observed that the S/G values obtained with the thioacidolysis method were in most cases higher than those calculated from CuO results (average CuO = *ca.* 80% of thioacidolysis). The interesting and large variation in the S/G ratio in the silver birch xylem material is explained by the fact that the material presented the large natural variation within a natural birch population. It would be of interest to see whether this is reflected in the pulping properties of the wood material. However, the total lignin content (Klason + acid-soluble lignin) showed very little variation, and most samples fell from 23% to 25% of dry weight. The sum of the analytical monomers released by the CuO method reached 25% to 55% of the Klason lignin. In earlier accounts, thioacidolysis covers *ca.* 20%–40% of the weight of the lignin presumably present [20].

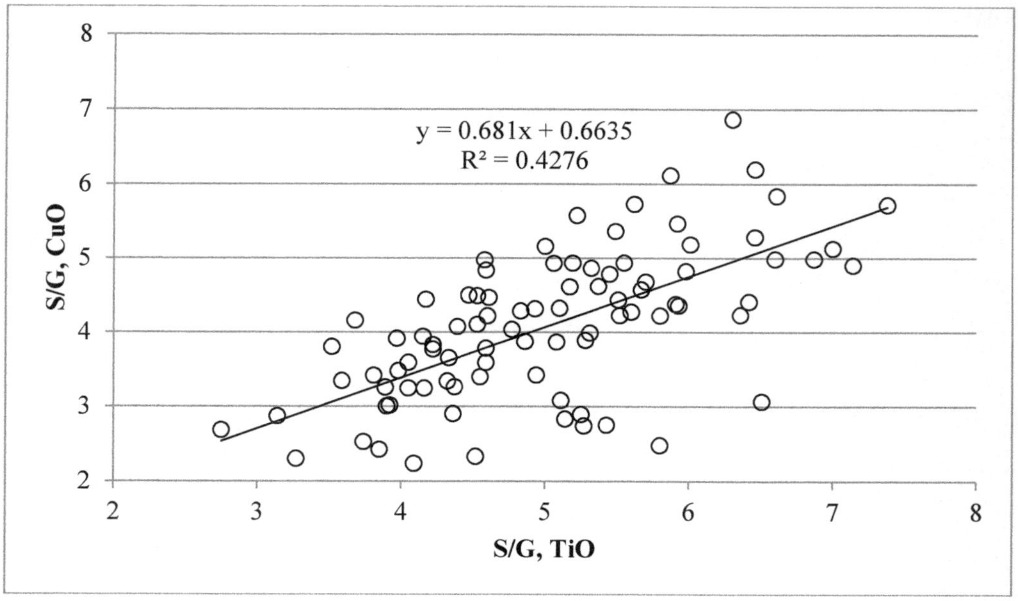

Figure 7. Comparisons of the S/G ratios obtained from the CuO (S/G, CuO) and thioacidolysis (S/G, TiO) analysis results. The wood material originated altogether from 88 trees of 22 different genotypes presenting natural variation in silver birch.

3. Experimental Section

For the different tissues of the silver birch trunk (*Betula pendula* Roth), we felled three birch trees at the experimental field of the Viikki Campus at the University of Helsinki. The trees were of the clone No. V5834 created from material of the Punkaharju experimental site of the Finnish Forest Research Institute. The trees were 13 years old, *ca.* 10 m in height and their diameter *ca.* 11 cm at a 1.5-m height. Longitudinal sections (20 µm in thickness) were cut in the tangential direction in whole birch trunk pieces with a cryomicrotome and divided into five fractions—1, cork cambium; 2, non-conductive phloem; 3, conductive phloem, cambium and differentiating xylem; 4, lignified xylem; 5, previous year's xylem—to show differences in lignin amounts and quality between the tissues. For microscopy, the sections were stained with 1% safranin and then 1% Alcian blue, which stain lignified cell walls red and unlignified cell walls blue, respectively.

A field experiment in Punkaharju, southeastern Finland (61°48' N, 29°18' E), was established in 1999 for long-term monitoring of within-stand differences between silver birch (*Betula pendula* Roth) genotypes in growth phenomena. The experiment was planted on abandoned former agricultural land with a fine sandy till soil and had a randomized complete block design with 6 replicate blocks. All blocks contained four replicate trees of the 22 genotypes at a planting distance of 2 × 2 m. The genotypes were micro-propagated from trees selected randomly in one stand (one hectare in size) of *B. pendula* that was naturally regenerated after logging operations in 1979. In 2008, one tree per each genotype in four randomly selected blocks was felled, altogether 88 trees. Sample discs were cut at breast height (1.3 m). Three–four of the outermost annual rings were removed from two cardinal directions (south and north) in order to avoid any fungal infection and decay near the pith. The samples were further cut into 2 × 2-mm sticks.

3.1. Quantitative Analysis of Lignin

3.1.1. Klason and Acid-Soluble Lignin Determination

Samples were ground frozen with a blade-mill (Polymix PX-A10). The dry solids content of the milled wood samples was determined at 103 °C. The samples of air-dried wood powders (3 g) were extracted with acetone using a Soxhlet apparatus for 6 h [21]. After evaporation of the solvents, the residues were dried at 103 °C, allowed to cool in a desiccator and then weighed. The amount of acid-insoluble lignin was determined by the Klason method [5]. The samples of the extracted wood powders (300 mg) were treated with 3 cm^3 of 72% sulfuric acid under vacuum for 1 h. The mixtures were diluted with about 82-cm^3 portions of water and autoclaved at 125 °C for 1 h. The precipitates were collected with glass fiber filter SS GF 52 × 47 mm by suction filtration and washed with water. The filters with the acid-insoluble lignin (Klason lignin) were dried at 103 °C, cooled in the desiccator and weighed. In order to determine the amount of acid-soluble lignin, the filtrates were diluted with water to 250 cm^3. Absorption of the acid solutions with the dissolved lignin was measured at 203 nm using sulfuric acid of the same concentration as a blank.

The absorbance readings were obtained with a Shimadzu UV-2401 PC UV-VIS Recording spectrophotometer. The total lignin (Klason lignin + acid-soluble lignin) content was calculated from the unextracted wood as follows: Klason lignin% = p(100 − u)/m, in which p = precipitate (g), u = extractives (%) and m = calculated dry weight of extracted sample (g). The acid-soluble lignin content was calculated using a lignin absorptivity of 128 $L \cdot g^{-1} \cdot cm^{-1}$ and corrected because of the absorption of carbohydrates according to [21]. The total lignin content of each sample was determined as the mean of the duplicate measurements; this is referred to as the measured lignin content of the sample.

3.1.2. AcBr-Based Determination

The powder was either lyophilized (birch tissues) or dried at 60 °C for 48 h (birch clones), and 5- or 10-mg samples were extracted with acetone to remove soluble extractives. Lignin contents were determined using the acetyl bromide method [22,23]. A sample of 5 or 10 mg of phloem and inner bark tissue was weighed into a 10-mL screw-capped test tube and sonicated with 5 mL of acetone for 30 min. The extract was pipetted off, and the extractive free sample was dried. It was redissolved in 5 mL 20% (v/v) AcBr-acetic acid solution (containing 100 μL 70% perchloric acid, a hazardous chemical) and the sample was kept in a block heater at 50 °C for 3 h with regular shaking [22]. After treatment, the sample was frozen at −20 °C for 15 min in order to stop the reaction. The melt solution was transferred to a 50-mL volumetric flask containing 5 mL 2 M NaOH and 12 mL 100% acetic acid. The solution was diluted into 50 mL with acetic acid. The UV spectrum was measured with a Shimadzu UV-2401 spectrometer at 280 nm. Lignin content was calculated using the following expression:

$$\text{Lignin\%} = 100(As - Ab)V/aW$$

As = absorbance of sample
Ab = absorbance of blank
V = volume of solution

W = weight of sample
a = the absorptivity of a lignin standard calculated for each analysis series

AcBr determination was calibrated with the total lignin content (Klason + acid-soluble) of a silver birch wood sample (lignin standard).

3.2. Qualitative Analysis of Lignin

3.2.1. Thioacidolysis: Method, TiO

The syringyl:guaiacyl (S/G) determinations of the wood lignin were made by the modified method of thioacidolysis [12]. An extractive-free wood sample (10–30 mg) was mixed with 3 mL of freshly prepared thioacidolysis reagent (2 M BF_3 etherate in an 8.75:1(v/v) dioxane/ethanethiol mixture). The thioacidolysis proceeded at 103 °C for 4 h. After cooling, the reaction mixtures were rinsed with water in the reaction tube containing methylene chloride (CH_2Cl_2). The lignin fragments were extracted from the aqueous phase thrice with 2 mL CH_2Cl_2. The organic fractions were combined, dried by the addition of anhydrous Na_2SO_4 and evaporated under nitrogen. Before the GC-MS analysis, the dried samples were silylated with 0.5 mL 20% TMSI-pyridine mixture (TMSI = 1 − (trimethylsilyl)imidazole) at 60 °C for 1 h, and then, the silylation was continued at room temperature overnight. The GC-MS analyses were performed using a HP 6890 GC-system equipped with a Mass Selective Detector 5873 and an HP-5 capillary column (30 m × 0.25 mm i.d., 0.25 μm film thickness). Helium was used as a carrier gas, with a flow of 1.5 mL/min.

The mass spectral characterizations were done according to Rolando *et al.* [23]. S/G ratios were defined as peak area ratios of syringyl and guaiacyl monomeric erythro/threo isomers, assuming that they have similar response factors, which is the case [11].

3.2.2. Alkaline Cupric (II) Oxidation: Method, CuO

Alkaline cupric (II) oxidations were carried out according to the procedure developed by Hedges and Ertel [16]. Some modifications, including those introduced by Goni and Montgomery [24], were made. Oxidations were performed using a microwave digestion system MSD-2000 and a liquid phase hydrolysis accessory set (CEM Corporation). A known amount of sample, typically 50 mg, 15 mL 2 M NaOH (N_2-spurged), 500 mg CuO, 50 mg $Fe(NH_4)_2(SO_4)_2 \cdot 6H_2O$ and 50 μL of internal standard solution (2 mg/mL of cinnamic acid) were added into the reaction vessel. The reaction vessel (120 mL) covers were sealed, and the closed reaction vessels were connected to each other via the Teflon tubing and loaded into the microwave oven. The fiber-optic temperature probe (Thermo-Optic) was connected to the first reaction vessel and the pressure/vacuum line to the last reaction vessel. Air was removed from the reaction vessels and was replaced by N_2-gas. The microwave-assisted CuO-treatment of samples was done at 150 °C for 1.5 h in a N_2 atmosphere.

After oxidation, the cooled solutions were acidified, and the CuO reaction products were extracted with 4 mL of ethyl acetate three times. The water present in the combined ethyl acetate extract was removed by the addition of Na_2SO_4. After evaporation, the extracts were silylated with N,*O*-bis(trimethyl-silyl)trifluoroacetamide (BSTFA) with 1% trimethylchlorosilane (TMCS). The

GC-MS detection of trimethylsilyl derivatives of CuO oxidation products was done with the same GC-MS instrument as was used in the thioacidolysis method.

The phenolic compounds derived by CuO oxidation were divided into four different structural groups; guaiacyls (G), syringyls (S), *p*-hydroxyphenyls (H) and hydroxycinnamyls (C). The vanillyl, syringyl and *p*-hydroxyl groups consist of corresponding aldehyde, ketone and carboxylic acid. The hydroxycinnamyl group includes ferulic acid and *p*-coumaric acid. Quantification of individual compounds was performed using the relative response factors adjusted to the peak area of internal standard and individual lignin phenols. The response factors were calculated from the analysis results of standard solutions. The amounts of different structural groups, as well as the total lignin content were counted as the sum of individual compounds.

4. Conclusions

Lignin is a complex macromolecule, and hence, it has been stated many times that there is no single method for the accurate determination of the quality or quantity of lignin [8,25]. Because of the high molecular weight and insolubility, knowledge of the composition of wood lignin is obtained indirectly, e.g., by the identification of lignin degradation products. This can be done with CuO and thioacidolysis methods, which give valuable information on the monolignol composition of lignin. Oxidation with CuO induces cleavage of β-*O*-4 ether bonds in lignin and retains the three carbons of the side chain of the phenol, and the lignin macromolecule is hydrolyzed.

As we have shown, CuO and thioacidolysis give different degradation products and, depending on how they are included in the calculations, give different S/G ratios. Another possible reason is that in the CuO method, monomers are also released from other units, such as β-5. The reason for the differences may be because the CuO method is perhaps also capable of releasing monomers from other units, such as ß-5. In addition, direct comparisons of S/G-ratio lignin data generated from the CuO and thioacidolysis methods are difficult, as different lignin subunits were analyzed in each process; however, the components are analogous. Hence, it is vital that for comparisons, the S/G-ratios should be the results of the same analytical method.

Downscaling of the sample size presents another problem: representative analysis and results are challenging if the sample size is on a mg scale as, e.g., vessels and fibers behave differently in grinding and may result in uneven samples. This is more accentuated in samples containing other tissues than xylem as, e.g., phloem fibers differ from xylem in the S/G ratio. Both the CuO and thioacidolysis results show clearly that silver birch phloem sclereids have a much lower S/G ratio than xylem. Similar results have been found in poplar tissues by thioacidolysis [19].

In summary, we can say that the different silver birch tissues showed marked differences in lignin amount and S/G ratio, and the results differed among the methods tested. The S/G ratios calculated from the thioacidolysis and CuO results correlated positively, but with an R^2 value of 0.43. This means that the S/G-ratios, as well as other qualitative lignin parameters should always be determined with the same method if comparisons between tissues and species are made. However, this work indicates clearly that the CuO method, which has been used earlier mainly for soil analyses, is a good alternative to study the monomeric composition and S/G ratio of wood lignins without the drawbacks of thioacidolysis.

Acknowledgments

The project was funded by a grant from the Academy of Finland (Grant No. 264621) to K.V.F. We wish to thank our technician, Eija Rinne, for the cryotome sectioning of the silver birch woody material, and Irmeli Luovula and Hanna Kautto for the lignin analyses.

Author Contributions

K.V.F., P.S., J.I., J.A.A.S. and K.N. designed the experiments, T.T. conducted the lignin analyses and drew the graphs, J.I. felled the trees for anatomical sections. K.V.F., P.S. and T.T. wrote the manuscripts which all the other authors commented.

Conflicts of Interest

The authors declare no conflict of interest.

References

1. Ralph, J.; Brunow, G.; Boerjan, W. Lignins. In *Encyclopedia of Life Sciences*. Available online: http://www.els.net (accessed on 15 February 2015).
2. Vanholme, R.; Morreel, K.; Darrah, C.; Oyarce, P.; Grabber, J.H.; Ralph, J.; Boerjan, W. Metabolic engineering of novel lignin in biomass crops. *New Phytol.* **2012**, *196*, 978–1000.
3. Eudes, A.; Liang, Y.; Mitra, P.; Loque, D. 2014: Lignin bioengineering. *Curr. Opin. Biotechnol.* **2014**, *26*, 189–198.
4. Chiang, V.L.; Puumala, R.J.; Takeuchi, H.; Ecker, R.E. Comparison of softwood and hardwood kraft pulping. *Tech. Assoc. Pulp Pap. Ind. J.* **1988**, *71*, 173–176.
5. Dence, C.W. The determination of lignin. In *Methods in Lignin Chemistry*; Lin, S.Y., Dence, C.W., Eds.; Springer-Verlag: Heidelberg, Germany, 1992; pp. 33–61.
6. Ibáñez, A.B.; Bauer, S. Downscaled method using glass microfiber filters for the determination of Klason lignin and structural carbohydrates. *Biomass Bioenergy* **2014**, *68*, 75–81.
7. Hatfield, R.; Fukushima, R.S. Can lignin be accurately measured? *Crop Sci.* **2005**, *45*, 832–839.
8. Brinkmann, K.; Blaschke, L.; Polle, A. Comparison of different methods for lignin determination as a basis for calibration of near-infrared reflectance spectroscopy and implications of lignoproteins. *J. Chem. Ecol.* **2002**, *28*, 2483–2501.
9. Moreira-Vilar, F.C.; de Cassia Siqueira-Soares, R.; Finger-Teixeira, A.; de Oliveira, D.M.; Ferro, A.P.; da Rocha, G.J.; de Lourdes, L.; Ferrarese, M.; dos Santos, W.D.; Ferrarese-Filho, O.; *et al.* The acetyl bromide method is faster, simpler and presents best recovery of lignin in different herbaceous tissues than klason and thioglycolic acid methods. *PLoS ONE* **2014**, *9*, e110000.
10. Johnson, D.B.; Moore, W.E.; Zank, L.C. The spectrophotometric determination of lignin in small wood samples. *Tappi* **1961**, *44*, 793–798.
11. Lapierre, C.; Monties, B.; Rolando, C. Preparative thioacidolysis of spruce lignin: Isolation and identification of main monomeric products. *Holzforschung* **1986**, *40*, 47–50.

12. Lapierre, C. Determining lignin structure by chemical degradations. In *Lignin and Lignans. Advances in Chemistry*; Heitner, C., Dimmel, D.R., Schmidt, J.A., Eds.; CRC Press Taylor Francis Group: Boca Raton, FL, USA, 2010; pp. 11–48.
13. Thevenot, M.; Dignac, M.-F.; Rumpel, C. Fate of lignins in soils: A review. *Soil Biol. Biochem.* **2010**, *42*, 1200–1211.
14. Spielvogel, S.; Prietzel, J.; Kögel-Knabner, I. Changes of lignin phenols and neutral sugars in different soil types of a high-elevation forest ecosystem 25 years after forest dieback. *Soil Biol. Biochem.* **2007**, *39*, 655–668.
15. Kuo, L.J.; Louchouarn, P.; Herbert, B.E. Fate of CuO-derived oxidation products during plant combustion: Application to the evaluation of char input to soil organic matter. *Org. Geochem.* **2008**, *39*, 1522–1536.
16. Hedges, J.I.; Ertel, J.R. Characterization of lignin by capillary gas chromatography of cupric oxide oxidation products. *Anal. Chem.* **1982**, *54*, 174–178.
17. Kaiser, K.; Brenner, R. Characterization of lignin by gas chromatography and mass spectrometry using a simplified CuO oxidation method. *Anal. Chem.* **2012**, *84*, 459–464.
18. Ralph, J. Hydroxycinnamates in lignification. *Phytochem. Rev.* **2010**, *9*, 65–83.
19. Christiernin, M. Lignin composition in cambial tissues of poplar. *Plant Physiol. Biochem.* **2006**, *44*, 700–706.
20. Yue, F.; Lu, F.; Sun, R.S.; Ralph, J. Synthesis of lignin-derived thioacidolysis monomers and their uses as quatitation standards. *J. Agric. Food Chem.* **2012**, *60*, 922–928.
21. KCL. Massan ja Puun Kokonaisligniinipitoisuus. (Total lignin content of wood and pulp). *KCL (Finnish Pulp and Paper Research Institute) Reports*; KCL: Espoo, Finland, 1982; 115b, p. 3.
22. Hatfield, R.D.; Grabber, J.; Ralph, J.; Brei, K. Using the acetyl bromide assay to determine lignin concentrations in herbaceous plants: Some cautionary notes. *J. Agric. Food Chem.* **1999**, *47*, 628–632.
23. Rolando, C.; Monties, B.; Lapierre, C. *Methods in Lignin Chemistry*, Springer Series in Wood Science; Springer-Verlag: Berlin, Germany, 1992; pp. 334–349.
24. Goñi, M.A.; Montgomery, S. Alkaline CuO oxidation with a microwave digestion system: Lignin analyses of geochemical samples. *Anal. Chem.* **2000**, *72*, 3116–3121.
25. Davin, L.B.; Patten, A.M.; Jourdes, M.; Lewis, N.G. Lignins: A twenty-first century challenge. In *Biomass Recaltritrance. Deconstructing the Plant Cell Wall for Bioenergy*; Himmel, M.E., Ed.; Blackwell Publishing Ltd.: Oxford, UK, 2008; pp. 213–305.

14

Measuring the Mechanical Properties of Plant Cell Walls

Hannes Vogler [1,*], Dimitrios Felekis [2], Bradley J. Nelson [2] and Ueli Grossniklaus [1]

[1] Institute of Plant Biology and Zurich-Basel Plant Science Center, University of Zurich, Zollikerstrasse 107, CH 8008 Zürich, Switzerland; E-Mail: grossnik@botinst.uzh.ch

[2] Institute of Robotics and Intelligent Systems, ETH Zurich, Tannenstrasse 3, CH 8092 Zürich, Switzerland; E-Mails: dfelekis@ethz.ch (D.F.); bnelson@ethz.ch (B.J.N.)

* Author to whom correspondence should be addressed; E-Mail: hannes.vogler@botinst.uzh.ch

Academic Editor: Philip J. Harris

Abstract: The size, shape and stability of a plant depend on the flexibility and integrity of its cell walls, which, at the same time, need to allow cell expansion for growth, while maintaining mechanical stability. Biomechanical studies largely vanished from the focus of plant science with the rapid progress of genetics and molecular biology since the mid-twentieth century. However, the development of more sensitive measurement tools renewed the interest in plant biomechanics in recent years, not only to understand the fundamental concepts of growth and morphogenesis, but also with regard to economically important areas in agriculture, forestry and the paper industry. Recent advances have clearly demonstrated that mechanical forces play a crucial role in cell and organ morphogenesis, which ultimately define plant morphology. In this article, we will briefly review the available methods to determine the mechanical properties of cell walls, such as atomic force microscopy (AFM) and microindentation assays, and discuss their advantages and disadvantages. But we will focus on a novel methodological approach, called cellular force microscopy (CFM), and its automated successor, real-time CFM (RT-CFM).

Keywords: cell wall; cytomechanics; turgor pressure; Young's modulus; pollen tube; cellular force microscope

---

1. Why Should We Study Cytomechanics?

Cytomechanical studies gained increasing interest over the last years after having been neglected for a long time. Mechanical forces have been recognized to influence many biological processes in animal biology, such as the differentiation of stem cells [1–4], embryo development [5–7] or tumor invasion, to name just a few [8–10]. In plant sciences, special attention has been paid to the mechanical properties of the cell wall, which plays an important role in plant stability and the resistance against pathogens. Several comprehensive reviews describing available methods in plant mechanics were published recently [11–15] and will only briefly be summarized here.

In the first part of this review, we will provide a few examples of well-known agricultural and industrial problems that are related to the mechanical properties of the plant cell wall. However, the main focus will be on pollen tubes and the effect of alterations in the biochemical composition of their cell wall on its mechanical properties. In the second part, we will discuss different approaches to assess these properties, with a more extended description of the cellular force microscope (CFM), a novel and flexible system to measure a wide range of forces exerted by living cells. A short outlook for new directions and developments in cytomechanical studies will be given at the end.

Modern agriculture has to deal with diverse problems related to the composition and stability of the cell wall. Crop lodging, the permanent displacement of plant shoots, is a world-wide problem that causes severe harvest losses. For the U.K., it was reported that every 3–4 years, on average, 15%–20% of the wheat area lodges [16]. In individual fields, however, the lodged area can be much larger. Countermeasures to reduce the risk of lodging involve the production of dwarf varieties on the breeders' side [17] and the application of plant growth regulators to shorten crops on the farmers' side [18]. Unfortunately, these approaches lead to a yield reduction, even in years with favorable weather conditions when lodging is not a problem. Knowledge about the correlation between the biochemical composition and the mechanical stability of the cell wall could help to produce crops that reach the full yield potential, while being more resistant to lodging.

The mechanical properties of the leaf directly affect plant-herbivore interactions, resistance against pathogens and nutrient recycling via litter decomposition [19]. Plants regulate the interactions with insects by modulating their surfaces to facilitate or limit insect access [20]. Modification of leaf toughness alters palatability or digestibility and, thus, can be considered a defense strategy against herbivory [21]. Similar mechanisms restrict the access of microorganisms to plant tissue [22].

While increased leaf toughness provides protection against herbivorous pests, it poses a problem for digestion when used as animal feed. A reduction in lignin content, as it occurs in the brown midrib mutants of maize and sorghum [23–25], leads to better digestibility and consequently higher milk production in dairy cattle [26]. However, this comes at the cost of lower yield, mostly due to lodging. Nevertheless, the use of brown midrib maize for silage on dairy farms is economically profitable where normal weather conditions are such that lodging only poses a minor problem [26].

Many industrial processes depend on an optimal cell wall composition. Removal of lignin, for example, is a cost-intensive process in paper production, and the paper industry would benefit tremendously from reduced lignin contents in the processed wood. However, lignin is important for tree stability and cannot be reduced to arbitrary levels without knowing how to compensate for the loss of stability. Similarly, the bottleneck in cost-effective biofuel production is the efficient degradation of cell walls into fermentable sugars [27]. Therefore, knowledge about the contribution of individual compounds to the overall mechanical properties of the cell wall is crucial to optimize cell wall composition and to reduce the costs in these industries.

To understand the complex interactions between expanding cells within a tissue or an entire organism, we first need a profound understanding of the mechanical processes during growth at the single-cell level. Fundamentally, changes in the shape and size of a cell during morphogenesis result from the deformation and modification of the existing cell wall, as well as from the secretion and deposition of newly-synthesized wall materials.

2. The Mechanical Cell Wall Properties Are Defined by Its Composition

The mechanical properties of the plant cell wall are defined by its biochemical composition and the specific interactions between individual wall polymers. The primary cell wall is a complex compound, assembled from cellulose microfibrils, hemicelluloses, pectic polysaccharides and cell wall proteins. It is thought that individual microfibrils are linked via xyloglucan (XyG), the most abundant hemicellulose in dicotyledonous plants, to build a load-bearing network. In one of the cell wall models, XyG chains tether neighboring, well-spaced microfibrils by direct crosslinking, and the resulting cellulose/XyG network is embedded in a pectin matrix [28]. A second possibility under consideration is that the microfibril crosslinking is indirect. In this case, XyG polymers bind covalently to the pectin network, and the so-embedded XyG chains connect cellulose microfibrils that are in close contact via hydrogen bonds. The close vicinity of the fibrils renders the XyG between them inaccessible for endoglucanases [29]. If the first model were true, endoglucanase digestion of heat-inactivated cell walls should provoke cell wall extension in an extensometer assay. In cucumber cell walls, however, this effect was not found, which favors the second model [29], at least in this case.

Cellulose microfibrils act as the main load-bearing constituents of the cell wall. In a typical elongating cell, the microfibrils are arranged transversally around the cell, thereby providing reinforcement against growth in width and allowing preferential growth in length [30]. A passive reorientation from a transverse to a longitudinal pattern occurs during the elongation process [31,32]. However, in the tip-growing pollen tubes, the levels of cellulose are quite low, and despite their nearly perfectly cylindrical shape, transversally-oriented cellulose microfibrils have not been found [33,34]. This finding indicates that cellulose cannot be the only compound that counteracts the transverse stress, which is twice as high as the longitudinal stress [35].

Arabidopsis thaliana plants lacking xyloglucan xylosyltransferases (XXT) have a strong reduction of xyloglucans in their cell walls. However, even double mutants for *xxt1* and *xxt2* show only a weak growth phenotype, suggesting that xyloglucan cannot be the main load-bearing tether that holds the cellulose microfibrils together [29]. Interestingly, the same double mutants show a severe pollen tube growth phenotype leading to the loss of cell wall integrity and eventual bursting [36]. Overexpression of xyloglucan endotransglucosylase/hydrolases (XTHs), on the other hand, decreases the wall yield threshold. Since the amount of xyloglucans was not increased in XTH overexpressing plants, this may be an indirect effect, caused, for example, by alterations in cellulose deposition and crystallinity, due to changes in the cell wall matrix near nascent microfibrils [37].

Mutants that are deficient in linear pectic arabinans display increased stiffness and reduced strain in epidermal cells of *Arabidopsis* stems. Despite a reduction in pectic arabinan of 54% in the stem of *arabinan-deficient–1* (*arad1*) mutants, these plants show no altered growth phenotype [38]. The above examples illustrate that compensation mechanisms may suppress some effects of the changes in cell wall composition and that phenotypes become only visible in specific cell types or under unfavorable growth conditions. Moreover, such compensation mechanisms suggest the existence of a cell wall sensing system, which relays information from the cell wall to the interior of the cell, so it can react with regard to the synthesis of new cell wall components [39–41].

To get a clearer picture of the mechanical consequences caused by altered cell wall composition, it is important to quantify the resulting mechanical parameters.

3. Measuring Turgor Pressure

Plant cell growth is a fine-tuned interplay between turgor-driven cell expansion and the resistance provided by the cell wall. The importance of turgor pressure for plant growth has been known for a long time. Measuring it, however, was only possible via the plasmolytic method, where the osmolality of the external medium was increased until the protoplast started to detach from the cell wall [42,43]. It was only in 1968 when the first direct measurement method became available. The pressure probe is an oil-filled capillary that is directly inserted into the vacuole of the cell to be studied [44]. Initially, this technique was used to measure turgor in algae with very large cells [45,46], but continuous development made it possible to use the pressure probe also for small cells, such as stomatal guard cells [47,48].

The pressure probe is a versatile instrument, not only for measuring, but also for manipulating turgor pressure within a cell by injecting or removing a known volume of liquid. Determining the volumetric modulus in this way allows conclusions about the average cell stiffness [49–51]. Despite the versatile information it can provide, the pressure probe has some disadvantages. It is invasive and, therefore, not useful for repeated measurements on the same cell. Furthermore, its handling is rather complicated, which limits the number of measurements one can make in a day.

Ball tonometry was developed to overcome these difficulties. This method is not invasive, albeit indirect, and is based on the compression of individual plant cells with a large diameter sphere (50–500 μm) [52,53]. Due to the large diameter of the indenter, which is much bigger than the thickness of the cell wall, the elasticity of the cell wall itself can be neglected. Thus, turgor pressure is determined by the ratio between the known load and the contact area between the sphere and the cell surface. Turgor values measured with this method agree well with the direct pressure probe measurements.

4. Measuring Mechanical Cell Wall Properties

The plant cell wall is a remarkable structure that must, at the same time, be strong enough to withstand turgor pressure, yet flexible enough to allow controlled cell expansion. To solve this dilemma, a perfect coordination between the deformation and modification of the existing wall and the secretion and deposition of new cell wall material is required. The secret of these mechanical properties lies within the compound structure of the cell wall. Although the main components are well described [54], there is a lack of knowledge concerning their individual contributions to the mechanical properties of the cell wall. In the past, numerous methods were developed to assess cell wall stiffness [12]. Extensometer assays provided insight into the elasticity of isolated cell walls or entire tissues [55–59]. However, it is extremely difficult to perform experiments on individual living cells, and measuring local differences in mechanical cell wall properties is not possible using this approach.

A new, interesting approach to measure mechanical properties of living cells uses custom-made microfluidic chips, which are especially useful to study growing cells, such as pollen tubes. By modulating channel shape and diameter, as well as the stiffness of the surrounding walls, it is possible to draw conclusions about the pushing force and the elasticity of the elongating pollen tube [60–62]. As with extensometer assays, however, it is not possible to get information about local differences in the mechanical properties of the cell wall at high resolution. Techniques, such as optical or magnetic tweezers, provide extremely highly-resolved force and position data down to the single molecule level for animal cells [63–65]. However, their force range is limited to piconewtons, which is too low to assess the mechanical properties of plant cell walls.

A solution to overcome these limitations was provided by nano- and micro-indentation assays. A cell is placed on a stiff surface, *i.e.*, a glass slide, and probed with an end-effector that has a well-defined tip geometry. Cellular stiffness can be determined by monitoring the forces that are necessary to create small, local deformations of the cell wall. Several factors influence the measured apparent stiffness. Among these are cell parameters, such as cell wall elasticity, cell wall thickness, cell geometry and turgor pressure, as well as system parameters, for example indenter geometry or contact angle [66–68].

Atomic force microscopy (AFM) is frequently used in nanoindentation studies, allowing the combined acquisition of highly-resolved topographical and stiffness information. AFM measurements with sharp indenters and very shallow indentations allow the direct extraction of the elastic modulus of the cell wall from experimental data without the influence of turgor pressure. Using such a setup, it was shown that the outer cell wall of the *Arabidopsis* shoot apex was more elastic at the periphery than at the tip [69].

However, the forces that can be applied by an AFM are usually in the nanonewton range. Thus, they are too small to deform the walls of turgid cells enough to measure in-plane elasticity, which is influenced by cellulose microfibrils. Instead, these measurements are sensitive to the elastic modulus normal to the cell wall. This provides information about the gel matrix component of the cell wall, which contains mainly pectins and hemicelluloses [54]. Peaucelle and colleagues circumvented this problem by gluing a 5-μm glass bead to a stiff cantilever, which extended the range of the AFM to microindentation dimensions. Using this approach, the authors could show that cell softening in the incipient leaf primordium correlated with pectin methylesterase (PME) activity [70].

Using a microindentation approach combined with finite element method (FEM) modeling, Forouzesh and colleagues were not only able to characterize the elastic and viscoelastic properties of *Arabidopsis* leaf cell walls, but they could also extract turgor pressure values and get a rough estimate of the cell wall thickness [71]. Table 1 gives an overview of the advantages and disadvantages of selected methods used to measure cytomechanical properties.

Table 1. Advantages and disadvantages of mechanical characterization methods

Method	Force Resolution	Force Range	Position Resolution	Position Range	Specimen aspect ratio	High throughput	Non- destructive
Magnetic or Optical tweezers	✓	✗	✗	✗	✗	✓	✗
Extensometer	✓	✓	✗	✗	✗	✓	✗
Microfluidic chip	✓	✗	✗	✗	✗	✓	✗
AFM	✓	✗	✓	✓	✗	✗	✓
CFM	✓	✓	✓	✓	✓	✗	✓
RT-CFM	✓	✓	✓	✓	✓	✓	✓

5. Cellular Force Microscopy

A complete mechanical characterization of plant cells requires the application of multi-scale loads from nanonewtons to millinewtons, requirements that are satisfied by the cellular force microscope (CFM). The CFM is a mechatronic system, combining micropositioners and force-sensing devices with standard light microscopes. The core of the system is a microelectromechanical systems (MEMS)-based capacitive microforce sensor [72], which can be equipped with probes of different geometries, depending on whether global or local properties are to be characterized. Tungsten probes with diameters ranging from 100 nm to 5 μm and lengths ranging from micrometers to millimeters are available. Extensions with various geometries (e.g., spheres or pyramids) can also be attached to the sensor tip. The force sensor is mounted on a three-axis positioning system to allow fast and precise positioning over a range of 27 mm with a resolution of 5 nm. The entire system is integrated with an inverted light microscope, to which a CCD camera is attached to facilitate the initial positioning of the sensor probe and to follow the course of the experiment. A schematic representation of the CFM is shown in Figure 1.

The CFM was first described by Felekis and colleagues. SI-traceable calibration and thorough uncertainty analysis characterization make the MEMS microforce sensors an ideal tool for the investigation of the micromechanical properties of living plant cells [73]. A careful study of epidermal cells from onion peel revealed the influence of turgor pressure and indentation angle on CFM measurements. The large travel range of the positioning system allowed raster scans on large areas that cover several cell files, resulting in combined topography and stiffness maps. Only the use of long, thin probes made it possible to assess the depressions between two adjacent cells. By using an FEM-based modeling approach, the authors could identify local differences in mechanical cell wall properties [67].

CFM experiments on growing lily pollen tubes confirmed previous results, which described that the tip of the pollen tube appears to be softer than the more distal regions. At first glance, this agrees well with the fact that the biochemical composition of the cell wall in the apical region is different from that in the rest of the pollen tube. At the tip, pectins are methylesterified, whereas in the shank, they become cross-linked by calcium (Ca^{2+}). Crystalline cellulose and callose are absent from the apical region. Based on these differences, it was concluded that the tip region of the pollen tube must be softer than the shank, which made sense, since growth occurs exclusively in the tip region. Vogler and colleagues [68] combined the CFM measurements with an FEM-based modeling approach, which allowed them to extract the Young's modulus of the cell wall, as well as turgor pressure, using data on the apparent stiffness and shrinkage of the pollen tube upon plasmolysis, respectively. For the FEM model, uniform material properties were assumed, *i.e.*, no differences between pollen tube tip and shank. Nevertheless, the simulated apparent stiffness curves closely reflected the measurements, showing lower apparent stiffness in the tip region. This means that the observed decrease in stiffness towards the tip can be explained by the geometry of the pollen tube alone and, despite biochemical differences in composition, is not dependent on different mechanical properties of the cell wall material in tip and shank.

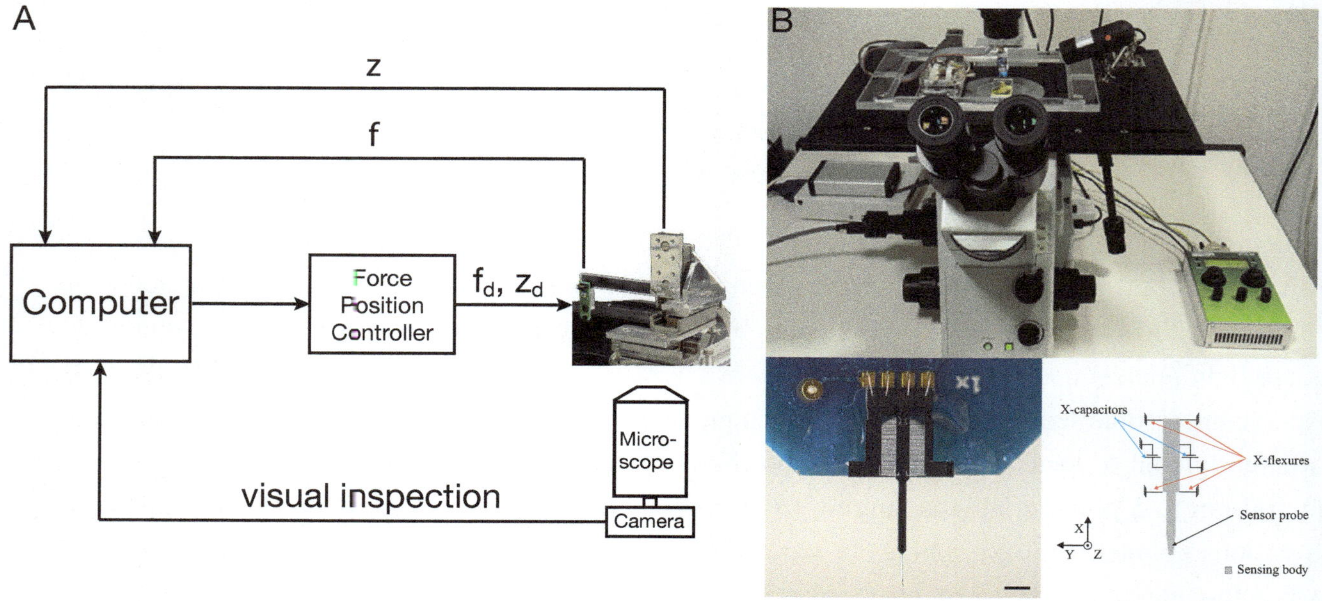

Figure 1. The cellular force microscope (CFM). (**A**) Schematic representation of the CFM. f and z are the measured values and f_d, z_d are the desired values of force and position, respectively. (**B**) The CFM setup on an inverted microscope. The lower panels show an image of the MEMS-based force sensor on the left and a schematic on the right side. Scale bar = 1 mm. Images in (B) taken from [68].

The influence of cell geometry on the apparent stiffness measurements illustrates one of the problems arising in CFM experiments or indentation experiments in general. Since available sensors are only capable of measuring forces parallel to the axis of the probe, every contact angle between the sample and the sensor probe >0 will affect the measured forces and, thus, the apparent stiffness [15]. Therefore, it will be important to produce 2D sensors capable of measuring vertical and lateral forces at the same time or to develop systems that allow the rotation of either the sample or the sensor in order to measure normal to the cell wall surface.

The CFM has been instrumental in determining the effects of specific cell wall mutations on the mechanical properties of the wall. For instance, the *Arabidopsis xxt1 xxt2* double mutant displays a severe reduction in XyG. This leads to an altered morphology and, eventually, the bursting of pollen tubes. A very similar phenotype was found in the *xyloglucanase113* (*xeg113*) mutant. This mutant lacks functional extensins, which are a class of cell wall remodeling proteins. CFM measurements, in combination with FEM modeling, provided insights into the mechanical properties of such mutant pollen tubes. XyG deficiency weakened the cell wall, represented by a decrease in the elastic modulus (Young's modulus) of the cell wall, while turgor pressure was unaffected in comparison to wild-type pollen tubes. In contrast, in the *xeg113* mutant, cell wall elasticity was not altered in comparison to the wild type, but turgor pressure was increased [36].

Another set of CFM experiments was performed to evaluate the effect of Ca^{2+} homeostasis on the apparent stiffness of *Arabidopsis* pollen tubes. Ca^{2+} homeostasis plays an important role in cell wall integrity. In an NADPH oxidase-deficient double mutant (*rbohH rbohJ*), which displays a pollen tube bursting phenotype, it was shown that the tip-focused Ca^{2+} gradient was dissipated [74,75]. CFM measurements revealed that *rbohH rbohJ* double mutants have a lower apparent stiffness than wild-type pollen tubes. Moreover, increasing the Ca^{2+} concentration in the growth medium reduced bursting and increased the apparent stiffness of the pollen tube to wild-type levels [76].

These findings exemplify that CFM data in combination with an FEM-based modeling approach is a tool that allows the precise characterization of the fine-tuned interplay between turgor pressure as a driving force of growth and the flexibility of the cell wall to control cell extension and expansion.

6. Real-Time Cellular Force Microscopy

Due to its flexibility, the CFM can be used to study the mechanical properties of multiscale biological specimens. However, taking measurements of fast growing cells is tedious. For example, following the growing tip of a pollen tube to measure mechanical cell wall properties at the same position over time requires manual positioning of the sensor probe for every single measurement, which also includes the repositioning of the microscope stage. Because of the variation between biological specimens, when measuring their mechanical properties, it is very important to measure a high number of samples in a given experiment. Furthermore, the goal is not only to measure, but also to correlate mechanical properties with phenotypes among different cell wall mutants. When cells with either a limited lifetime or rapid change of their mechanical properties are studied, high throughput and speed is required.

Automation is a key factor for high-throughput experiments, and to this end, the real-time CFM (RT-CFM) was developed [77,78]. By automating procedures, such as sample selection and mechanical loading, the experiment time is drastically reduced, which allows hundreds of measurements a day, whereas the CFM is limited to a few tens of measurements. For the sample selection routine, computer vision techniques were employed, which allow the user to define an area of interest (AoI) of the sample. The RT-CFM automatically creates an interactive image based on overlapping snapshots of the AoI. On the resulting interactive image, the user then selects single point(s), a mesh of points with defined characteristics (number of points, point-to-point distance), an arbitrary area along with a number of points, a point-to-point distance or a combination of the above. The exact measurement points are calculated, and the sensor tip is positioned automatically at each point using visual servoing techniques (vision-based robot control). A computer vision algorithm calculates the image distance between the sensor location and the target point and transforms this distance to positioner coordinates, and the sample is moved until the image distance is minimized.

Moreover, a functionality for cell tracking is included in the RT-CFM. The output of the tracking module can be a distinctive feature of a cell, e.g., the tip of a growing pollen tube. The coordinates of the tracked feature can be used as input to the visual servoing module, and thus, measurements can be performed at a localized point with respect to the feature. This functionality is very useful, for instance for conducting repeated measurements at a given distance from the pollen tube tip.

A further improvement of the RT-CFM lies in its positioning capabilities and acquisition architecture. The experiments with the CFM revealed a need for sub-micrometer positioning with minimum errors, due to the size of the cells being at the microscale. Thus, a dual-stage configuration is used that consists of a combination of positioners mounted on top of each other. In the dual stage, a coarse XY positioner for large range, but slow and low-resolution movements is used, integrated with a fine XYZ positioner for fast and high-resolution, but small range movements. The dual-stage configuration benefits from the advantages of both positioners and achieves a position resolution of 2 nm over a range of 25 mm. With the parallel kinematic design of the current system, cumulative guiding errors and backlash are avoided. At the same time, non-linearities, poor accuracy and off-axis motions are compensated for by a parallel metrology approach [77,78].

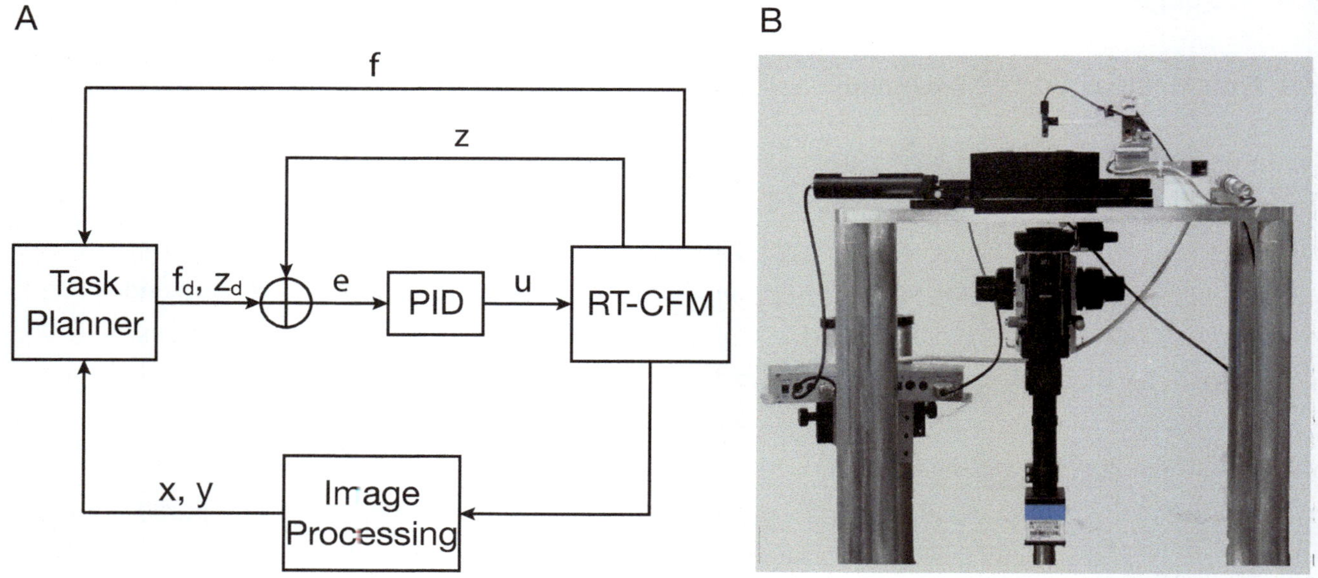

Figure 2. The real-time cellular force microscope (RT-CFM). (**A**) Schematic representation of the RT-CFM. f and z are the measured values, and f_d, z_d are the desired values of force and position, respectively. e is the error between the desired and the measured values of force and position, and u is the control signal that is fed into the RT-CFM positioner. x and y are the coordinates of the measurement location to which the system has to be positioned. The task planner switches the system between measurement, positioning, calibration and post-processing tasks. The proportional-integral-derivative (PID) is the control scheme used to control to the positioner. (**B**) Custom inverted RT-CFM setup. Block diagram in (A) adapted from [78].

The actuation control input and the position readout in the RT-CFM are interfaced with a real-time computer. This configuration allows one to acquire real-time, deterministic experimental data and implement real-time control with force and position feedback. Due to the faster dynamics of the piezoelectric stack actuators in combination with the real-time control, the execution time required for experiments is drastically reduced, allowing for faster measurements. A schematic representation of the RT-CFM is shown in Figure 2.

The abilities of the RT-CFM were demonstrated by characterizing the apparent stiffness over the entire area of a pollen tube as it grows, in contrast to measurements along the pollen tube longitudinal axis that have been previously obtained [68,73,79]. Felekis and colleagues took measurements over an area of 60 μm × 50 μm, starting at the pollen tube tip with maximum forces reaching 1.5 μN. The measured stiffness maps were combined with topography maps of growing pollen tubes. To study the effect of cell wall growth on the mechanical properties of the cell wall, a 14 μm × 5 μm area on the pollen tube was measured at three different time points during its growth. The visual servoing functionality allowed the accurate localization of the sensor tip on the growing cell and the comparison of successive experiments. The experimental time was 30 seconds and the time interval between experiments 2.5 minutes [77].

In addition to the mechanical characterization of the pollen tube, it would be interesting to understand how external mechanical stresses are perceived, influence behavior and are, in turn, modified by the cell. In combination with fluorescence microscopy, the RT-CFM is a powerful tool for quantifying the physiological responses of cells to mechanical stimulation. For instance, Felekis and colleagues applied precise mechanical loading while monitoring the changes in intracellular Ca^{2+} levels [77].

7. Outlook

The RT-CFM is a versatile, highly automated system, which allows high-throughput mechanical measurements on growing cells. In combination with fluorescence microscopy, it will allow well-defined mechanical stresses to be applied to individual cells and the physiological responses to be followed in real time. Such a setup would, for example, allow mechanical triggers to be set in different parts of growing pollen tubes that express a Ca^{2+}-sensing construct and observe the effects on the homeostasis of intracellular Ca^{2+}. Moreover, RT-CFM has the potential to be used in combination with multi-well plates to compare different growth conditions in parallel or in combination with microfluidic channels, which would allow researchers to manipulate conditions and measure the changes in real time.

Despite the advantages of the CFM and RT-CFM, there are remaining problems that need to be addressed. It is not possible to measure cell wall stiffness directly due to the influence of turgor pressure. FEM-based models are required to extract the Young's modulus and turgor pressure from the measured apparent stiffness values, which involves the measurement of additional parameters, such as pollen tube geometry and cell wall thickness. Another disadvantage is that measurements can only be taken perpendicular to the cell surface, and thus, it is not possible to get data from curved regions of cells, for example the pollen tube tip. This problem could be overcome by the development of 2D-MEMS force sensors, which allow force measurements along the Z- and X-axis. While prototypes of such 2D-MEMS force sensors exist [80,81], they are not yet commercially available.

Although not yet applied to living cells, further technological developments, such as multi-frequency atomic force microscopy (MF-AFM), will open completely new, exciting possibilities [82–85]. Improved cell wall models based on CFM or RT-CFM data can be used to identify regions where rapid changes in stiffness parameters occur. The application of MF-AFM on these areas will deliver high resolution data for cell topography, force, elasticity, dissipative energy loss as a measure of viscosity, and much more.

Another method with potential for studying biophysical parameters in plant cell growth is photoacoustic microscopy (PAM). It was successfully used to determine viscoelasticity in animal tissues, especially tumors [86,87]. With a lateral resolution of tens of microns, PAM was limited to measuring differences in viscoelasticity at the tissue level until recently. Newer setups, however, provide a resolution at the submicron level [88,89], which opens the possibility to precisely determine the subcellular regions where cell expansion occurs.

Acknowledgments

We apologize to our colleagues whose work could not be included in this review due to space limitations. Research on pollen tube cytomechanics in our laboratories is supported by the University of Zurich, the ETH Zurich and the Research, Technology and Development Projects "Plant Growth in a Changing Environment" and "MechanX", supported by SystemsX.ch, the Swiss Initiative in Systems Biology (to Bradley J. Nelson and Ueli Grossniklaus).

Conflicts of Interest

The authors declare no conflict of interest.

References

1. Kasper, G.; Glaeser, J.; Geissler, S.; Ode, A.; Tuischer, J.; Matziolis, G.; Perka, C.; Duda, G. Matrix metalloprotease activity is an essential link between mechanical stimulus and mesenchymal stem cell behavior. *Stem Cells* **2007**, *25*, 1985–1994.
2. Khani, M.; Tafazzoli-Shadpour, M.; Rostami, M.; Peirovi, H.; Janmaleki, M. Evaluation of mechanical properties of human mesenchymal stem cells during differentiation to smooth muscle cells. *Ann. Biomed. Eng.* **2014**, *42*, 1373–1380.
3. Shah, N.; Morsi, Y.; Manasseh, R. From mechanical stimulation to biological pathways in the regulation of stem cell fate. *Cell Biochem. Funct.* **2014**, *32*, 309–325.
4. Strauss, S.; Dudziak, S.; Hagemann, R.; Barcikowski, S.; Fliess, M.; Israelowitz, M.; Kracht, D.; Kuhbier, J.; Radtke, C.; Reimers, K.; *et al.* Induction of osteogenic differentiation of adipose derived stem cells by microstructured nitinol actuator-mediated mechanical stress. *PLoS One* **2012**, *7*, e51264.
5. Fernandez-Sanchez, M.; Serman, F.; Ahmadi, P.; Farge, E. Mechanical induction in embryonic development and tumor growth integrative cues through molecular to multicellular interplay and evolutionary perspectives. *Methods Cell Biol.* **2010**, *98*, 295–321.
6. Piccolo, S. Mechanics in the embryo. *Nature* **2013**, *504*, 223–225.
7. Pouille, P.A.; Ahmadi, P.; Brunet, A.C.; Farge, E. Mechanical signals trigger myosin II redistribution and mesoderm invagination in *Drosophila* embryos. *Sci. Signal.* **2009**, *2*, ra16.
8. Lokody, I. Tumour-promoting tissue mechanics. *Nat. Rev. Cancer* **2014**, *14*, 296.

9. Menon, S.; Beningo, K. Cancer cell invasion is enhanced by applied mechanical stimulation. *PLoS One* **2011**, *6*, e17277.
10. Suresh, S. Biomechanics and biophysics of cancer cells. *Acta Biomater.* **2007**, *3*, 413–438.
11. Burgert, I.; Keplinger, T. Plant micro- and nanomechanics: Experimental techniques for plant cell-wall analysis. *J. Exp. Bot.* **2013**, *64*, 4635–4649.
12. Geitmann, A. Experimental approaches used to quantify physical parameters at cellular and subcellular levels. *Am. J. Bot.* **2006**, *93*, 1380–1390.
13. Milani, P.; Braybrook, S.; Boudaoud, A. Shrinking the hammer: Micromechanical approaches to morphogenesis. *J. Exp. Bot.* **2013**, *64*, 4651–4662.
14. Moulia, B. Plant biomechanics and mechanobiology are convergent paths to flourishing interdisciplinary research. *J. Exp. Bot.* **2013**, *64*, 4617–4633.
15. Routier-Kierzkowska, A.; Smith, R. Measuring the mechanics of morphogenesis. *Curr. Opin. Plant Biol.* **2013**, *16*, 25–32.
16. Berry, P.; Sylvester-Bradley, R.; Berry, S. Ideotype design for lodging-resistant wheat. *Euphytica* **2007**, *154*, 165–179.
17. Flintham, J.E.; Börner, A.; Worland, A.J.; Gale, M.D. Optimizing wheat grain yield: Effects of *Rht* (gibberellin-insensitive) dwarfing genes. *J. Agric. Sci.* **1997**, *128*, 11–25.
18. Crook, M.J.; Ennos, A.R. The effect of nitrogen and growth regulators on stem and root characteristics associated with lodging in two cultivars of winter wheat. *J. Exp. Bot.* **1995**, *46*, 931–938.
19. Onoda, Y.; Westoby, M.; Adler, P.; Choong, A.; Clissold, F.; Cornelissen, J.; Diaz, S.; Dominy, N.; Elgart, A.; Enrico, L.; *et al.* Global patterns of leaf mechanical properties. *Ecol. Lett.* **2011**, *14*, 301–312.
20. Whitney, H.; Federle, W. Biomechanics of plant-insect interactions. *Curr. Opin. Plant Biol.* **2013**, *16*, 105–111.
21. Dimarco, R.; Nice, C.; Fordyce, J. Family matters: Effect of host plant variation in chemical and mechanical defenses on a sequestering specialist herbivore. *Oecologia* **2012**, *170*, 687–693.
22. Goriely, A.; Tabor, M. Estimates of biomechanical forces in *Magnaporthe grisea*. *Mycol. Res.* **2006**, *110*, 755–759.
23. Barrière, Y.; Argillier, O. *Brown-midrib* genes of maize: A review. *Agronomie* **1993**, *13*, 865–876.
24. Jorgenson, L. *Brown midrib* in maize and its linkage relations. *Agron. J.* **1931**, 549–557.
25. Mechin, V.; Laluc, A.; Legee, F.; Cezard, L.; Denoue, D.; Barriere, Y.; Lapierre, C. Impact of the *brown-midrib bm5* mutation on maize lignins. *J. Agric. Food Chem.* **2014**, *62*, 5102–5107.
26. Sattler, S.E.; Funnell-Harris, D.L.; Pedersen, J.F. *Brown midrib* mutations and their importance to the utilization of maize, sorghum, and pearl millet lignocellulosic tissues. *Plant Sci.* **2010**, *178*, 229–238.
27. Pauly, M.; Keegstra, K. Cell-wall carbohydrates and their modification as a resource for biofuels. *Plant J.* **2008**, *54*, 559–568.
28. Pauly, M.; Albersheim, P.; Darvill, A.; York, W. Molecular domains of the cellulose/xyloglucan network in the cell walls of higher plants. *Plant J.* **1999**, *20*, 629–639.

29. Park, Y.; Cosgrove, D. A revised architecture of primary cell walls based on biomechanical changes induced by substrate-specific endoglucanases. *Plant Physiol.* **2012**, *158*, 1933–1943.
30. Schopfer, P. Biomechanics of plant growth. *Am. J. Bot.* **2006**, *93*, 1415–1425.
31. Anderson, C.; Carroll, A.; Akhmetova, L.; Somerville, C. Real-time imaging of cellulose reorientation during cell wall expansion in *Arabidopsis* roots. *Plant Physiol.* **2010**, *152*, 787–796.
32. Preston, R. The case for multinet growth in growing walls of plant cells. *Planta* **1982**, *155*, 356–363.
33. Anderson, J.R.; Barnes, W.S.; Bedinger, P. 2,6-Dichlorobenzonitrile, a cellulose biosynthesis inhibitor, affects morphology and structural integrity of petunia and lily pollen tubes. *J. Plant Physiol.* **2002**, *159*, 61–67.
34. Aouar, L.; Chebli, Y.; Geitmann, A. Morphogenesis of complex plant cell shapes: The mechanical role of crystalline cellulose in growing pollen tubes. *Sex. Plant Reprod.* **2010**, *23*, 15–27.
35. Castle, E.S. Membrane tension and orientation of structure in the plant cell wall. *J. Cell. Comp. Physiol.* **1937**, *10*, 113–121.
36. Draeger, C.; Vogler, E.; Ndinyanka, T.; Weber, A.; Knox, J.; Felekis, D.; Smith, R.; Nelson, B.; Grossniklaus, U.; Ringli, C. Synergistic influence of cell wall xyloglucan and extensins on the growth and mechanical properties of *Arabidopsis* pollen tubes. Submitted for publication, 2015.
37. Miedes, E.; Suslov, D.; Vandenbussche, F.; Kenobi, K.; Ivakov, A.; van der Straeten, D.; Lorences, E.; Mellerowicz, E.; Verbelen, J.; Vissenberg, K. Xyloglucan endotransglucosylase/hydrolase (XTH) overexpression affects growth and cell wall mechanics in etiolated *Arabidopsis* hypocotyls. *J. Exp. Bot.* **2013**, *64*, 2481–2497.
38. Verhertbruggen, Y.; Marcus, S.; Chen, J.; Knox, J. Cell wall pectic arabinans influence the mechanical properties of *Arabidopsis thaliana* inflorescence stems and their response to mechanical stress. *Plant Cell Physiol.* **2013**, *54*, 1278–1288.
39. Cheung, A.; Wu, H. THESEUS 1, FERONIA and relatives: A family of cell wall-sensing receptor kinases? *Curr. Opin. Plant Biol.* **2011**, *14*, 632–641.
40. Lindner, H.; Muller, L.; Boisson-Dernier, A.; Grossniklaus, U. CrRLK1L receptor-like kinases: Not just another brick in the wall. *Curr. Opin. Plant Biol.* **2012**, *15*, 659–669.
41. Wolf, S.; Hofte, H. Growth control: A saga of cell walls, ROS, and peptide receptors. *Plant Cell* **2014**, *26*, 1848–1856.
42. Beck, W.A. Determining the osmotic value at incipient plasmolysis. *Trans. Am. Microsc. Soc.* **1929**, *48*, 204–208.
43. De Vries, H. *Eine Methode zur Analyse der Turgorkraft*; Bernstein: Berlin, Germany, 1884.
44. Green, P. Growth physics in *Nitella*: A method for continuous *in vivo* analysis of extensibility based on a micro-manometer technique for turgor pressure. *Plant Physiol.* **1968**, *43*, 1169–1184.
45. Tomos, A.; Leigh, R. The pressure probe: A versatile tool in plant cell physiology. *Annu. Rev. Plant Physiol. Plant Mol. Biol.* **1999**, *50*, 447–472.
46. Zimmermann, U.; Steudle, E. *Physical Aspects of Water Relations of Plant Cells*; Woolhouse, H., Ed.; Elsevier: London, UK, 1979; Volume 6; pp. 45–117.
47. Franks, P. Use of the pressure probe in studies of stomatal function. *J. Exp. Bot.* **2003**, *54*, 1495–1504.

48. Husken, D.; Steudle, E.; Zimmermann, U. Pressure probe technique for measuring water relations of cells in higher plants. *Plant Physiol.* **1978**, *61*, 158–163.
49. Franks, P.; Buckley, T.; Shope, J.; Mott, K. Guard cell volume and pressure measured concurrently by confocal microscopy and the cell pressure probe. *Plant Physiol.* **2001**, *125*, 1577–1584.
50. Hukin, D.; Doering-Saad, C.; Thomas, C.; Pritchard, J. Sensitivity of cell hydraulic conductivity to mercury is coincident with symplasmic isolation and expression of plasmalemma aquaporin genes in growing maize roots. *Planta* **2002**, *215*, 1047–1056.
51. Proseus, T.; Ortega, J.; Boyer, J. Separating growth from elastic deformation during cell enlargement. *Plant Physiol.* **1999**, *119*, 775–784.
52. Lintilhac, P.; Wei, C.; Tanguay, J.; Outwater, J. Ball tonometry: A rapid, nondestructive method for measuring cell turgor pressure in thin-walled plant cells. *J. Plant Growth Regul.* **2000**, *19*, 90–97.
53. Wei, C. An insight into cell elasticity and load-bearing ability. Measurement and theory. *Plant Physiol.* **2001**, *126*, 1129–1138.
54. Cosgrove, D. Growth of the plant cell wall. *Nat. Rev. Mol. Cell Biol.* **2005**, *6*, 850–861.
55. Chanliaud, E.; Gidley, M.J. *In vitro* synthesis and properties of pectin/*Acetobacter xylinus* cellulose composites. *Plant J.* **1999**, *20*, 25–35.
56. Cosgrove, D. Wall extensibility: Its nature, measurement and relationship to plant cell growth. *New Phytol.* **1993**, *124*, 1–23.
57. Edge, S.; Steele, D.; Chen, A.; Tobyn, M.; Staniforth, J. The mechanical properties of compacts of microcrystalline cellulose and silicified microcrystalline cellulose. *Int. J. Pharm.* **2000**, *200*, 67–72.
58. Kutschera, U. Cessation of cell elongation in rye coleoptiles is accompanied by a loss of cell-wall plasticity. *J. Exp. Bot.* **1996**, *47*, 1387–1394.
59. Wei, C.; Lintilhac, L.; Lintilhac, P. Loss of stability, pH, and the anisotropic extensibility of *Chara* cell walls. *Planta* **2006**, *223*, 1058–1067.
60. Agudelo, C.G.; Sanati, Nezhad, A.; Ghanbari, M.; Naghavi, M.; Packirisamy, M.; Geitmann, A. TipChip: A modular, MEMS-based platform for experimentation and phenotyping of tip-growing cells. *Plant J.* **2013**, *73*, 1057–1068.
61. Sanati Nezhad, A.; Naghavi, M.; Packirisamy, M.; Bhat, R.; Geitmann, A. Quantification of the Young's modulus of the primary plant cell wall using Bending-Lab-On-Chip (BLOC). *Lab Chip* **2013**, *13*, 2599–2608.
62. Sanati Nezhad, A.; Naghavi, M.; Packirisamy, M.; Bhat, R.; Geitmann, A. Quantification of cellular penetrative forces using lab-on-a-chip technology and finite element modeling. *Proc. Natl. Acad. Sci. USA* **2013**, *110*, 8093–8098.
63. Hawes, C.; Osterrieder, A.; Sparkes, I.; Ketelaar, T. Optical tweezers for the micromanipulation of plant cytoplasm and organelles. *Curr. Opin. Plant Biol.* **2010**, *13*, 731–735.
64. Huang, C.C.; Wang, C.F.; Mehta, D.S.; Chiou, A. Optical tweezers as sub-pico-newton force transducers. *Opt. Commun.* **2001**, *195*, 41–48.
65. Zlatanova, J.; Leuba, S. Magnetic tweezers: A sensitive tool to study DNA and chromatin at the single-molecule level. *Biochem. Cell Biol.* **2003**, *81*, 151–159.

66. Bolduc, J.; Lewis, L.; Aubin, C.; Geitmann, A. Finite-element analysis of geometrical factors in micro-indentation of pollen tubes. *Biomech. Model. Mechanobiol.* **2006**, *5*, 227–236.
67. Routier-Kierzkowska, A.; Weber, A.; Kochova, P.; Felekis, D.; Nelson, B.; Kuhlemeier, C.; Smith, R. Cellular force microscopy for *in vivo* measurements of plant tissue mechanics. *Plant Physiol.* **2012**, *158*, 1514–1522.
68. Vogler, H.; Draeger, C.; Weber, A.; Felekis, D.; Eichenberger, C.; Routier-Kierzkowska, A.; Boisson-Dernier, A.; Ringli, C.; Nelson, B.; Smith, R.; *et al.* The pollen tube: A soft shell with a hard core. *Plant J.* **2013**, *73*, 617–627.
69. Milani, P.; Gholamirad, M.; Traas, J.; Arneodo, A.; Boudaoud, A.; Argoul, F.; Hamant, O. *In vivo* analysis of local wall stiffness at the shoot apical meristem in *Arabidopsis* using atomic force microscopy. *Plant J.* **2011**, *67*, 1116–1123.
70. Peaucelle, A.; Braybrook, S.; Le Guillou, L.; Bron, E.; Kuhlemeier, C.; Höfte, H. Pectin-induced changes in cell wall mechanics underlie organ initiation in *Arabidopsis*. *Curr. Biol.* **2011**, *21*, 1720–1726.
71. Forouzesh, E.; Goel, A.; Mackenzie, S.; Turner, J. *In vivo* extraction of *Arabidopsis* cell turgor pressure using nanoindentation in conjunction with finite element modeling. *Plant J.* **2013**, *73*, 509–520.
72. Sun, Y.; Wan, K.; Roberts, K.; Bischof, J.; Nelson, B. Mechanical property characterization of mouse zona pellucida. *IEEE Trans. Nanobiosci.* **2003**, *2*, 279–286.
73. Felekis, D.; Muntwyler, S.; Vogler, H.; Beyeler, F.; Grossniklaus, U.; Nelson, B. Quantifying growth mechanics of living, growing plant cells *in situ* using microrobotics. *IET Micro Nano Lett.* **2011**, *6*, 311–316.
74. Boisson-Dernier, A.; Lituiev, D.; Nestorova, A.; Franck, C.; Thirugnanarajah, S.; Grossniklaus, U. ANXUR receptor-like kinases coordinate cell wall integrity with growth at the pollen tube tip via NADPH oxidases. *PLoS Biol.* **2013**, *11*, e1001719.
75. Lassig, R.; Gutermuth, T.; Bey, T.; Konrad, K.; Romeis, T. Pollen tube NAD(P)H oxidases act as a speed control to dampen growth rate oscillations during polarized cell growth. *Plant J.* **2014**, *78*, 94–106.
76. Franck, C.; Vogler, H.; Grossniklaus, U. University of Zurich, Zürich, Switzerland. Unpublished data, 2014.
77. Felekis, D.; Vogler, H.; Mecja, G.; Muntwyler, S.; Nestorova, A.; Huang, T.; Sakar, M.; Grossniklaus, U.; Nelson, B. Real-time, automated characterization of 3D morphology and mechanics of developing plant cells, *Int.l J. Rob. Res.* **2015**, doi:10.1177/0278364914564231.
78. Felekis, D.; Vogler, H.; Mecja, G.; Muntwyler, S.; Sakar, M.S.; Grossniklaus, U.; Nelson, B.J. High-throughput analysis of the morphology and mechanics of tip growing cells using a microrobotic platform. In Proceedings of the 2014 IEEE/RSJ International Conference on Intelligent Robots and Systems (IROS 2014), Chicago, IL, USA, 14–18 September 2014.
79. Geitmann, A.; Parre, E. The local cytomechanical properties of growing pollen tubes correspond to the axial distribution of structural cellular elements. *Sex. Plant Reprod.* **2004**, *17*, 9–16.

80. Beyeler, F.; Muntwyler, S.; Nagy, Z.; Graetzel, C.; Moser, M.; Nelson, B. Design and calibration of a MEMS sensor for measuring the force and torque acting on a magnetic microrobot. *J. Micromech. Microeng.* **2008**, *18*, ARTN 025004.
81. Beyeler, F.; Muntwyler, S.; Nelson, B. A six-axis MEMS force-torque sensor with micro-Newton and nano-Newtonmeter resolution. *J. Microelectromech. Syst.* **2009**, *18*, 433–441.
82. Garcia, R.; Herruzo, E. The emergence of multifrequency force microscopy. *Nat. Nanotechnol.* **2012**, *7*, 217–226.
83. Pukhova, V.; Banfi, F.; Ferrini, G. Energy dissipation in multifrequency atomic force microscopy. *Beilstein J. Nanotechnol.* **2014**, *5*, 494–500.
84. Sebastian, A.; Shamsudhin, N.; Rothuizen, H.; Drechsler, U.; Koelmans, W.; Bhaskaran, H.; Quenzer, H.; Wagner, B.; Despont, M. Note: Micro-cantilevers with AlN actuators and PtSi tips for multi-frequency atomic force microscopy. *Rev. Sci. Instrum.* **2012**, *83*, 096107.
85. Shamsudhin, N.; Rothuizen, H.; Despont, M.; Lygeros, J.; Sebastian, A. Micro-cantilever design and modeling framework for quantitative multi-frequency AFM. In Proceedings of the 2012 12th IEEE Conference on Nanotechnology (IEEE-NANO), Birmingham, UK, 20–23 August 2012.
86. Gao, G.; Yang, S.; Xing, D. Viscoelasticity imaging of biological tissues with phase-resolved photoacoustic measurement. *Opt. Lett.* **2011**, *36*, 3341–3343.
87. Zhao, Y.; Yang, S.; Chen, C.; Xing, D. Simultaneous optical absorption and viscoelasticity imaging based on photoacoustic lock-in measurement. *Opt. Lett.* **2014**, *39*, 2565–2568.
88. Zhang, C.; Maslov, K.; Wang, L. Subwavelength-resolution label-free photoacoustic microscopy of optical absorption *in vivo*. *Opt. Lett.* **2010**, *35*, 3195–3197.
89. Zhang, C.; Maslov, K.; Hu, S.; Chen, R.; Zhou, Q.; Shung, K.; Wang, L. Reflection-mode submicron-resolution *in vivo* photoacoustic microscopy. *J. Biomed. Opt.* **2012**, *17*, 020501.

Permissions

All chapters in this book were first published in Plants, by MDPI; hereby published with permission under the Creative Commons Attribution License or equivalent. Every chapter published in this book has been scrutinized by our experts. Their significance has been extensively debated. The topics covered herein carry significant findings which will fuel the growth of the discipline. They may even be implemented as practical applications or may be referred to as a beginning point for another development.

The contributors of this book come from diverse backgrounds, making this book a truly international effort. This book will bring forth new frontiers with its revolutionizing research information and detailed analysis of the nascent developments around the world.

We would like to thank all the contributing authors for lending their expertise to make the book truly unique. They have played a crucial role in the development of this book. Without their invaluable contributions this book wouldn't have been possible. They have made vital efforts to compile up to date information on the varied aspects of this subject to make this book a valuable addition to the collection of many professionals and students.

This book was conceptualized with the vision of imparting up-to-date information and advanced data in this field. To ensure the same, a matchless editorial board was set up. Every individual on the board went through rigorous rounds of assessment to prove their worth. After which they invested a large part of their time researching and compiling the most relevant data for our readers.

The editorial board has been involved in producing this book since its inception. They have spent rigorous hours researching and exploring the diverse topics which have resulted in the successful publishing of this book. They have passed on their knowledge of decades through this book. To expedite this challenging task, the publisher supported the team at every step. A small team of assistant editors was also appointed to further simplify the editing procedure and attain best results for the readers.

Apart from the editorial board, the designing team has also invested a significant amount of their time in understanding the subject and creating the most relevant covers. They scrutinized every image to scout for the most suitable representation of the subject and create an appropriate cover for the book.

The publishing team has been an ardent support to the editorial, designing and production team. Their endless efforts to recruit the best for this project, has resulted in the accomplishment of this book. They are a veteran in the field of academics and their pool of knowledge is as vast as their experience in printing. Their expertise and guidance has proved useful at every step. Their uncompromising quality standards have made this book an exceptional effort. Their encouragement from time to time has been an inspiration for everyone.

The publisher and the editorial board hope that this book will prove to be a valuable piece of knowledge for researchers, students, practitioners and scholars across the globe.

List of Contributors

Suyun Zhang
Plant Biodynamics Laboratory, Biology Institute Leiden, Leiden University, Sylviusweg 72, Leiden 2333, BE, The Netherlands

Bert van Duijn
Plant Biodynamics Laboratory, Biology Institute Leiden, Leiden University, Sylviusweg 72, Leiden 2333, BE, The Netherlands
Fytagoras, Sylviusweg 72, Leiden 2333, BE, The Netherlands

Seiji Takeda
Graduate School of Life and Environmental Sciences, Kyoto Prefectural University, Sakyo-ku, Kyoto 606-8522, Japan
Biotechnology Research Department, Kyoto Prefectural Agriculture Forestry and Fisheries Technology Center, Seika, Kyoto 619-0244, Japan

Akira Iwasaki
Laboratory of Plant Organ Development, National Institute for Basic Biology, Okazaki, Aichi 444-8585, Japan
Department of Botany, Graduate School of Science, Kyoto University, Kyoto 606-8502, Japan

Kiyoshi Tatematsu
Laboratory of Plant Organ Development, National Institute for Basic Biology, Okazaki, Aichi 444-8585, Japan

Kiyotaka Okada
Laboratory of Plant Organ Development, National Institute for Basic Biology, Okazaki, Aichi 444-8585, Japan
National Institute of Natural Sciences, Tokyo 105-0001, Japan

Tyler McCubbin
Department of Plant Sciences, University of California, Davis, CA 95616, USA

Elias Bassil
Department of Plant Sciences, University of California, Davis, CA 95616, USA

Shiqi Zhang
Department of Plant Sciences, University of California, Davis, CA 95616, USA

Eduardo Blumwald
Department of Plant Sciences, University of California, Davis, CA 95616, USA

Rita Brendel
Botanical Institute, Karlsruhe Institute of Technology, Karlsruhe 76131, Germany

Katharina Svyatyna
Botanical Institute, Karlsruhe Institute of Technology, Karlsruhe 76131, Germany

Yusuke Jikumaru
RIKEN Plant Science Center, Tsurumi, Yokohama, Kanagawa 230-0045, Japan

Michael Reichelt
Department of Biochemistry, Max Planck Institute for Chemical Ecology, Jena 07745, Germany

Axel Mithöfer
Department of Bioorganic Chemistry, Max Planck Institute for Chemical Ecology, Jena 07745, Germany

Makoto Takano
Department of Plant Physiology, National Institute of Agrobiological Sciences, Tsukuba, Ibaraki 305-8602, Japan

Yuji Kamiya
RIKEN Plant Science Center, Tsurumi, Yokohama, Kanagawa 230-0045, Japan

Peter Nick
Botanical Institute, Karlsruhe Institute of Technology, Karlsruhe 76131, Germany

Michael Riemann
Botanical Institute, Karlsruhe Institute of Technology, Karlsruhe 76131, Germany

Joseph K. E. Ortega
Bioengineering Laboratory, Mechanical Engineering Department, University of Colorado Denver, Denver, CO 80217, USA

Jason T. Truong
Bioengineering Laboratory, Mechanical Engineering Department, University of Colorado Denver, Denver, CO 80217, USA

Cindy M. Munoz
Bioengineering Laboratory, Mechanical Engineering Department, University of Colorado Denver, Denver, CO 80217, USA

David G. Ramirez
Bioengineering Laboratory, Mechanical Engineering Department, University of Colorado Denver, Denver, CO 80217, USA

Ingeborg Lang
Cell Imaging and Ultrastructure Research, University of Vienna, Althanstrasse 14, A-1090 Vienna, Austria

Stefan Sassmann
Cell Imaging and Ultrastructure Research, University of Vienna, Althanstrasse 14, A-1090 Vienna, Austria

Brigitte Schmidt
Cell Imaging and Ultrastructure Research, University of Vienna, Althanstrasse 14, A-1090 Vienna, Austria

George Komis
CR-Hana, Palacký University Olomouc, Šlechtitelů 586/11, 783 71 Olomouc-Holice, Czech Republic

David S. Domozych
Department of Biology and Skidmore Microscopy Imaging Center, Skidmore College, Saratoga Springs, NY 12866, USA

Hui Peng
Food Quality Laboratory, Beltsville Agricultural Research Center, Agricultural Research Service, United States Department of Agriculture, Beltsville, MD 20705, USA
College of Life Sciences, Guangxi Normal University, Guilin 541004, China

Tianbao Yang
Food Quality Laboratory, Beltsville Agricultural Research Center, Agricultural Research Service, United States Department of Agriculture, Beltsville, MD 20705, USA

Wayne M. Jurick II
Food Quality Laboratory, Beltsville Agricultural Research Center, Agricultural Research Service, United States Department of Agriculture, Beltsville, MD 20705, USA

Hyacinthe Le Gall
EA3900-BIOPI, Biologie des Plantes et Innovation, Université de Picardie Jules Verne, 80039 Amiens, France

Florian Philippe
EA3900-BIOPI, Biologie des Plantes et Innovation, Université de Picardie Jules Verne, 80039 Amiens, France

Jean-Marc Domon
EA3900-BIOPI, Biologie des Plantes et Innovation, Université de Picardie Jules Verne, 80039 Amiens, France

Françoise Gillet
EA3900-BIOPI, Biologie des Plantes et Innovation, Université de Picardie Jules Verne, 80039 Amiens, France

Jérôme Pelloux
EA3900-BIOPI, Biologie des Plantes et Innovation, Université de Picardie Jules Verne, 80039 Amiens, France

Catherine Rayon
EA3900-BIOPI, Biologie des Plantes et Innovation, Université de Picardie Jules Verne, 80039 Amiens, France

Ann-Iren Kittang Jost
Centre for Interdisciplinary Research in Space (CIRiS), NTNU Samfunnsforskning AS, Dragvoll Allé 38, Trondheim NO-7491, Norway

Takayuki Hoson
Department of Biology, Graduate School of Science, Osaka City University, Sumiyoshi-ku, Osaka 558-8585, Japan

Tor-Henning Iversen
Department of Biology, The Norwegian University of Science and Technology (NTNU), Realfagbygget, Trondheim NO-7491, Norway

Hyangju Kang
Department of Life Sciences, Pohang University of Science and Technology, Pohang 790-784, Korea

Inhwan Hwang
Department of Life Sciences, Pohang University of Science and Technology, Pohang 790-784, Korea
Division of Integrative Biosciences and Biotechnology, Pohang University of Science and Technology, Pohang 790-784, Korea

Yusuke Tagashira
Graduate School of Agricultural and Life Sciences, The University of Tokyo, Bunkyo-ku, Tokyo 113-8657, Japan

Tomoe Shimizu
Graduate School of Agricultural and Life Sciences, The University of Tokyo, Bunkyo-ku, Tokyo 113-8657, Japan

Masanobu Miyamoto
Graduate School of Agricultural and Life Sciences, The University of Tokyo, Bunkyo-ku, Tokyo 113-8657, Japan

Sho Nishida
Graduate School of Agricultural and Life Sciences, The University of Tokyo, Bunkyo-ku, Tokyo 113-8657, Japan

Kaoru T. Yoshida
Graduate School of Agricultural and Life Sciences, The University of Tokyo, Bunkyo-ku, Tokyo 113-8657, Japan

Kurt V. Fagerstedt
Department of Biosciences, Division of Plant Biology, P.O. Box 65, University of Helsinki, FI-00014 Helsinki, Finland

Pekka Saranpää
Natural Resources Institute Finland, Jokiniemenkuja 1, Fi-01300 Vantaa, Finland

Tarja Tapanila
Natural Resources Institute Finland, Jokiniemenkuja 1, Fi-01300 Vantaa, Finland

Juha Immanen
Department of Biosciences, Division of Plant Biology, P.O. Box 65, University of Helsinki, FI-00014 Helsinki, Finland
Institute of Biotechnology, P.O. Box 65, Helsinki University, FI-00014 Helsinki, Finland

Juan Antonio Alonso Serra
Department of Biosciences, Division of Plant Biology, P.O. Box 65, University of Helsinki, FI-00014 Helsinki, Finland

Kaisa Nieminen
Natural Resources Institute Finland, Jokiniemenkuja 1, Fi-01300 Vantaa, Finland

Hannes Vogler
Institute of Plant Biology and Zurich-Basel Plant Science Center, University of Zurich, Zollikerstrasse 107, CH 8008 Zürich, Switzerland

Dimitrios Felekis
Institute of Robotics and Intelligent Systems, ETH Zurich, Tannenstrasse 3, CH 8092 Zürich, Switzerland

Bradley J. Nelson
Institute of Robotics and Intelligent Systems, ETH Zurich, Tannenstrasse 3, CH 8092 Zürich, Switzerland

Ueli Grossniklaus
Institute of Plant Biology and Zurich-Basel Plant Science Center, University of Zurich, Zollikerstrasse 107, CH 8008 Zürich, Switzerland